METHODS IN MOLECULAR BIOLOGY

Series Editor
John M. Walker
School of Life and Medical Sciences
University of Hertfordshire
Hatfield, Hertfordshire, UK

For further volumes:
http://www.springer.com/series/7651

For over 35 years, biological scientists have come to rely on the research protocols and methodologies in the critically acclaimed *Methods in Molecular Biology* series. The series was the first to introduce the step-by-step protocols approach that has become the standard in all biomedical protocol publishing. Each protocol is provided in readily-reproducible step-by step fashion, opening with an introductory overview, a list of the materials and reagents needed to complete the experiment, and followed by a detailed procedure that is supported with a helpful notes section offering tips and tricks of the trade as well as troubleshooting advice. These hallmark features were introduced by series editor Dr. John Walker and constitute the key ingredient in each and every volume of the *Methods in Molecular Biology* series. Tested and trusted, comprehensive and reliable, all protocols from the series are indexed in PubMed.

Cancer Cell Culture

Methods and Protocols

Edited by

Dania Movia and Adriele Prina-Mello

Trinity Translational Medicine Institute, Trinity Centre for Health Sciences, Dublin, Ireland

Humana Press

Editors
Dania Movia
Trinity Translational Medicine Institute
Trinity Centre for Health Sciences
Dublin, Ireland

Adriele Prina-Mello
Trinity Translational Medicine Institute
Trinity Centre for Health Sciences
Dublin, Ireland

ISSN 1064-3745 ISSN 1940-6029 (electronic)
Methods in Molecular Biology
ISBN 978-1-0716-3055-6 ISBN 978-1-0716-3056-3 (eBook)
https://doi.org/10.1007/978-1-0716-3056-3

© The Editor(s) (if applicable) and The Author(s), under exclusive license to Springer Science+Business Media, LLC, part of Springer Nature 2023, Corrected Publication 2023
This work is subject to copyright. All rights are solely and exclusively licensed by the Publisher, whether the whole or part of the material is concerned, specifically the rights of translation, reprinting, reuse of illustrations, recitation, broadcasting, reproduction on microfilms or in any other physical way, and transmission or information storage and retrieval, electronic adaptation, computer software, or by similar or dissimilar methodology now known or hereafter developed.
The use of general descriptive names, registered names, trademarks, service marks, etc. in this publication does not imply, even in the absence of a specific statement, that such names are exempt from the relevant protective laws and regulations and therefore free for general use.
The publisher, the authors, and the editors are safe to assume that the advice and information in this book are believed to be true and accurate at the date of publication. Neither the publisher nor the authors or the editors give a warranty, expressed or implied, with respect to the material contained herein or for any errors or omissions that may have been made. The publisher remains neutral with regard to jurisdictional claims in published maps and institutional affiliations.

Cover image courtesy of Despina Bazou.

This Humana imprint is published by the registered company Springer Science+Business Media, LLC, part of Springer Nature.
The registered company address is: 1 New York Plaza, New York, NY 10004, U.S.A.

Preface

In the past ten years, cancer research has been advancing at a fast pace with the discoveries of the influence of the tumor microenvironment on tumor growth and progression, of new associated biomarkers, and the subsequent development of new in vitro experimental research models. These have enabled the investigation of biochemical and molecular pathways in a mechanistic way with accuracy and precision previously unknown, thus enabling the discovery of new targeted drugs and immunotherapies. A continuous need for standardized in vitro methods for cancer cell culture has however emerged, thus thriving to achieve precision targeting, accelerated drug development, and predictive efficacy assessment of new treatments.

The intention behind this book is to provide researchers with a collection of cell models that can be used for preclinical cancer research. After an initial introductory overview, the methodological chapters describe both conventional two-dimensional (2D) and more innovative three-dimensional (3D) culturing techniques, and how these can be deployed to address specific cancer types, or subtypes, or to recapitulate the interaction complexity between cancer cells and treatments.

This book is therefore divided into two parts. The first part includes introductory chapters, with historical and chronological mini-review type chapters which present the state-of-the-art in preclinical cancer research, from the basics of 2D cancer cell culture, to the cell models at the air-liquid interface, and finally moving on to the most recent advancements in the development of 3D complex spheroid models and dedicated disease animal models.

Following that, the second part of the book is dedicated to the technical chapters, which illustrate step-by-step methodologies for specific cancer cell culture methods. These chapters have been written by contributors with extensive expertise, and who use these models in their laboratories for preclinical cancer research. Methods range from the generation of isogenic cancer cell lines, and the use of serum-free growth conditions, to the formation of 3D cell cultures and the use of specific assays for the efficacy assessment of new anticancer therapies.

In conclusion, the overall value of this book on in vitro cancer cell methods lies on the practical experimental step-by-step knowledge transfer provided by the contributing authors, their extensive knowledge sharing in the presented models, and the ability to connect with them for assistance going forward in the adoption and use of any of the presented models.

Therefore, with such invaluable contributions, we hope that every specific contribution here presented can represent the starting point for further improved and advanced models to be developed by the community of researchers involved in creating a breakthrough in cancer research, whether through the development of new treatments or through the creation of new models for getting insight on unknown cancer pathophysiology mechanisms.

Dublin, Ireland

Dania Movia
Adriele Prina-Mello

Acknowledgments

We are grateful to the authors of the various chapters, for their expert input and constructive cooperation in the timely preparation of this book. We also extend our thanks to Prof. John M. Walker, as Series Editor.

We gratefully acknowledge the partial financial support of Science Foundation Ireland and the Irish Research Council (SFI-IRC Pathway Programme to D.M.), and European Union under the Horizon 2020 (NoCanTher (grant 685795), Safe-N-Medtech (grant 814607), and Expert (grant 825828) projects) and Horizon-Europe (INSPIRE project) programs.

It is with genuine gratitude and warm regard that we dedicate this book in memory of Prof. Yuri Volkov, immunologist, cancer researcher, and mentor, who has meant and continues to mean so much to us. Although he is no longer among us, his memory continues to inspire our research.

Dania Movia
Adriele Prina-Mello

Contents

Contributors

REBECCA AMET • *The John Durkan Leukaemia Laboratory, Trinity Translational Medicine Institute, Trinity College and St James's Hospital, Dublin, Ireland*
MARTIN P. BARR • *Thoracic Oncology Research Group, Trinity St James's Cancer Institute, Trinity Translational Medicine Institute, St James's Hospital & Trinity College Dublin, Dublin, Ireland*
DESPINA BAZOU • *Department of Haematology, Mater Misericordiae University Hospital, Dublin, Ireland; School of Medicine, University College Dublin, Dublin 4, Ireland*
SARAH BROPHY • *The John Durkan Leukaemia Laboratory, Trinity Translational Medicine Institute, Trinity College and St James's Hospital, Dublin, Ireland*
AOIFE CANNON • *Department of Surgery, Trinity St. James's Cancer Institute, Trinity Translational Medicine Institute, Dublin, Ireland*
ALINE CHARY • *Environmental Research and Innovation (ERIN) Department, Luxembourg Institute of Science and Technology (LIST), Esch-sur-Alzette, Luxembourg*
MELISSA J. CONROY • *Cancer Immunology Research Group, Department of Physiology, School of Medicine, Trinity College Dublin, Dublin, Ireland*
LUISANA DI CRISTO • *D3 PharmaChemistry, Nanoregulatory Group, Italian Institute of Technology, Genoa, Italy*
PAUL DOWLING • *Department of Biology, Maynooth University, National University of Ireland, Kildare, Ireland*
JAMES DREW • *CRUK Beatson Institute, Glasgow, UK*
KATIE DUNPHY • *Department of Biology, Maynooth University, National University of Ireland, Kildare, Ireland*
STEPHEN P. FINN • *Thoracic Oncology Research Group, Trinity Translational Medicine Institute, St. James's Hospital, Dublin, Ireland; Department of Clinical Medicine, Trinity College Dublin, Dublin, Ireland; Department of Histopathology and Morbid Anatomy, Sir Patrick Dun Translational Research Lab, St. James's Hospital, Dublin, Ireland; Department of Histopathology, Labmed Directorate, St. James's Hospital, Dublin, Ireland; Cancer Molecular Diagnostics, Labmed Directorate, St. James's Hospital, Dublin, Ireland*
HAYLEY FOY-STONES • *Cryobiology Laboratory Stem Cell Facility, St James's Hospital, Dublin, Ireland*
SARA FUOCHI • *Universität Bern, Experimental Animal Center, Bern, Switzerland*
VIOLA GALLIGIONI • *Netherlands Institute for Neuroscience - KNAW, Amsterdam, The Netherlands*
NICOLA GARDINER • *Cryobiology Laboratory Stem Cell Facility, St James's Hospital, Dublin, Ireland*
STEVEN G. GRAY • *Thoracic Oncology Research Group, Trinity Translational Medicine Institute, St. James's Hospital, Dublin, Ireland; Department of Clinical Medicine, Trinity College Dublin, Dublin, Ireland*
PRIYANKA GUPTA • *Centre for 3D models of Health and Disease, Division of Surgery and Interventional Science, University College London, London, UK*
GARY HANNON • *Laboratory of Biological Characterization of Advanced Materials (LBCAM), Trinity Translational Medicine Institute, Trinity College Dublin, Dublin,*

Ireland; Trinity St. James's Cancer Institute, St. James's Hospital, Trinity College Dublin, Dublin, Ireland

DAVID C. HAY • *Centre for Regenerative Medicine, University of Edinburgh, Edinburgh, UK*

SARAH HOLMES • *Laboratory of Biological Characterization of Advanced Materials (LBCAM), Trinity Translational Medicine Institute, Trinity College, Dublin, Ireland*

MANTAS JONAITIS • *Centre for Regenerative Medicine, University of Edinburgh, Edinburgh, UK*

ALVILE KASARINAITE • *Centre for Regenerative Medicine, University of Edinburgh, Edinburgh, UK*

NIAMH LYNAM-LENNON • *Department of Surgery, Trinity St. James's Cancer Institute, Trinity Translational Medicine Institute, Dublin, Ireland*

JOANNE LYSAGHT • *Cancer Immunology and Immunotherapy Group, Department of Surgery, Trinity Translational Medicine Institute, St James's Hospital, Dublin, Ireland*

ELAINE MA • *CRUK Beatson Institute, Glasgow, UK; Institute of Cancer Sciences, University of Glasgow, Glasgow, UK*

LAURA M. MACHESKY • *CRUK Beatson Institute, Glasgow, UK; Institute of Cancer Sciences, University of Glasgow, Glasgow, UK*

STEPHEN G. MAHER • *Department of Surgery, Trinity St. James's Cancer Institute, Trinity Translational Medicine Institute, Dublin, Ireland*

LAURE MARIGNOL • *Translational Radiobiology and Oncology Group, Applied Radiation Therapy Trinity Research Group, Trinity College Dublin, Dublin, Ireland*

ANTHONY M. MCELLIGOTT • *The John Durkan Leukaemia Laboratory, Trinity Translational Medicine Institute, Trinity College and St James's Hospital, Dublin, Ireland; Trinity St. James's Cancer Institute, Trinity College and St James's Hospital, Dublin, Ireland*

DANIA MOVIA • *Applied Radiation Therapy Trinity (ARTT), Discipline of Radiation Therapy, Trinity Centre for Health Sciences, Trinity College Dublin, Dublin, Ireland; Laboratory for Biological Characterisation of Advanced Materials (LBCAM), Trinity Translational Medicine Institute, Trinity Centre for Health Sciences, Trinity College Dublin, Dublin, Ireland; Trinity St. James's Cancer Institute, St. James's Hospital, Trinity College Dublin, Dublin, Ireland*

LANCE L. MUNN • *Edwin L. Steele Laboratory, Department of Radiation Oncology, Massachusetts General Hospital, Harvard Medical School, Boston, MA, USA*

EIMEAR MYLOD • *Cancer Immunology and Immunotherapy Group, Department of Surgery, Trinity Translational Medicine Institute, St James's Hospital, Dublin, Ireland*

EZGI ONER • *Department of Pharmaceutical Biotechnology, Faculty of Pharmacy, Izmir Katip Celebi University, Balatcik, Izmir, Turkey; Thoracic Oncology Research Group, Trinity Translational Medicine Institute, St. James's Hospital, Dublin, Ireland; Department of Clinical Medicine, Trinity College Dublin, Dublin, Ireland*

ADRIELE PRINA-MELLO • *Laboratory for Biological Characterization of Advanced Materials (LBCAM), Trinity Translational Medicine Institute, Trinity Centre for Health Sciences, Trinity College Dublin, Dublin, Ireland; Nanomedicine and Molecular Imaging Group, Trinity Translational Medicine Institute, (TTMI), School of Medicine, Trinity College Dublin, Dublin, Ireland; Trinity St. James's Cancer Institute, St. James's Hospital, Trinity*

College Dublin, Dublin, Ireland; Advanced Materials and Bioengineering Research (AMBER) Centre, CRANN Institute, Trinity College Dublin, Dublin, Ireland

STEFANIA SABELLA • *D3 PharmaChemistry, Nanoregulatory Group, Italian Institute of Technology, Genoa, Italy*

MELISSA ANNE TUTTY • *Laboratory for Biological Characterization of Advanced Materials (LBCAM), Trinity Translational Medicine Institute, Trinity Centre for Health Sciences, Trinity College Dublin, Dublin, Ireland; Nanomedicine and Molecular Imaging Group, Trinity Translational Medicine Institute (TTMI), School of Medicine, Trinity College Dublin, Dublin, Ireland*

EIRINI G. VELLIOU • *Centre for 3D models of Health and Disease, Division of Surgery and Interventional Science, University College London, London, UK*

Part I

Overview of the Existing Methods for Cancer Cell Culture

Chapter 1

Cancer Cell Culture: The Basics and Two-Dimensional Cultures

Melissa Anne Tutty, Sarah Holmes, and Adriele Prina-Mello

Abstract

Despite significant advances in investigative and therapeutic methodologies for cancer, 2D cell culture remains an essential and evolving competency in this fast-paced industry. From basic monolayer cultures and functional assays to more recent and ever-advancing cell-based cancer interventions, 2D cell culture plays a crucial role in cancer diagnosis, prognosis, and treatment. Research and development in this field call for a great deal of optimization, while the heterogenous nature of cancer itself demands personalized precision for its intervention. In this way, 2D cell culture is ideal, providing a highly adaptive and responsive platform, where skills can be honed and techniques modified. Furthermore, it is arguably the most efficient, economical, and sustainable methodology available to researchers and clinicians alike.

In this chapter, we discuss the history of cell culture and the varying types of cell and cell lines used today, the techniques used to characterize and authenticate them, the applications of 2D cell culture in cancer diagnosis and prognosis, and more recent developments in the area of cell-based cancer interventions and vaccines.

Key words Cancer cell culture, Cell line, Primary cells, Cell culture analysis, Characterization, Cancer research

1 Introduction

1.1 Cancer Cell Lines

Cell culture is a century-old and vitally important tool in cancer research, widely used by biomedical researchers in hospitals, academic settings, and industry laboratories for many applications [1]. The first citations of cell culture go back 125 years ago, to a publication from Roux et al., who successfully cultured tissue from chick embryos for several days in saline [2, 3]. Following this, Ljunggren showed in 1898 that it was possible for the human skin to survive in ascitic fluid ex vivo [4], and in 1907, Harrison showed that frog embryo tissue in frog lymph clots could not only survive but also sprout fibers [5]. Losee and Ebeling (1914) were the first to culture cancer cells [6], with William Earle producing the first continuous rodent cell lines at the National Cancer

Dania Movia and Adriele Prina-Mello (eds.), *Cancer Cell Culture: Methods and Protocols*,
Methods in Molecular Biology, vol. 2645, https://doi.org/10.1007/978-1-0716-3056-3_1,
© The Author(s), under exclusive license to Springer Science+Business Media, LLC, part of Springer Nature 2023

Table 1
Key terms used in cancer cell culture

Term	Definition
Authenticated/validated cell line	Cell line whose identity, purity, sterility, and functionality is confirmed
Cell bank	Repository of cell lines
Certificate of analysis	
Cell culture	Growth and maintenance of tissue explants in vitro
Cell line	Cells subcultured beyond their initial primary culture
Continuous cell line	Cell line that has unlimited number of doublings; immortal cell culture
Clone	Cells derived from one single origin cell
Confluent	Cells covering all substrate
Doubling time	Time taken for cells to double
Finite cell line	Cell line with limited lifespan; undergoes senescence after a certain point
Lag growth phase	Initial slow growth phase: Cells subculture here
Log growth phase	Exponential rapid growth phase
Passage/subculture	Expansion of cells from one flask to another
Plateau hase	Slow growth phase; occurs when cells become confluent
Primary culture	Initial culture of dissociated cells from tumors or extracted from blood
Substrate	Matrix/material on which cells grow
Tissue bank/biobank	Repository of human tissue samples
Tissue culture	Growth and maintenance of dissociated cells in vitro

Adapted from Ref. Langdon [9]

Research Institute in 1943 [7]. Earle was closely followed by George Gay, who in 1951 produced the first human continuous cell line, HeLa, from the cancer patient Helen Lane (discussed in more detail below) [3]. To date, HeLa cells are still one of the most popular and widely used cells. In the decades following these events, there has been a huge expansion in the use of cell lines, and there are now thousands available, from a large number of varying cell banks (Table 1), including commercial ones, where cell lines can be readily purchased, research-based cell banks, and nonprofit ones. Cell culture not only assists in the discovery of new molecules but also elucidates their functions. It is also used for the time-effective and high-throughput toxicity screening of new therapeutic compounds. Cell culture is not only undertaken with cancer cells but also healthy ones, and it is now commonplace to obtain non-neoplastic cells such as endothelial or inflammatory cells for various studies. Cancer cells derived from tumorous tissues are extracted from samples and passaged many times in appropriate supplemented media to promote growth.

1.1.1 Definitions

There are many definitions associated with cell culture [3]. The phrase "cell culture" itself refers to the maintenance of disaggregated cells in vitro, that is, outside the human body. In contrast to this, the term "organ culture" relates to intact tissues. The phrase "tissue culture" can be used to discuss both. Other common terms in cell culture include "primary," "immortalized," "passage," and/or "subculture." A primary cell culture is the initial culture, and it is this culture that undergoes many subcultures or passages (both terms have similar meaning) to form a cell line. Cell lines can be either "finite" or "continuous," with continuous cell lines being "transformed." Some of the most common definitions relating to cell culture are listed in Table 2. The application of cell lines, that is, in vitro cell-based cultures, is widely used in a variety of different fields within medical research, specifically in both basic cancer research and drug discovery. Cancer cell lines are useful as they provide an indefinite source of biological material for many different experimental purposes. If cultured correctly, validated and confirmed cell lines, that is, cell lines whose identity, purity, sterility, and functionality are confirmed, can retain the genetic properties of the cancer they come from, making them critical tools in cancer biology. Cell cultures can be one of two types—adherent or suspension. Adherent cells grow via attachment and are anchorage-dependent. Adherent cells are normally derived from tissues of organs. On the other hand, suspension cells do not require attachment for growth and are anchorage-independent cells, floating in culture medium in "suspension." Most suspension cells are isolated from blood, with a small number derived from tissues like hepatocytes or intestinal cells [8, 9].

1.2 Primary Versus Immortalized Cell Lines

There are two key cell types with regard to cell culture: primary and immortalized cells. While primary cells are isolated directly from human or animal donors and have their original characteristics highly preserved, immortalized cells are modified or passaged to allow them to proliferate indefinitely, making them a cost-effective and convenient, albeit not wholly biologically relevant, tool for cancer biology [10]. There are many stark differences between both these cell types, shown in Table 3 and also in more detail below.

1.2.1 Primary Cell Culture

Primary cells are cells isolated directly from animal or human sources, using mechanical or enzymatic extraction methods. Once cells are isolated, they are cultured in artificial environments with appropriate growth factors/nutrients, which support their proliferation and growth. While primary cells are at a disadvantage due to

Table 2
Main cell banks, their country of origin, respective websites, and contents [137]

Bank	Website	Country	Contents	Status
American tissue culture collection (ATCC)	www.atcc.org	USA	Over 950 cancer cell lines. Large collection of immortalized cell lines from human tissues	Nonprofit
European collection of animal cell culture (ECACC)	www.hpacultures.org.uk	UK	Over 1100 cell lines from over 45 species	Nonprofit
HyperCLDB	//		bioinformatics.istge.it/hypercldb	Italy
Database of hyperlinks from Interlab project; Facilitates rapid searches	Unconfirmed			
Deutsche Sammlung von Mikroorganismen und Zelkulturen GmbH (German Collection of Microorganisms and Cell Cultures)	www.dsmz.de	Germany	652 human and animal cell lines; Collection of 586 human leukemia-lymphoma cell lines	Unconfirmed
Banca Biologica e Cell Factory (Interlab cell line collection)	www.iclc.ie	Italy	Over 268 cell lines	Unconfirmed
Asterand	www.asterand.com	USA	Commercial supplier of human primary cells and cell lines	Commercial
Japanese Collection of Research Bioresources	Cellbank.nibio.go.jp	Japan	Over 1000 human and animal cell lines	Commercial
RIKEN gene bank	www.brc.riken.go.jp	Japan	Over 1000 human and animal cell lines	Nonprofit

their limited lifespan, they are advantageous in other ways [11]. Historically, researchers have employed immortalized cell lines for studying tissue function and response; however, using these cell lines that contain chromosomal abnormalities and gross mutations may be a poor indicator of normal phenotype and progression of disease. When using primary cells, there is also the opportunity to study the donor and not just the cells being used, with factors including medical history, age, race, and sex, all factors that can be considered in the model. This is also useful with regard to the growing trend of personalized medicine. Here, primary cells must be used to avoid critical issues such as lack of tissue complexity and donor variability associated with immortalized cell lines. Human primary cells are often used to study intercellular and

Table 3
Key differences between primary and immortalized cell lines

Cell characteristics	Primary cells	Immortalized cell line
Origin	Isolated from human or animal donor	Isolated from tumors
Lifespan	Finite	Unlimited if cultured correctly
Biological properties	Bears close resemblance to in vivo	De-differentiation, not close to in vivo situations
Genetics	Bears close resemblance to in vivo	Artificial, can be severe genetic changes once immortalized
Proliferation	Generally low (depends on cell type/conditions)	High
Function	Closely resembles cell function	Can lack appropriate functions/ function lost
Maintenance	If passaged/maintained no longer considered primary cells	Can be maintained long term in appropriate conditions
Donor characteristics	Available	Not available
In vivo model	Yes	No
Uses	Inflammation, immunology, toxicology, vaccine studies, or when a close resemblance to in vivo situation is needed	Studying cancer/tumors; in situations where primary not available
Examples	Primary endothelial cells, hepatocytes, fibroblasts, and stem cells	HepG2, Huh7, A549, LLC-PK1, HUVEC, JURKAT

Adapted from Ref. [13]

intracellular communication, widely used in developmental biology, and also for studying the mechanisms of various diseases like diabetes, Parkinson's disease, and cancer, making them a more advanced and biologically relevant model for recapitulating tissue cell type [12]. Various primary cell types can be isolated from a wide array of tissues, as seen in Fig. 1, and they include primary (1) endothelial cells, used in wound healing, cancer studies, and high-content toxicology screening; (2) fibroblasts, used in tissue engineering and regeneration; (3) primary immune cells like peripheral blood mononuclear cells (PBMCs), used in cell-based assays; (4) keratinocytes, useful in the study of skin cancer and psoriasis; (5) melanocytes, which can be used in wound healing, toxicity, and melanoma studies and in cosmetic research; (6) primary smooth muscle cells (SMCs), used to model hypertension fibrosis; and (7) primary stem cells, for modeling disease states [13].

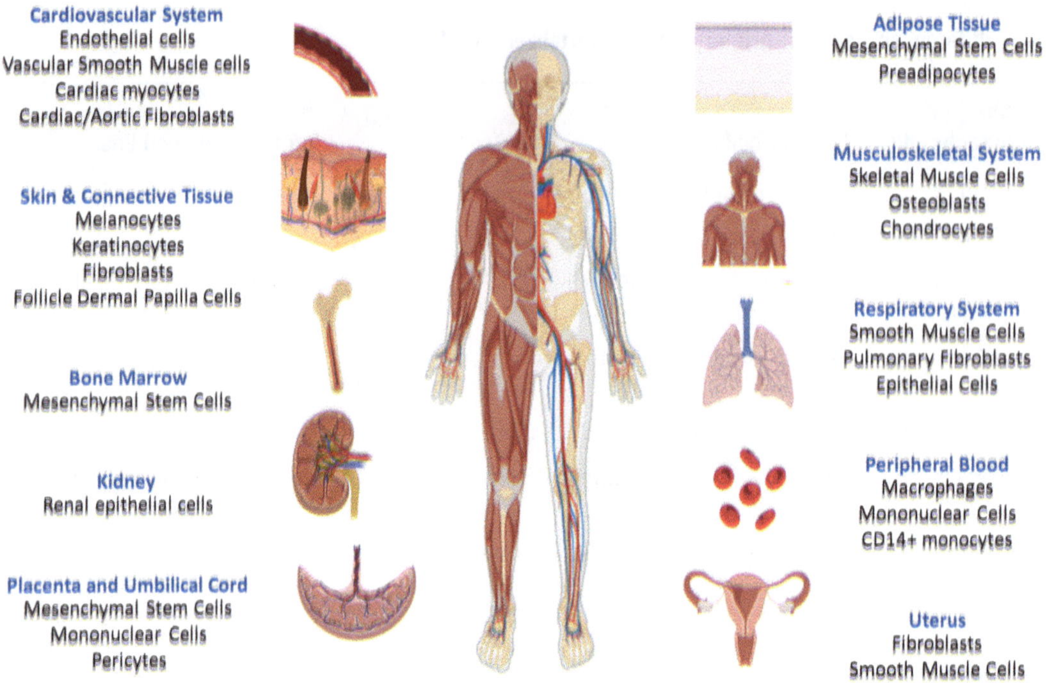

Fig. 1 Sources of primary cells in the human body. Primary are isolated from many different tissues in the body, ranging from skin and connective and adipose tissues to the kidney and respiratory systems. (Adapted from Ref. [13])

1.2.2 Immortalized Cell Lines

Immortalized cell lines are often used in cancer research instead of primary cells due to the many advantages they offer such as unlimited supply, cost-effectiveness, ease of use, and lack of ethical concerns associated with them compared to animal and human tissue [13]. Cell lines generate a pure population of cells, which is important as it provides a constant cell population with reproducible results. Immortalized cell lines have revolutionized cancer research and to date are used in many areas including drug toxicity and metabolism screening; generation of artificial tissues, for example, artificial skin; vaccine and antibody production; gene function studies; and the synthesis of biological compounds. Immortalized cell lines are extremely popular, something which is confirmed by the approximate 3600 cell lines, from over 150 different species from the American Type Culture Collection (ATCC) Cell Biology (Virginia, USA). One of the most important scientific discoveries of the last century has been the discovery of the first immortalized human cell line, HeLa, a cervical carcinoma cell line obtained by Dr. Georgy Gey, a Johns Hopkins researcher, in 1951 during the treatment of Henrietta's cancer [6]. These cells have never been commercially distributed or sold and to date are still offered freely for scientific research by Johns Hopkins, due to the fact they do not own the rights for them. In the past decades, this cell line has contributed to many scientific breakthroughs and discoveries, in varied areas ranging from the development of polio to leukemia studies, the study of the AIDS virus, the study of cancer, and most recently the development of the COVID-19 vaccines [14–16]. While many other cell lines are available today, HeLa cells are still an extremely popular choice in medical research. As previously mentioned, thousands of immortalized cell lines exist today. Some of the cell lines most used today for routine cancer research purposes are listed below, in Table 4 [17–19].

1.3 Techniques Used to Characterize/Authenticate Cancer Cell Lines

In order to carry out reliable and meaningful research through the use of cell-based assays, it is obviously of great importance to ensure that the cells used for experimentation have the correct origin (identity) and phenotype [20]. Due to the fact there are thousands of cell lines currently in regular use, it is essential to not only authenticate them but also characterize them regularly. This is vitally important for a variety of reasons. In order for a particular cellular model to have any value, the relationship of the cell line to its tissue of origin must be well established, and the cell line properties should reflect closely the properties of the cell type it was derived from [9]. For example, it is vital that HepG2, an immortalized liver cell line derived from human hepatocellular carcinoma, exhibits characteristics that are consistent with the liver and is epithelial in nature.

When a new cell line is developed, it is critically important to fully establish its origin, particularly cell lines from primary tumors, or those formed from effusions, ascites, and metastatic deposits. For example, in 1999 the SW626 colon cell line was redesignated after it was originally mislabeled as an ovarian cell line in 1974 [21]. In an ideal situation, a cell genetic makeup will remain constant; however, other features like expression levels may change, with cells also losing differentiation characteristics [22, 23]. As cells are cultured over time, clones may also emerge. For these reasons, it is crucial to assess culture purity and any possible cross-contaminations with other cell lines. The issue of cross-contamination has been long-standing problem in in vitro cell culture for decades and still poses a major issue today [24–26].

There are many notable cases in the history of in vitro experimentation whereby cell line "cross-contamination" has occurred. Most often, this has happened due to simple human errors, by accidental switching of cell cultures or mislabeling [27]. Cross-contamination has been prevalent since the 1970 and 1980s, when Gartler and Nelson-Rees demonstrated that one in three cell lines was either contaminated or was in fact completely replaced by other cell lines [28–31]. The largest amount of contamination was seen with HeLa. Established in 1951, HeLa was very quickly distributed following its inception; however, with this rapid growth and not only its mixing with other cell lines but also its tendency to outgrow other cell lines caused contamination to occur [17]. This

Table 4
Commonly used immortalized cell lines (most used), their source, gender and age, and potential applications

Cell line (s)	Source, gender/age	Applications
A549	Human adenocarcinoma alveolar basal epithelial cells, male, 58 years	Used for studying the metabolic processing of lung tissue; used in the identification of drug delivery mechanisms in tissue
BALB/ 3T3	Mouse fibroblasts from disaggregated BALB/c mouse embryos, 14–17 day	Uses in tumorigenicity studies, and for studying contact inhibition and viral transformation
Caco-2	Human epithelial cells from colorectal adenocarcinoma, male, 72 years	Primarily used as a model of the intestinal epithelial barrier
CHO	Chinese hamster ovary, female	Extensively used in the pharmaceutical industry to generation stable cell lines for mass production of therapeutic proteins; up to 3–10 grams of recombinant protein can be produced per liter of culture
HeLa	Human cervical carcinoma, epithelial, female, 31 years	Isolated in 1951, first immortal human cells to be grown in culture. Have contributed to many medical breakthroughs
HepG2	Human hepatocellular carcinoma, male, 15 years	Model system for studying liver metabolism and toxicity of xenobiotics; used in detection of cytotoxicity and genotoxicity agents; drug targeting studies; form bio-artificial livers and used in complex liver models
HEK293	Human embryonic kidney, female, fetal	Used in biotechnology industry to produce therapeutic proteins and viruses for gene therapy; used in the safety of various chemicals
HUVEC	Human umbilical vein endothelial cells, male and female cells	Commonly used for investigating various physiological and pharmacological events, such as blood coagulation, macromolecule transport, angiogenesis, and fibrinolysis
JURKA T	Human T lymphocyte cells, male, 14 years	Used to study acute T cell leukemia, T cell signaling, and the expression of various chemokine receptors susceptible to viral entry, particularly HIV
LLC-PK1	Porcine kidney, male, 3–4 weeks	Used a model for epithelial tissue; used in wide spectrum of pharmacologic and metabolic research investigations
MCF7	Human epithelial breast tissue from with metastatic adenocarcinoma, female, 69 years	In vitro breast studies due to fact they have retained several characteristics particular to mammary epithelium, including processing of estrogen, in the form of estradiol, via estrogen receptors (ER) in the cell cytoplasm

(continued)

Table 4
(continued)

Cell line (s)	Source, gender/age	Applications
THP-1	Human monocyte cells extracted from peripheral blood from an acute monocytic leukemia patient	Extensively for studying monocyte/ macrophage functions, drug and nutrient transport, signaling pathways and mechanisms. A common model for estimating the modulation of monocyte and macrophage activity

Adapted from Ref. [13]

problem has still not disappeared, and contamination is still widespread and being detected routinely by DNA fingerprinting [24, 25, 32].

A further issue is not only cross-contamination between the same species but also cross-contamination between other species, which is also common. For these reasons, it is crucial to not only have scientific methods that confirm a cell lines unique nature (these methods will be explained in detail below from Subheading 1.3.1) but also the animal species from which they were originally derived [9]. A variety of methods exist for characterizing and authenticating cell lines, as observed in Fig. 2, and they not only assess the general features of cells but also underpin the species of origin and also unique characteristics. These methods include PCR analysis, DNA profiling and fingerprinting, flow cytometry, morphological assessment, growth and cell cycle analysis, species-specific antibody detection, cytogenetics, karyotyping, and some key methods, which will be discussed in more detail below.

1.3.1 Cytogenetics

One useful way for distinguishing between individual cell types in populations for many years has been cytogenetics, and while DNA profiling has been a more cost-effective and easy method for many years, cytogenetics is a good complimentary technique for defining and characterizing cancer cell lines [33]. When compared to profiling, cytogenetics does have some advantages. Firstly, observing specific changes in chromosomes gives clues as to the changes in biology in the disease being studies, that is, p53 mutations in cancer or the Philadelphia chromosome in chronic myeloid leukemia. Furthermore, using microscopic techniques means deviations in subgroups that may not be found while sampling the whole population can be closely monitored, with various user-friendly techniques such as fluorescence in situ hybridization (FISH) having been developed, which allow easy analysis [34].

Conventional cytogenetics techniques include staining techniques, which identify chromosomes and their modifications, such as trypsin Giemsa (G), reverse Giemsa (R), constitutive

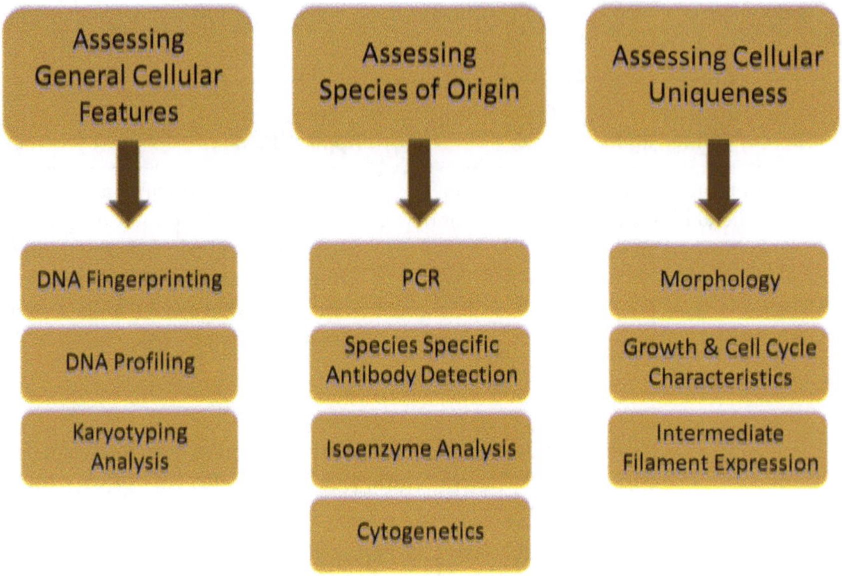

Fig. 2 Methods of cell culture analysis. Methods vary from those which assess general features, to identifying species of origin, to unique cellular characteristics. (Adapted from Ref. Langdon et al. [9])

heterochromatin (C), and quinacrine fluorescent (Q) staining, whereby specific bands are formed. G-banding or staining is used widely [35]. Here, trypsin is used to digest chromosomal proteins, with strong band staining being observed following Giemsa staining. R-banding (from reverse Giemsa) provides a banding pattern, which is different but as informative as traditional G staining, and C-band staining highlights the presence of heterochromatin. Finally, Q staining, that is, quinacrine (mustard or dihydrochloride), intercalates DNA and produces bands, which result from differential quenching of fluorescence. This type of staining produces a pattern different to G-staining. In more recent decades, the development of FISH has allowed investigation at all levels from whole chromosomes the whole way down to single, individual genes. FISH was developed during the late 1980s, and unlike conventional in situ hybridization, it was found to be more accurate and less time-consuming [34]. FISH also has more advantages over other conventional cytogenetic techniques, with its resolution being more superior to other banding analysis methods and the fact that it can be undertaken independently of cell cycle as signals are viewed in interphase nuclei. It is also possible using FISH to detect numerous targets at the same time. To date, there have been many different applications of FISH developed, including spectral karyotyping (SKY) [36], combined binary ratio labeling (COBRA) [37], color-changing karyotyping, and multiplex-FISH (M-FISH) [9, 38].

1.3.2 DNA Fingerprinting/Profiling

DNA fingerprinting and profiling are two other valuable tools for characterizing cell lines, with their applications including not only correctly identifying cell lines but also comparing difference sources of the same cell type. Developed in 1985 by Alec Jeffries, today it is also extensively used in forensics and works via exploiting the variability, which is found in "noncoding" regions of the human genome, representing around 90% of DNA bases [39]. Ten percent of the human genome encodes genes, and in this 10%, genetic variation between people is relatively small; however, in the other 90%, large differences are observed in individuals. It is assumed that this is due to the fact these regions are less important for survival [9]. It has also become even more apparent in recent decades that the vast areas of this noncoding DNA are orientated into repeating sequences called "variable number of tandem repeats," or VNTR. Two different types of repeating sequences have been identified, (1) minisatellites, which are 10–100 bp in length, and (2) microsatellites, or short tandem repeats (STRs), which are 2–5 bp in length. Most originally identified VNTR areas had many base pairs, that is, 20–50 in each repeat, with repeats reaching from 50 to many hundred; therefore, each region may be anywhere from 1000 bp up to over 10,000, making both the number of repeats and their lengths very characteristic. In a variable VNTR locus, >95% of the population has alleles of differing lengths; therefore, the chance of two unrelated people having the same length of allelic combination at a specific VNTR is extremely low, at <1%.

To date, many different strategies have been designed to profile these variable regions, with each divided into either multi-locus or single-locus approaches. In a multi-locus approach, multi-locus probes interact with many loci in a genome, and fingerprints can be observed via Southern blot analysis. By doing so, a single or double band can be produced. DNA profiling previously used larger VNTR regions, which were subjected to restriction fragment length polymorphism (RFLP) analysis with restriction fragments, produced following digestion with restriction enzymes, and with separation by electrophoresis and the bands identified by Southern blot analysis; however, this method was limited in that the restriction fragments may often are not separated, making it difficult to distinguish fragments that are similar, but have different lengths.

More current DNA profiling techniques look at STR regions, in combination with polymerase chain reaction (PCR) DNA amplification. This method is not only much faster but also far less labor-intensive than the RFLP method, due to the fact many key steps are automated. As well as this, far less material is needed for PCR-based analysis, where only approximately 1 nanogram of material is needed, in comparison to RFLP, which requires up to 20 nanograms of DNA. DNA fingerprinting has been used by many cell banks, with the ECACC using probes designed by Jeffreys to confirm cell line identity and compares being closely related cell lines. The DSMZ have also used the PCR-based methodology, in conjunction with RFLP for fingerprinting.

1.3.3 Flow Cytometry

Flow cytometry is a more modern but still well-established method with many applications, including for identifying cell populations in solution. Flow cytometry measures a variety of parameters, including cell size, granularity, DNA content, gene expression, surface receptors, and intracellular proteins, and can easily distinguish between varying cell types in mixtures based on the expression of various biomarkers or proteins. This methodology is advantageous as it can perform these measurements at rapid speeds, with the potential to measure between three to six varying properties in a single sample, cell by cell, for approximately 10,000 cell events in a minute.

1.4 Methods to Verify Species of Origin

1.4.1 Antibody Detection

A common laboratory technique, immunofluorescence or fluorescent antibody staining, can also be used to verify cell species origin [40]. This simple method involves using species-specific antiserum that binds to target cells, which is further coupled to a fluorescently labelled anti-globulin. Fluorescently labeled cells are then studied under fluorescent microscopy. One common dye used in this technique is FITC, that is, fluorescein isothiocyanate. Antibody detection, an area that is still continually advancing today, is highly sensitive and is said to detect even one contaminating cell in a cell population of over 10,000 [25].

1.4.2 Polymerase Chain Reaction (PCR)

Previously mentioned above, polymerase chain reaction (PCR) is a widely used methodology for rapidly making millions to billions of copies, or partial copies, of specific pieces of DNA, by taking small fragments and amplifying it enough so that it can be studied in detail. Invented in 1983 by Kary Mullis, PCR is a fundamental to genetic research and testing and is used for many applications in both forensics and biomedical research [41]. Due to the fact that PCR machines are common in most laboratories, it is a very convenient methodology for determining species origin of cells. One specific cell bank, the German Collection of Microorganisms and Cell Cultures, at the Leibniz Institute (DSMZ), uses PCR to verify their cell lines, using primers that amplify rDNA, alu, and β-globin. After amplification of the isolated genomic material, gel electrophoresis is used to separate amplified DNA fragments by size, showing differences between cell lines from different species.

1.4.3 Isoenzyme Analysis

Various enzymes exist as isoforms, which have different structural configurations, also called isoenzymes. Each species has a specific isoenzyme profile, with these isoforms also having differing abilities to move electrophoretically [42]. This allows them to be analyzed using gel electrophoresis. Today, various enzymes have been shown to be useful for determining species of origin, including aspartate aminotransferase, lactate dehydrogenase (LDH), nucleoside phosphorylase (NP), glucose-6-phosphate dehydrogenase (G6PD), and peptidase B [43]. In isoenzyme analysis, lysates extracted from cells ran on agarose gels during electrophoresis and then allowed to react with MTT dye ((3-(4,5-dimethylthiazol-2-yl)-2,5-diphenyl-2H-tetrazolium bromide) in the presence of phenazine methosulfate and a particular substrate to form a formazan band on the gel. To determine species origin, the band migration ratios are then compared against libraries of standards. This method is relatively sensitive and can detect contamination if the contamination level is above 10% [44].

1.5 General Characteristics of Cancer Cell Lines

1.5.1 Morphology

Perhaps the most obvious feature of a cell line is their appearance under the microscope. Various cell lines have very characteristic appearances in terms of population growth, cell size, or cell shape [6]. For example, fibroblasts have a string-like shape, elongated morphology "spindle," compared to epithelial cells, which are characteristically rounded (these factors however do depend on the adhesion of surfaces) [45]. Cells like HepG2 tend to cluster together in culture colonies, while other cells remain spread and distributed, and others grow in tighter colonies. A549 cells also form confluent epithelial monolayers that also have elongated morphology. Other cell lines are grown in suspension, such as HeLa or non-small cell lung cancer (NSCLC), whereby cells tend to adhere to surfaces, such as flasks, plates, and petri dishes. Between cell lines, ability to adhere or be removed from common cell culture flasks can vary massively, also including the presence of mycoplasma infection, which is influencing adherence. Cell morphology can be found within the cell supplier general information, and deviation from classical characteristics like a change in adhesion may be indicative of a contamination or an impeding issue, which can impact the cellular normal growth [46].

1.5.2 Colony Forming Efficiencies (CFE) and Cell Growth Rate

As stated, each cell line will have its own characteristic growth rate and density provided by the cell bank supplier. Cells are cultured under normal growth conditions using consistent recommended good-quality reagents, as indicated by the cell bank supplier or established cell culture protocol. Therefore, any deviation from these regular characteristics can be indicative of issues with cultures. Under define conditions, cells can be grown either in colonies or in monolayer on surface. The percentage of cells, which can produce colonies in either culture matrices like agar gel or substrates like plastic, is also a characteristic of specific cell lines and determines the cell growth rate [33].

1.5.3 Cell Cycle Characteristics/Ploidy

Flow cytometry is a widely used laboratory technique today for a variety of applications. One such application is for observing cell DNA content and the cell cycle [47]. By staining cells with DNA stains like propidium iodide (PI), nuclear DNA content can be analyzed, including the ploidy content of the cells and also the number of cells that are in varying stages of the cell cycle, that is, G0, G1, S, and G2/M phases [48].

1.5.4 Expression Profiles

Various proteins are extremely helpful for characterizing cellular origins, with these proteins easily measured using common techniques like western blotting, immunocytochemistry, or FACS. Intermediate filaments like desmin, vimentin, cytokeratin, and neurofilament proteins are useful for characterizing myogenic cells, mesenchymal cells, epithelial cells, and neurons, respectively [49]. Cell lines also can be characterized by more specific proteins. For example, liver cells like HepG2 strongly express the ATP-binding cassette transporter protein MRP2 (multidrug resistance protein 2), which plays important roles of detoxification [50]; the breast cancer cell line MCF-7, which expresses high levels of the estrogen receptor [51]; and A549 cells, which expresses P-glycoprotein (P-gp), a ubiquitous efflux transport protein, which is associated with variability in drug disposition and interindividual variability in drug treatment response [52], for instance. This information is also often provided by the cell supplier to enable the recognition and checking of the cell quality.

1.6 Important Culturing Considerations

There are specific criteria for selecting an appropriate cell line for the experiments you may want to undertake, as described below. These include but are not limited to the following:

1.6.1 Species Type

Both non-primate and nonhuman cell lines can be considered as they have fewer limitations and ethical considerations than human-sourced cells. However, the type of experiment being undertaken dictates what type of cell source is needed [10].

1.6.2 Continuous or Finite Cell Lines

An important consideration is whether cell type needs to remain in culture continuously, or for a finite period [10]. While immortalized, continuous cell types are easier and more predictable with regard to cloning and maintenance, finite cell types, like the liver cell line HepaRG, are more accurate with regard to functionality. This is an important consideration for metabolism studies [53].

1.6.3 Functional Characteristics

Endpoints must be carefully considered before a cell type is chosen, and the cell type must fit the purpose of the experiment [10]. For example, liver- or kidney-derived cell lines like HepG2 or LLC-PK1 are preferrable for drug toxicity screening [54, 55], and A549 cells are the gold standard for answering key questions in lung cancer biology and for modeling the lung [56].

1.6.4 Growth Characteristic and Conditions

Considerations must be made to specific requirements of cells with regard to cloning efficiency, density, growth rate, and ease of maintenance/growth, for example, for gathering large amounts of genetic material or recombinant proteins, a cell line may be chosen that grows in suspension with a fast cell growth rate (replication) [10].

1.6.5 Transformed or Normal Cell Type

Cell lines that are transformed normally have an increased growth rate, are continuous, and require less serum in culture medium and higher plating efficiency; however, they have undergone permanent changes in their phenotype due to genetic transformation [8]. Examples of the most commonly used transformed cell cancer lines include CaCo-2 and HT-29.

1.6.6 Other Considerations

Several considerations must be accounted when using cell lines, either immortalized or primary. If using immortalized cell line, it is expected that it is well characterized and from a validated source. If a finite cell line is used, it is important to plan to have sufficient amounts to complete all experiments. It is important to check that the cell line remains stable over time, and it is easy to culture, or specialist materials are needed. Make sure the cell line is easy to expand and can generate sufficient stocks to freeze down a batch as a backup, if possible. If an abnormal cell line is used, it is important to make sure that normal equivalent cell line can be used as a control to check on the abnormal pattern of the investigated cell line, for instance, the use of specific cell line for breast cancer research (e.g., SK-BR-3 as hypertriploid human cell line) and the comparison with adenocarcinoma and non-cancerogenic cell line (e.g., MCF-7 and MCF-10). These are all factors that must be considered [57].

2 Two-Dimensional (2D) Cell Culture in Cancer Diagnosis

There are a variety of two-dimensional (2D) methodologies in the diagnosis of cancer, including various laboratory and imaging tests, liquid biopsies, smears, and DNA screening. These will be discussed in more detail below [23, 58].

2.1 Liquid Biopsies/Blood Analysis

There are a wide array of in vitro laboratory tests, such as urine and blood tests, which can identify specific abnormalities in the body that be indicative of cancer [59]. With regard to 2D cell-based methodologies used in the diagnosis of cancer, one of the most popular methodologies is looking at circulating tumor cells, or CTCs [60]. CTC tests are commonly used to monitor cancers such as metastatic breast, prostate, and colorectal cancers and work on the principle that cancer cells detach from solid tumors and filter into the bloodstream. From here, they can be captured, identified, and counted in various ways.

One such method involves extracting CTCs from whole blood for DNA and RNA analysis on a 2D microwell chip, via staining using a variety of detection stains like DAPI, CD45-APC, CD16, and CK-PE [61]. CTC tests are often used prior to beginning therapy or during treatment. Another 2D slide and blood-based methodology is complete blood counts, or CBC [62]. Here, blood samples can be taken and the varying types of cells in the sample observed, with too few or too many of a certain type of cell, or abnormal cells being a marker of cancer. More specifically, this test looks at levels of red blood cells, white blood cells, and platelets, while also measuring hemoglobin. CBCs can detect various cancers including leukemia and are also useful for monitoring patients throughout treatment [63].

2.2 Imaging and Histology

Microscopic imaging is a widely used and vitally important tool in the diagnosis of cancer [64]. Histology is the study of the microscopic anatomy of various cells and tissues in organisms, with this type of analysis performed by examining very thin slices, that is, sections, of tissues under a light or electron microscopes. Today, histology is regarded as the "gold standard" in the clinical diagnosis of cancer and is widely used in the detection and grading of cancers including breast, prostate, cervical, and lung [65]. Histological type is determined by examining suspect tissues excised via biopsy or surgery, and if this varies from what is normally found, it can be indicative of cancer, which has spread from a primary site, that is, metastasis. Histology also can identify specific cell types, for example, the difference between different skin cancers such as melanoma, basal cell carcinoma, and squamous cell carcinoma. Histology works in the following manner. Following collection of tissue sample from a surgical procedure, sample is fixed as soon as possible to preserve it (fixation method depends on imaging and analysis, which will be performed). After fixation, tissues must be supported in some manner, for example, frozen or embedded in a solid wax or paraffin mold, to allow sufficiently thin sections to be sliced using a microtome. The tissues are dehydrated, cleared, and then infiltrated, with the embedded tissue finally cut into thin sections using a microtome, approximately 5 μm for light microscopy and from 80 to 100 nm for electron microscopy. Following cutting, sections are floated in warm water to smooth any wrinkles and to stop the thin sections curling. The sections are then mounted on glass slides via an adhesive or by heating. They are then ready for microscopy. Slides can be stained with a variety of stains, depending on the required endpoint. Each dye binds to a certain cellular structure, with intensity of the color of stain varying across different tissue structures. The most common staining for light microscopy of tissue samples is hematoxylin and eosin (H&E), with hematoxylin staining the nuclear components of cells a dark blue color and eosin staining the cytoplasmic organelles varying shades of pink or red. Various other "special stains" also exist for specific diagnostic needs, such as Masson's trichrome, popular for its ability to stain collagen fiber blue and Prussian blue stain, used for identifying iron stored in the body [66]. Following staining, glass or plastic coverslips are added to stained slides to protect the tissue during storage and to provide better visual quality for microscopy.

2.3 Biopsies

Often to confirm a cancer diagnosis, a portion of cells will be surgically removed from the tumor tissue and the cells tested in a laboratory. This is known as a biopsy. Depending on the type of suspected cancer, a specific biopsy may be performed [58]. These include the following:

1. Fine needle aspiration (FNA): Here, a small, hollow, thin needle removes cells and fluid from a tumor mass. For tumors deep inside the body, a CT scan or ultrasound is used to guide the needle [67].
2. Core biopsy: Core biopsies are similar to an FNA procedure; however, here a larger needle is used [68].
3. Excisional biopsy: During an excisional biopsy, the area is numbed, the skin is cut, and the entire tumor is removed [69].
4. Incisional biopsy: Similar to an excisional biopsy, however a smaller portion of tumor is removed [70].
5. Laparoscopic biopsy: A small incision is made and a laparoscope (a small surgical device with a camera) inserted to see inside the abdomen and then obtain tissue samples [71].
6. Endoscopic biopsy: An endoscope, that is, a flexible, slim tube with a camera and light attached at the end is inserted into the body, through various areas like the mouth, nose, bladder, throat, or lungs. Here, small medical tools can be passed down through the tube and cells or tissue samples removed [72].
7. Skin biopsy: Various types of skin biopsies are used today. One method, a punch biopsy, is done with a small device that removes a small portion of tissue, including several layers of skin (epidermis, dermis, and subcutaneous fat). Another type, a shave biopsy, removes only the uppermost layers of the skin, that is, the epidermis and part of the dermis. These tests are used for diagnosing basal cell or squamous cell skin cancers [73].

2.4 Cytology/Smears

In comparison to a biopsy, whereby tissue from a certain area of the body is removed and analyzed for cancer, a cytology test removes and studies a smaller number of cells [74]. In a cytology test, the cytological morphology of the collected cells is studied under a microscope following some sort of staining. The examined cells are extracted from the body using a variety of methods, including scraping tissue surface (e.g., during a cervical smear) or by collecting body fluids like phlegm. In comparison to tissue biopsies, specimens for cytological analysis are generally easier to acquire, are less likely to cause complications for patient and cause less discomfort, and cost less, with cytology tests not only used for diagnosis but also in screening [75]. As previously mentioned, while histology looks at entire blocks of tissue, cytology differs in that it only looks at single cell types.

2.5 DNA/RNA Screening and Genetic Testing

DNA screening and genetic testing also help estimate a person's chance of developing cancer in their lifetime. It works by looking for specific changes, that is, mutations, that happen in chromosomes, genes, or proteins, therefore predicting the risk of developing a specific form of cancer [76]. It also allows patients to determine if they may pass particular mutations, which may pose a risk to their children and overall can provide information of a patient's overall health. DNA screening is widely used for cancers such as ovarian, breast, prostate, pancreatic, colon, thyroid, kidney, and stomach cancer, as well as sarcoma and melanoma [77–79]. Genetic testing is extremely useful for patients who have a hereditary risk to cancer, that is, patients who may have inherited certain genes and have a family history of certain cancers.

3 Two-Dimensional (2D) Cell Culture in Cancer Prognosis

Two-dimensional cell culture is also useful for determining prognosis once a patient is diagnosed with cancer [23]. The most common way of doing so is with tumor profiling. Tumor profiling relates to the use of laboratory tests, which can check for specific genes or gene mutations, as well as various biomarkers and proteins in tumor tissue samples. Tumor profiling is an extremely useful tool in personalized cancer care, as it is not only used to help plan cancer treatments but can also predict whether cancer will metastasize to other parts of the body or come back in the future [80].

Conventional cancer therapy works via the inhibition of various cellular processes in the body; however, newer treatments like targeted therapy work by attacking specific pathways or genes, which are responsible for tumor growth. Once a tumor tissue sample has been obtained from a patient through a core biopsy (samples can also be taken directly from the body during surgery), tumor DNA is extracted from the cells, before being purified and sent to the lab. This DNA is then sequenced to look for specific gene mutations, which may make a tumor unresponsive to certain treatments [81]. Today, tumor profiling is widely used to determine if the pathways of a patient's specific tumors will match up with targeted treatments that are available. It is used for many types of cancers but more specifically in lung and colon cancer and sarcoma and melanoma, due to the fact these cancers are more often related to genetic mutations, which respond well to targeted treatments [82].

Despite being useful, tumor profiling is not without its limitations, however, and while it has the power to underpin hundreds of genes in the blueprint of a tumors DNA, tumor profiling has only yet been developed for a relatively small number of these genes. In addition, even when there is a targeted drug available for a specific tumor gene, there is no guarantee that the tumor will respond to treatment. Even if responsive, this result may only be temporary, and as cancer is a dynamic disease, constantly growing and changing, tumor profiles that are taken even a year apart may yield very different results. However, despite being in its infancy, tumor profiling and targeted therapy are very exciting and viable treatment strategies for assessing cancer prognosis.

4 Two-Dimensional (2D) Cell Culture in Drug Screening

As previously indicated, in vitro 2D cell cultures are widely used tools for not only improving our understanding of cellular biology and tissue morphology but also for helping us understand diseases and the action of drug compounds in greater detail [23]. They are also a critical part of cancer research. Up until the 1980s, the in vivo mouse models of L1210V leukemia or P388 were used by the National Cancer Institute (NCI) for the systematic pre-screening of drugs [83]. Despite being highly stable and convenient in terms of set up and data interpretation, as well as relatively low cost, these models were significantly limited by their inability to identify therapeutic agents aimed for treating solid tumors. Due to this limitation, in the late 1980s, a comprehensive panel of 60 different human cells lines was developed, originating for various tumors including lung, colon, breast, kidney, prostate, melanoma, and leukemia, called NCI60 [84]. In recent decades, this panel has been greatly extended, and cell lines now exist from a large array of tissue types including, as well as the ones above, liver, bone, brain, cervical, esophageal, leukemia, lymphoma, myeloma, ovarian, pancreatic, prostate, and thyroid.

Choosing an appropriate culture method for the study of cancer allows us to understand tumor biology better, as well as helping in the development and optimization of new chemotherapeutic drugs or treatment strategies for cancer therapy [83]. The majority of research with regard to cancer biology is based upon experimentation conducted using 2D cultures; however, in more recent decades, the use of 3D cultures has also emerged, which will be discussed later in this book in Chap. 5. Regarding 2D cell cultures, they can be carried out in suspension, or in adherent conditions, where cells are attached to solid plastic or glass substrates. 2D cultures are advantageous as they are simple to undertake and relatively low-cost to maintain. There are also a very large number of functional assays available today based upon 2D cultures. These include cytotoxicity and viability assays, like MTT and LDH, which measure the toxic potential or efficacy of various materials used either as cancer interventions or directly in the treatment of cancer; cell proliferation assays, which detect changes in cell division changes in a cell population; clonogenic assays, which track cell survival and colony forming efficiency, cell migration, and invasion; and adhesion assays, which can be used for the fast screening of therapeutic chemotherapy drugs and for assaying new materials, which are involved in these processes in cancer cells [85]. Each of these assays is discussed in more detail below.

4.1 Clonogenic Assay

Clonogenic assays, or colony formation assays, are in vitro cell survival assays, which are based on the ability of single cells to form colonies of at least 50 cells. Clonogenic assays test the ability of a cell population to undergo "unlimited" division, and they are widely used to determine cell reproductive death after radiotherapy treatment, as well as for determining the effectiveness of various agents including cytotoxic drugs, nanomaterials, and genes [86]. Following incubation with these materials, 100 cells are seeded in a 6-well plate or petri dish. Plates are then left for a specific incubation period, normally from 1 to 3 weeks depending on the cell line used. At least six cell divisions must form sufficiently for large colonies, which are visible with the naked eye. Colonies are the fixed, stained, and counted [87].

4.2 Cell Proliferation and Cytotoxicity Assays

Cell viability can be defined as the number of cells that are healthy in a sample, and a key marker or vital indicator of cell viability and indeed the mechanisms of action of various proteins, genes, or pathways that are involved in cell survival or death following exposure to a cytotoxic agent is cell proliferation [88]. Cell proliferation assays have many applications in cancer research ranging from the screening of chemotherapeutics to analyzing cell activity. These assays typically detect changes in cell populations or changes occurring in numbers of cells dividing. These assays can be broken down into varying methodologies, that is, metabolic/enzyme activity assays, cell membrane permeability assays, cell proliferation marker assays, ATP quantification assays, cell adherence assays, and DNA synthesis assays. The methodology chosen will depend on a variety of parameters, including cell type being studied, the protocol being used, and the information required, and the specific research question should be considered in detail. Irrespective of the type of assay used, it is important to know at the end of the experiment how many viable cells remain. Some key proliferation and cytotoxicity assays will be discussed in more detail below.

4.2.1 Metabolic and Enzyme Activity: MTT, LDH, and Alamar Blue

The MTT assay is a colorimetric assay that assesses cell metabolic activity. Based on the conversion of MTT ((3-[4,5-dimethylthiazol-2-yl]-2,5 diphenyl tetrazolium bromide) to insoluble formazan crystals, via mitochondrial NADPH-dependent oxidoreductase enzymes released in living cells, the MTT assay allows us to determine the number of viable cells in a sample by measuring mitochondrial activity, which is proportional to the amount of formazan crystals formed [89]. Cells that are proliferating exhibit greater amounts of MTT conversion, in comparison to dying or dead cells, which exhibit low metabolism, which in turn yields low MTT conversion. At the completion of the assay, the formazan crystals formed are solubilized with DMSO or SDS, and measurements read using a spectrophotometer between 540 and 720 nm. The MTT assay is most commonly used for detecting cytotoxicity of drug compounds under varying conditions and concentrations, and it is also possible to determine IC50 values, that is, the 50% half maximal inhibitory concentration of drugs using this assay, if appropriate controls are used. It is very useful in in vitro pre-clinical assessment for predicting dosing prior to in vivo experimentation and also in the study of drug resistance in cell lines. However, it is not useful for any follow-up studies as the cells are killed during the protocol, and it is impossible to distinguish between cytostatic and cytotoxic drugs using MTT [90].

The LDH assay, or lactate dehydrogenase, is an assay that allows the determination of cytotoxicity in cells via the release of the cytoplasmic enzyme LDH into cell culture medium following membrane disruption in damaged and dying cells. This two-step procedure works in the following way: (1) LDH catalyzes the conversion of lactate to pyruvate via the reduction of NAD+ to NADH. Following this, there is the reduction of a tetrazolium salt (INT) to a red formazan product, via diasphorase enzymes using NADH. Using a spectrophotometer, the tetrazolium salt can then be measured colorimetrically, at 490–520 nm by a spectrophotometer. The amount of formazan produced is directly proportional to the amount of LDH that is released in the culture medium. LDH is widely used due to its easy and fast protocol; however, it is possible that the amount of LDH in cell culture medium may be contaminated by endogenous LDH activity. This would cause false-positive results and undermine the integrity of the assay. LDH half-life in medium is also limited, as is the assay sensitivity [91].

The Alamar Blue, or Resazurin, assay is another assay that determines numbers of proliferating cells. This nontoxic cell permeable vital dye enters cells and becomes reduced to resofurin, which fluoresces red due to the reducing environment in the cytosol. In proliferating cells, there is a higher rate of resazurin to resorufin, and in dead cells, a much lower rate is observed [92]. Dead or dying cells create a much lower fluorescence, since there is no active metabolism occurring, and this is how cell viability is determined. Fluorescent readouts from this assay are measured at

570 and 630 nm. It is to be noted that this method is preferred to others as it induces less toxicity than other methodologies and due to the fact that the dye can be used in chronic cytotoxicity assessment, that is, to conduct long-term experiments, as it does not kill the cells. Resazurin is also water-soluble, non-radioactive, and very stable in cell culture, making it a useful reagent [93]. However, much optimization is needed when using it, which may decrease its applicability and usefulness for rapid testing. It is also possible that interference during fluorescence readouts may occur, causing false results.

4.2.2 ATP Quantification: CellTiter-Glo™

ATP quantification or ATP assays are extremely useful markers of the toxic effects of compounds. Adenosine 5′-triphosphate, or ATP, is the major energy source for living organisms, with ATP mostly produced in the mitochondria, with ATP and the phosphate bond attached to it supplying the energy needed by the body to complete cellular process. By measuring the total ATP in a sample, it is possible to elucidate the toxicity of a compound, or indeed lack of toxicity of these agents on cells [94]. ATP quantification assays are based upon the monitoring of ATP levels via luminescence. Here, the natural enzyme luciferase reacts with ATP in cells, producing light that can be measured using a luminescent plate reader, with the amount of light produced directly proportional to cellular viability. One of the most well-known commercial ATP assays is the CellTiter-Glo™ assay. This assay exists in two forms: a normal assay used for 2D cultures and also a 3D assay kit, for use with organoids and cell spheroids [95]. This assay is specifically designed for use with multi-well formats and is ideal for use in high-throughput experiments. ATP assays provide many advantages over other assay formats in that they are quick and easy to undertake, are highly sensitive and cost-effective, and have large usage areas. As the endpoint signal is also present for a sustained time due to the stability of the assay, measurements are also easier to undertake. ATP assays have also been shown to be extremely accurate in cancer research [96], that is, they have been more accurate in predicting both sensitivity and resistance of platinum-resistant epithelial ovarian cancer tumors to other cytotoxic agents [97]. However, as also observed with the MTT assay, it is not possible to differentiate between cytostatic or cytotoxic effects of drugs, due to the fact the decrease in ATP levels could depend on a variety of parameters such as cell number, cell death, or various metabolic activities [90].

4.2.3 Membrane Integrity: G6PD Release, Trypan Blue

Another marker of toxicity in cells is any damage to cell membrane integrity. Various assays have been developed to assess membrane integrity, with two of the most common including the G6PD assay and Trypan Blue exclusion. The G6PD assay is a fluorescence-based microplate assay that quantifies cell death via the measurement of glucose-6-phosphate dehydrogenase, or G6PD, activity. This cytosolic enzyme leaches from cells once plasma membrane integrity is compromised, and in this assay, cell death is determined via correlating the activity of extracellular G6PD with the reduction of the compound resazurin to its fluorescent product, resorufin, via an enzyme-coupled reaction. In this reaction, rapid signal amplification from even very small amounts of G6PD is possible, making this assay advantageous over the conventional resazurin assay. The G6PD assay is also easy to undertake, nontoxic, and amenable to high-throughput screening, making it an extremely time and cost-effective marker of cytotoxicity [98].

One of the most well-known and widely used assays today is for assessing cell viability. Trypan blue, a blue azo dye, is impermeable for normal cell membranes and will only enter cells whose membranes are compromised. After being internalized by cells, trypan blue binds to the intracellular proteins and stains them blue. It is also commonly used for identifying and counting live and dead cells within cell populations. Here, viable cells are shown as small, round, refractive cells, in comparison to dead cells, which present as large, swollen, and dark blue [99].

4.3 Adhesion, Migration, Invasion, and Wound Healing Assays

Migration is a critical property of live cells, and it is crucial for normal cell development, immune responses, and various disease processes, including inflammation and the progression of cancer [100]. How the motility of a cell is affected by physical, chemical, and molecular features is a great challenge in understanding a cells migratory behavior. To date, there are various useful methods to study migration of cells, which is important in many areas of biomedical research including developmental biology, cell biology, immunology, vascular biology, immunology, and indeed cancer research. In vitro assays are an excellent approach to underpin in vivo situations and facilitate the study of a live cell's behavior. During the development of cancer, its progression, invasion, and metastasis occur when tumor cells spread from the primary site and move through the circulatory and lymphatic systems, invading basement membranes and endothelial walls, before finally entering distant organs [101]. Migration, invasion, and adhesion are crucial steps in this process; therefore, studying these processes in detail is critical in order to fight cancer. Some key assays routinely used for assessing cell adhesion and migration and their invasive and angiogenic potential are discussed briefly below and summarized in Table 5.

4.3.1 Adhesion Assays

Cell adhesion is an intricate mechanism that is involved in many processes ranging from cell migration and invasion, to embryogenesis, tissue remodeling, and wound healing. Here, cells have the ability to adhere to a variety of proteins in the ECM, such as collagen I and IV, laminin, fibronectin, and fibrinogen, forming complexes with different cytoskeletal components. They can also

Table 5
Common cell adhesion, migration, cell adhesion, migration, and invasion assays

Cellular property	Cellular activity	Format	Assay readout
Adhesion	Adhesion to extracellular matrix	48-well plate	Colorimetric or fluorometric
	Adhesion to endothelium or epithelium	96-well plate	Fluorometric
Migration	Random migration	24-well gap closure	Microscopy
	Migration toward immobilized ECM (haptotaxis)	24-well Boyden chamber	Colorimetric or Fluorometric
	Migration through endothelium toward a chemoattractant (transmigration)	24-well Boyden chamber	Fluorometric
	Migration toward a chemoattractant (chemotaxis)	24-well Boyden chamber	Colorimetric or fluorometric
		96-well Boyden chamber	Fluorometric
Invasion	Invasion through extracellular matrix	24-well Boyden chamber/ transwells	Colorimetric or fluorometric
		96-well Boyden chamber	Fluorometric
Wound healing	Migration across a wound field/gap	24-well gap closure	Light microscopy

Adapted from Ref. [138]

adhere to both epithelium and endothelium. Adhesion assays not only measure the contacts between cells and extracellular adhesion proteins but also allow us to gain more information about a variety of other cellular events [102]. The most common forms of adhesion assays test two things: (1) the ability of specific cell types or cell lines to adhere to various adhesive substrates and (2) to assess the sensitivity of specific cell-substrate interactions to inhibitors. Based on experimental conditions, the majority of cell adhesion assays fall into two categories, either static adhesion assays or dynamic, or flow, adhesion assays. Static assays assess the adhesion of common cell types like fibroblasts and epithelial cells to the ECM, whereas dynamic adhesion assays are commonly used for analyzing the adhesion of blood cells to endothelial cells, to themselves or to ECM proteins [103].

4.3.2 Migration and Invasion Assays

Migration is a fundamental property of living cells, and it is critical for typical development, immune responses, and disease processes including metastasis, inflammation, and cancer [100]. Examining the migratory potential of a cell is extremely useful in a wide array of areas in biomedical research, including vascular biology, immunology, cell and developmental biology, and indeed cancer biology. Both cell migration and invasion are very complex, multistep processes, which play critical roles in the progression of a variety of diseases, including atherosclerosis, arthritis, and cancer. The most widely known and well-accepted cell migration technique is the Boyden Chamber assay [104]. The transwell-based migration assay system utilizes a hollow plastic chamber, which is sealed at one end with a porous membrane. The chamber is suspended over a bigger well, which can contain cell culture medium and/or various chemo attractants. Cells are then placed inside this chamber and are allowed to migrate through the pores, to the underside of the membrane. The cells that have migrated are then stained and counted. The 2D Gap Closure assay is a further migration assay. Here, a defined area is created on a flat, 2D surface, and cells are seeded. The migration of cells is assessed in real time using microscopy. While the analysis for this assay is qualitative in nature, in recent years, various software and programs have been developed to qualify cell migration quantitatively. There are other types of assays that also assess migration, including the chemotaxis and haptotaxis assays, which describe cell migration toward chemo attractants in their surrounding environments or in immobilized ECM, respectively, and also the transmigration assay, which assesses the migration of cells through the vascular endothelium [105].

While migration assays determine regular cellular movement, invasion assays determine movement across the EMC and the ability to invade surrounding tissues. While migration assays do not use filters, invasion assays use filters with ECM molecules, including elastin and collagen [101]. As previously mentioned, while migration is crucial in the progression of various diseases, invasion encompasses cell migration through ECM degradation and proteolysis. Invasion assays can be undertaken in either 24 or 96 well format, using inserts that are pre-coated with basement membrane (ECM) matrixes, or Collagen or Laminin I.

4.3.3 Wound Healing/Scratch Assays

The in vitro wound healing, or scratch, assay is an easy and low-cost method of measuring the migration cells in vitro. Here, a cell monolayer is cultured before a portion of the cells are "scratched" and the migration that occurs as the cells close in on the scratch observed at regular intervals via image capture [106]. The rate of migration of the cells can then be determined. In comparison to other methods of assessing migration, the scratch assay is widely used for studying the effects cell-cell and cell-matrix interactions have on cell migration and is particularly useful as it is compatible with live cell imaging, and intracellular events can be monitored if desired. The wound healing assay can also be used to assess the migration of single cells at the edge of a scratch and is extremely useful due to the short amount of time it takes to run the assay, that is, a few hours to overnight [107].

4.4 Angiogenesis

Angiogenesis is a complex and strongly regulated process, which plays important roles in both physiological conditions like reproduction and wound healing and pathological conditions, including tumor growth and metastasis. Angiogenesis relates to the process of the formation of new blood vessels, that is, vascular sprouting, from pre-existing blood supplies, which can mature into stable vasculature, which can supply both oxygen and nutrients [108]. Angiogenesis is important in cancer due to the fact solid tumors require a blood supply if they are to grow beyond a few millimeters in size. Solid tumors are able to form this blood supply themselves, by producing various chemical signals that can stimulate angiogenesis. In turn, they can also stimulate normal cells nearby to release angiogenic signaling molecules. These newly formed blood vessels provide the growing tumors with the essential oxygen and nutrients they need to grow, allowing them to grow and subsequently allow cancer cells to invade neighboring tissues and translocate throughout the body. This in turn causes the formation of new colonies of cancerous cells, or metastasis [109].

There are many 2D in vitro angiogenic cell-based assays, which provide convenient systems for studying angiogenesis, including angiogenesis tube formation assays, microfluidic based angiogenesis assays, endothelial adhesion, invasion and migration assays (discussed previously), and wound healing assays. One of the most common and popular methods of studying cell migration and angiogenesis is the scratch wound healing assay. As discussed above, endothelial cells are grown on 2D surfaces until they are confluent, and a wound is produced by scraping cells with a pipette tip, cell scraper, or needle. The scratched space or "wounded area" will then fill cells via migration; however, cells in the cleared area will also eventually proliferate. The "wounded" area is photographed immediately after wounding and at specific time points thereafter. Cell migration is then quantified and expressed as average percentage of closure of the scratch area [107].

5 Two-Dimensional (2D) Cell Culture in Cellular-Based Cancer Interventions

It goes without saying the significant role 2D cell culture plays in the research and development of all cancer interventions (see Table 6 for a rudimentary outline of some available therapeutics) [110]. So too is its involvement in up-and-coming cell-based therapies. In recent decades, many cellular therapies have been designed to improve the body's ability to fight cancer. Manufacturing these cellular-based cancer interventions normally involves collecting specific cells from a patient or donor and modifying them to heighten their functionality, for the purpose of patient transfusion or vaccine production. With inherent antitumor responses, certain

Table 6
Various types of cancer therapy and approved drugs

Therapy type	Examples
Chemotherapy	5-Flurouracil Carboplatin
Hormone therapy	Abiraterone acetate Fulvestrant
Epigenetic modification	Azacitidine Decitabine
Immune stimulation and checkpoint inhibition	Aldesleukin Pembrolizumab
Angiogenesis inhibition	Bevacizumab Regorafenib
Cell signaling inhibition	Ibrutinib Imatinib Ceritinib
Adoptive immunotherapy	Anti-CD19 CAR-T cell therapy CART-Meso
Monoclonal antibody therapy	Cetuximab TDM-1 (Trastuzumab & DM1 chemo)
Oncolytic viruses	T-VEC (Imlygic™)
Vaccines	Sipuleucel-T (Provenge) DCVax®-L

immune cells (i.e., Dendritic Cells, M1 Macrophages, Natural Killer cells, $CD4^+$, and $CD8^+$ T cells) provide ideal therapeutic strategies for targeted and personalized cancer medicine [111]. Another cell type of interest, particularly in the area of hematological malignancy, is that of the stem cell and its associated reprogramming technology (i.e., induced pluripotent stem cells) [112]. Furthermore, cancer cells themselves can be harnessed for medicinal effect, for example, by way of vaccination [113]. To date, many different types of cellular therapies for treating cancer are being investigated, including tumor infiltrating lymphocytes (TIL), chimeric antigen receptor T (CAR T) cells, and other forms of modified T-cells, induced pluripotent stems cells (iPSCs), vaccines, natural killer (NK) cells, and various other immunotherapies.

5.1 Adoptive Cell Transfer and T Cells

The potent cytotoxic potential of certain T cell types is well known; coupled with their recognition of both tumor-specific antigens and cellular stress markers, their proficiency in killing cancer cells is unrivalled [111]. However, tumor-associated immuno-suppression, immuno-regulation, and immuno-evasion (e.g., increased

Tregs, decreased cancer cell MHC class I expression) can result in T cell depletion and blocking, thereby impeding this natural defense mechanism [114, 115]. The solution to this is adoptive T cell Transfer, whereby autologous or allogenic T cells are activated and expanded in large populations in vitro (10^{9-11}), before being infused into the patient [116]. Furthermore, this ex vivo approach provides the opportunity for both the patient and T cell population to be primed; the patient can be given treatment to promote an optimal in vivo environment for engraftment (e.g., non-myeloablative chemotherapy), while the T cells can be stimulated and/or modified to enhance their effector functions and specificity [117].

Variations of this concept have been developed to cater for differing patient-disease circumstance while still achieving the all-important tumor recognition. For example, T cells capable of tumor cell recognition can be cultured in vitro using a patient's tumor sample and T cell growth factor, IL-2. This method, however, is dependent upon the tumor sample containing sufficient tumor infiltrating lymphocytes with pre-existing tumor specificity [118]. Alternatively, co-cultures of nontumor T cells and antigen-presenting cells, pre-exposed to mutation peptides from patient tumor DNA, can also accomplish specificity [119]. A particularly innovative and popular approach in producing tumor recognition involves the genetic modification of peripheral T cells. This is done through vector-mediated insertion of antitumor receptor genes into the T cells, prior to expansion and patient infusion. The type of receptor introduced can either be a T cell receptor (MHC dependent), or a chimeric antigen receptor (CAR) (non-MHC dependent, but antigen target must be on cancer cell surface). A CAR is a manufactured receptor and, for the purposes of T cell modulation, combines an antibody-derived extracellular domain with intracellular T cell receptor signaling moieties (e.g., CD28, CD137) [120, 121].

CAR-Ts have seen great success in the last 5 years, with a number of FDA and EMA clinical approvals of anti-CD19 and anti-BCMA CAR-Ts for the treatment of B cell malignancies [120, 122]. As it stands, five generations of CAR-Ts have been developed to date [123], and they represent approximately 600 ongoing clinical trials for the treatment of various cancers [124]. Importantly, technological advances that have been fostered through adoptive T cell transfer have paved the way for its translation to other cell types, for example, NK cells, stem cells, and dendritic cells. Adoptive T cell transfer is not without its challenges; however, and with regard to cell culture, the main difficulty lies in growing the large but necessary number of cells, without loss of antigen specificity [125].

5.2 Induced Pluripotent Stem Cells (Leukemia)

Induced pluripotent stem cells, or iPSCs, can be described as an unlimited and self-renewing source of cells, which can further differentiate into a wide variety of cell types [126]. Here, iPSCs, which are derived from blood or skin cell,s are reprogrammed back into an embryonic-like pluripotent state, which allows them to produce an unlimited source of any type of human cell needed for a variety of therapeutic needs. Some of the primary and most interesting uses of iPSCs are in generating immune cells to fight cancer or generating novel iPSC cell lines from already existing cancer cell lines. Examples of these can be observed in Table 7, with many forms of cancer including prostate, lung, breast, and gastrointestinal cancers, as well as melanoma, glioblastoma, and various sarcomas [126].

Reprogramming human primary cancer cell lines into this interesting cell type has many potential biomedical applications. This transformative approach has, in recent years, become an invaluable tool in the study of cancers, which are derived from tissues or cell types with limited lifespans, or which are difficult to culture routinely or from samples that cannot be obtained easily from patients. It is possible that iPSCs can model tumors in situations where human cancer-related genes have no rodent counterpart, or where mutations are observed in human genes that are far too complex to engineer into the mouse genome [126]. It is also possible to preserve these unique genotypes using human-derived iPSCs by cell banking. However, even though there is a large amount of interest in iPSCs, they are not without their difficulties, and reprogramming cancer cells in this way can be complicated due to a variety of reasons including cancer-specific gene mutations, DNA damage accumulation, epigenetic modifications, and reprogramming-triggered cellular senescence. Despite these issues, however, many studies have succeeded in the generation of iPSCs, and various studies have reported generation of new, novel iPSC cell lines from common, existing cell lines.

5.3 NK Cells

While cancer cell loss of MHC class I can render T cells wholly ineffective, NK cells boast: (1) "missing-self" recognition, identifying the absence of an endogenous protein, and (2) "induced-self" recognition, identifying a marked increase of an endogenous protein. Although not as efficient as T cells, nor antigen-specific, NK cells possess powerful cytotoxic activity, making them promising candidates for cell-based cancer therapy [127]. In fact, some success has already been seen in this regard; the FDA has recently given Fast Track Designation to a human placental hematopoietic stem cell-derived NK cell therapy, CYNK-001, for the treatment of glioblastoma multiforme, which is now at Phase II clinical trials [128].

Also in Phase II clinical trials is oNKord®, an allogenic ex vivo cultured NK cell preparation from CD34+ umbilical cord blood

Table 7
Induced pluripotent cell (iPSC) lines derived from cancer

Cancer type	Primary cell/ immortalized cell line	Method of reprogramming	Cellular phenotype	References
Prostate cancer	PC-3	Retroviral mir-302s	Migratory in iPSCs reduced; cell cycle-related gene expression in iPSCs also reduced	[140]
Melanoma	R545		Melanocytes and melanoma cells reprogrammed via SOX2	
Hepatocellular carcinoma Pancreatic Esophageal Colorectal	PLC PANC-1/ AMIAPaCa-2 TE-10 DLD-1, HT-29	Combination of either retroviral or lentiviral OSKM, NANOG, LIN28, BCL2, KRAS, and shRNA for tumour suppressors optimized for each separate cell line	Acquired sensitivity to chemotherapies in embryoid body cells; Reduced invasive and tumorigenic potential of embryoid body cells; Increased expression of p16 and p53 of embryoid body cells in comparison to the parental cells	[141]
Lung	A549	Lentiviral OSNL and nondegradable HIFα	Tumorigenic properties of iPSCs in mice increased; A more aggressive and invasive iPSC	[142]
Breast	MCF-7	Retroviral O	MCF-7 iPSCs do not fully reprogram; Overexpression of Sox2; Cancer stem cell features such as high ALDH activity and CD44 expression observed	[143]
Pancreatic ductal adenocarcinoma (PDAC)	Patient-derived pancreatic ductal adenocarcinoma	Lentiviral O	Teratomas from PDAC iPSC-like cells undergo early to invasive stages of human cancer	[144]

Adapted from Ref. Jean J. Kim [126]

progenitors, which was granted Orphan Drug Designation by both the EMA and FDA for the treatment of Acute Myeloid Leukemia [129].

Despite this progress, however, obtaining sufficient quantities of NK cells while preserving their cytotoxic effects remains a serious challenge; striking the right balance with regard to cytokines and feeder cells during activation and expansion is of utmost importance and, perhaps even more crucial, is the NK cell source itself [130, 131].

5.4 Cell-Based Vaccines

5.4.1 Tumor Cell Vaccines

For this class of vaccine, autologous or allogenic tumor cells are irradiated and prepared ex vivo, followed by in vivo administration for an enhanced antitumor immune response [132]. To negate any pre-existing immunosuppressive properties of these cells, they can be mixed with an adjuvant or genetically modified to increase immunogenicity [133]; this was done in the case of the GVAX vaccine [134], which is currently in Phase II clinical trials for pancreatic and prostate cancer. The preferred use of autologous tumor cells is due to the conferred patient specificity, although surgery is required to collect a sufficient sample. On the other hand, allogenic vaccines are typically easier and less costly to produce [132].

5.4.2 Dendritic Cell Vaccines

As carriers and presenters of antigens, DCs are vital in eliciting strong adaptive immune responses to cancer, and it is for this reason they are a focus of cancer vaccine development [135]. Similar to that of other cell-based interventions, DC vaccines involve ex vivo cell preparation. Typically, they are loaded with tumor-specific peptides, recombinant proteins, or mRNA followed by their induced maturation and patient infusion [136, 137]. The DC vaccine, Sipuleucel-T (Provenge®), for the personalized treatment of prostate cancer, is the first and only approved therapeutic cancer vaccine currently on the market.

The success of this treatment type depends on a number of factors, including the quality of the chosen antigen, and the DCs' propensity for activation, induction of adaptive immunity, and tumor infiltration [132].

6 Overall Conclusion

Without a doubt, 2D cell culture is as relevant and effective today as it ever was; still an essential and evolving competency in a fast-paced industry, this efficient, economical, and accessible discipline remains an invaluable resource in the fight against cancer. From conventional and basic monolayer cultures and functional assays to more recent and ever-advancing cell-based cancer interventions,

2D cell culture is critical for both diagnosing cancer and underpinning the prognosis.

Having said this, in more recent decades, many three-dimensional (3D) in vitro models have been developed, to overcome some of the common issues seen with 2D models and to act as a steppingstone between 2D models and in vivo animal models. These 3D cultures of cancer cells have become more common for better recapitulating in vivo-like cellular environments and are now praised as being more scientifically accurate in the preclinical assessment of new drug candidates, prior to moving into more expensive and time-consuming in vivo animal models.

These models will be discussed in more detail in Chap. 5: 3D spheroids for cancer research.

References

1. Hudu SA, Alshrari AS, Syahida A, Sekawi Z (2016) Cell culture, technology: enhancing the culture of diagnosing human diseases. J Clin Diagn Res 10(3):De01-5. https://doi.org/10.7860/jcdr/2016/15837.7460
2. Roux W (1887) Beiträge zur Entwickelungsmechanik des Embryo. Arch Mikrosk Anat 29(1):157–212
3. Yao T, Asayama Y (2017) Animal-cell culture media: history, characteristics, and current issues. Reprod Med Biol 16(2):99–117. https://doi.org/10.1002/rmb2.12024
4. Ljunggren C (1898) On the safe survival of skin epithelial cells outside of the human organism with special reference to skin transplantation. Nordiskt Medicinskt Arkiv 31:1–10
5. Maehle AH (2011) Ambiguous cells: the emergence of the stem cell concept in the nineteenth and twentieth centuries. Notes Rec R Soc Lond 65(4):359–378. https://doi.org/10.1098/rsnr.2011.0023
6. Harnden DG (1977) Cell biology and cell culture methods—a review. In: Harkness RA, Cockburn F (eds) The cultured cell and inherited metabolic disease: monograph based upon proceedings of the fourteenth symposium of the society for the study of inborn errors of metabolism. Springer, Dordrecht, pp 3–15
7. Fan W, Lin CS, Potluri P, Procaccio V, Wallace DC (2012) mtDNA lineage analysis of mouse L-cell lines reveals the accumulation of multiple mtDNA mutants and intermolecular recombination. Genes Dev 26(4):384–394. https://doi.org/10.1101/gad.175802.111
8. Philippeos C, Hughes RD, Dhawan A, Mitry RR (2012) Introduction to cell culture. Methods Mol Biol 806:1–13. https://doi.org/10.1007/978-1-61779-367-7_1
9. Langdon SP (2004) Characterization and authentication of cancer cell lines: an overview. Methods Mol Med 88:33–42. https://doi.org/10.1385/1-59259-406-9:33
10. Segeritz CP, Vallier L (2017) Cell culture: growing cells as model systems in vitro. Basic Sci Methods Clin Res:151–172. https://doi.org/10.1016/B978-0-12-803077-6.00009-6. Epub 2017 Apr 7
11. Richter M, Piwocka O, Musielak M, Piotrowski I, Suchorska WM, Trzeciak T (2021) From donor to the lab: a fascinating journey of primary cell lines. Front Cell Dev Biol:9. https://doi.org/10.3389/fcell.2021.711381
12. Kaur G, Dufour JM (2012) Cell lines: valuable tools or useless artifacts. Spermatogenesis 2(1):1–5. https://doi.org/10.4161/spmg.19885
13. PromoCell (2019) Human primary cells and immortal cell lines: differences and advantages. promocell.com: PromoCell
14. Gilgenkrantz S (2014) Sixty years of HeLa cell cultures. Hist Sci Med 48(1):139–144
15. Stelzer-Braid S, Walker GJ, Aggarwal A, Isaacs SR, Yeang M, Naing Z et al (2020) Virus isolation of severe acute respiratory syndrome coronavirus 2 (SARS-CoV-2) for diagnostic and research purposes. Pathology 52(7):760–763. https://doi.org/10.1016/j.pathol.2020.09.012
16. Landry JJ, Pyl PT, Rausch T, Zichner T, Tekkedil MM, Stütz AM et al (2013) The genomic and transcriptomic landscape of a HeLa cell line. G3 (Bethesda) 3(8):1213–1224. https://doi.org/10.1534/g3.113.005777

17. Lucey BP, Nelson-Rees WA, Hutchins GM (2009) Henrietta Lacks, HeLa cells, and cell culture contamination. Arch Pathol Lab Med 133(9):1463–1467. https://doi.org/10.5858/133.9.1463
18. Gillet JP, Varma S, Gottesman MM (2013) The clinical relevance of cancer cell lines. J Natl Cancer Inst 105(7):452–458. https://doi.org/10.1093/jnci/djt007
19. Wilding JL, Bodmer WF (2014) Cancer cell lines for drug discovery and development. Cancer Res 74(9):2377–2384. https://doi.org/10.1158/0008-5472.Can-13-2971
20. Almeida JL, Cole KD, Plant AL (2016) Standards for cell line authentication and beyond. PLoS Biol 14(6):e1002476. https://doi.org/10.1371/journal.pbio.1002476
21. Furlong MT, Hough CD, Sherman-Baust CA, Pizer ES, Morin PJ (1999) Evidence for the colonic origin of ovarian cancer cell line SW626. J Natl Cancer Inst 91(15):1327–1328. https://doi.org/10.1093/jnci/91.15.1327
22. Yang Y-HK, Ogando CR, Wang See C, Chang T-Y, Barabino GA (2018) Changes in phenotype and differentiation potential of human mesenchymal stem cells aging in vitro. Stem Cell Res Ther 9(1):131. https://doi.org/10.1186/s13287-018-0876-3
23. Kapalczynska M, Kolenda T, Przybyla W, Zajaczkowska M, Teresiak A, Filas V et al (2018) 2D and 3D cell cultures – a comparison of different types of cancer cell cultures. Arch Med Sci 14(4):910–919. https://doi.org/10.5114/aoms.2016.63743
24. O'Brien SJ (2001) Cell culture forensics. Proc Natl Acad Sci 98(14):7656–7658. https://doi.org/10.1073/pnas.141237598
25. Stacey GN (2000) Cell contamination leads to inaccurate data: we must take action now. Nature 403(6768):356. https://doi.org/10.1038/35000394
26. Rojas A, Gonzalez I (2018) Cell line cross-contamination: a detrimental issue in current biomedical research. Cell Biol Int 42(3):272. https://doi.org/10.1002/cbin.10904
27. Stacey GB, Hawkins E, Hawkins JR (1999) DNA fingerprinting and characterisation of animal cell lines. In: Portner R (ed) Animal cell biotechnology: methods and protocols. Methods in biotechnology. Springer Nature
28. Nelson-Rees WA, Flandermeyer RR, Hawthorne PK (1974) Banded marker chromosomes as indicators of intraspecies cellular contamination. Science (New York, NY) 184(4141):1093–1096. https://doi.org/10.1126/science.184.4141.1093
29. Nelson-Rees WA, Daniels DW, Flandermeyer RR (1981) Cross-contamination of cells in culture. Science (New York, NY) 212(4493):446–452. https://doi.org/10.1126/science.6451928
30. Nelson-Rees WA, Flandermeyer RR (1976) HeLa cultures defined. Science (New York, NY) 191(4222):96–98. https://doi.org/10.1126/science.1246601
31. Gartler SM (1968) Apparent HeLa cell contamination of human Heteroploid cell lines. Nature 217(5130):750–751. https://doi.org/10.1038/217750a0
32. MacLeod RA, Dirks WG, Matsuo Y, Kaufmann M, Milch H, Drexler HG (1999) Widespread intraspecies cross-contamination of human tumor cell lines arising at source. Int J Cancer 83(4):555–563. https://doi.org/10.1002/(sici)1097-0215(19991112)83:4<555::aid-ijc19>3.0.co;2-2
33. Kaplan J, Hukku B (1998) Chapter 11: Cell line characterization and authentication. In: Mather JP, Barnes D (eds) Methods in cell biology. Academic, pp 203–216
34. Cui C, Shu W, Li P (2016) Fluorescence in situ hybridization: cell-based genetic diagnostic and research applications. Front Cell Dev Biol 4:89. https://doi.org/10.3389/fcell.2016.00089
35. Kannan TP, Zilfalil BA (2009) Cytogenetics: past, present and future. Malays J Med Sci 16(2):4–9
36. Guo B, Han X, Wu Z, Da W, Zhu H (2014) Spectral karyotyping: an unique technique for the detection of complex genomic rearrangements in leukemia. Transl Pediatr 3(2):135–139. https://doi.org/10.3978/j.issn.2224-4336.2014.01.02
37. Szuhai K, Tanke HJ (2006) COBRA: combined binary ratio labeling of nucleic-acid probes for multi-color fluorescence in situ hybridization karyotyping. Nat Protoc 1(1):264–275. https://doi.org/10.1038/nprot.2006.41
38. Henegariu O, Heerema NA, Bray-Ward P, Ward DC (1999) Colour-changing karyotyping: an alternative to M-FISH/SKY. Nat Genet 23(3):263–264. https://doi.org/10.1038/15437
39. Jeffreys AJ, Wilson V, Thein SL (1985) Hypervariable 'minisatellite' regions in human DNA. Nature 314(6006):67–73. https://doi.org/10.1038/314067a0
40. Nicholson WL, Comer JA, Sumner JW, Gingrich-Baker C, Coughlin RT, Magnarelli LA et al (1997) An indirect immunofluorescence assay using a cell culture-derived

antigen for detection of antibodies to the agent of human granulocytic ehrlichiosis. J Clin Microbiol 35(6):1510–1516. https://doi.org/10.1128/jcm.35.6.1510-1516.1997

41. Shampo MA, Kyle RA, Kary B (2002) Mullis—Nobel laureate for procedure to replicate DNA. Mayo Clin Proc 77(7):606. https://doi.org/10.4065/77.7.606

42. O'Brien SU, Kleiner G, Olson R, Shannon JE (1977) Enzyme polymorphisms as genetic signatures in human cell cultures. Science (New York, NY) 195(4284):1345–1348. https://doi.org/10.1126/science.841332

43. Klaus GS, Dörthe G, Hans GD (1995) Isoenzyme analysis as a rapid method for the examination of the species identity of cell cultures. In Vitro Cell Dev Biol Anim 31(2):115–119

44. Nims RW, Shoemaker AP, Bauernschub MA, Rec LJ, Harbell JW (1998) Sensitivity of isoenzyme analysis for the detection of interspecies cell line cross-contamination. In Vitro Cell Dev Biol Anim 34(1):35–39. https://doi.org/10.1007/s11626-998-0050-9

45. Fernandes IR, Russo FB, Pignatari GC, Evangelinellis MM, Tavolari S, Muotri AR et al (2016) Fibroblast sources: where can we get them? Cytotechnology 68(2):223–228. https://doi.org/10.1007/s10616-014-9771-7

46. Stacey GN (2011) Cell culture contamination. Methods Mol Biol 731:79–91. https://doi.org/10.1007/978-1-61779-080-5_7

47. Vignon C, Debeissat C, Georget M-T, Bouscary D, Gyan E, Rosset P et al (2013) Flow cytometric quantification of all phases of the cell cycle and apoptosis in a two-color fluorescence plot. PLoS One 8(7):e68425. https://doi.org/10.1371/journal.pone.0068425

48. Pekle E, Smith A, Rosignoli G, Sellick C, Smales CM, Pearce C (2019) Application of imaging flow cytometry for the characterization of intracellular attributes in Chinese hamster ovary cell lines at the single-cell level. Biotechnol J 14(7):e1800675. https://doi.org/10.1002/biot.201800675

49. Puts JJ, Vooijs GP, Huysmans A, van Aspert A, Ramaekers FC (1986) Cytoskeletal proteins as tissue-specific markers in cytopathology. Exp Cell Biol 54(2):73–79. https://doi.org/10.1159/000163346

50. Jedlitschky G, Hoffmann U, Kroemer HK (2006) Structure and function of the MRP2 (ABCC2) protein and its role in drug disposition. Expert Opin Drug Metab Toxicol 2(3):351–366. https://doi.org/10.1517/17425255.2.3.351

51. Pawlak KJ, Wiebe JP (2007) Regulation of estrogen receptor (ER) levels in MCF-7 cells by progesterone metabolites. J Steroid Biochem Mol Biol 107(3–5):172–179. https://doi.org/10.1016/j.jsbmb.2007.05.030

52. Cohen JS, Benton AS, Nwachukwu F, Ozedirne T, Teach SJ, Freishtat RJ (2010) P-glycoprotein transporter expression on a549 respiratory epithelial cells is positively correlated with intracellular dexamethasone levels. J Investig Med 58(8):991–994

53. Guillouzo A, Corlu A, Aninat C, Glaise D, Morel F, Guguen-Guillouzo C (2007) The human hepatoma HepaRG cells: a highly differentiated model for studies of liver metabolism and toxicity of xenobiotics. Chem Biol Interact 168(1):66–73. https://doi.org/10.1016/j.cbi.2006.12.003

54. Bandele OJ, Santillo MF, Ferguson M, Wiesenfeld PL (2012) In vitro toxicity screening of chemical mixtures using HepG2/C3A cells. Food Chem Toxicol 50(5):1653–1659. https://doi.org/10.1016/j.fct.2012.02.016

55. Mossoba ME, Sprando RL (2020) In vitro to in vivo concordance of toxicity using the human proximal tubule cell line HK-2. Int J Toxicol 39(5):452–464. https://doi.org/10.1177/1091581820942534

56. Swain RJ, Kemp SJ, Goldstraw P, Tetley TD, Stevens MM (2010) Assessment of cell line models of primary human cells by Raman spectral phenotyping. Biophys J 98(8):1703–1711. https://doi.org/10.1016/j.bpj.2009.12.4289

57. Fisher T (2022) Gibco cell culture basics: cell lines. Accessed 20 Apr 2022

58. Vaidyanathan R, Soon RH, Zhang P, Jiang K, Lim CT (2018) Cancer diagnosis: from tumor to liquid biopsy and beyond. Lab Chip 19(1):11–34. https://doi.org/10.1039/c8lc00684a

59. Lone SN, Nisar S, Masoodi T, Singh M, Rizwan A, Hashem S et al (2022) Liquid biopsy: a step closer to transform diagnosis, prognosis and future of cancer treatments. Mol Cancer 21(1):79. https://doi.org/10.1186/s12943-022-01543-7

60. Maly V, Maly O, Kolostova K, Bobek V (2019) Circulating tumor cells in diagnosis and treatment of lung cancer. In Vivo 33(4):1027–1037. https://doi.org/10.21873/invivo.11571

61. de Wit S, Zeune LL, Hiltermann TJN, Groen HJM, Dalum GV, Terstappen L (2018) Classification of cells in CTC-enriched samples by advanced image analysis. Cancers (Basel) 10(10). https://doi.org/10.3390/cancers10100377

62. Spell DW, Jones DV Jr, Harper WF, David BJ (2004) The value of a complete blood count in predicting cancer of the colon. Cancer Detect Prev 28(1):37–42. https://doi.org/10.1016/j.cdp.2003.10.002
63. Jaime-Pérez JC, García-Arellano G, Herrera-Garza JL, Marfil-Rivera LJ, Gómez-Almaguer D (2019) Revisiting the complete blood count and clinical findings at diagnosis of childhood acute lymphoblastic leukemia: 10-year experience at a single center. Hematol Transfus Cell Ther 41(1):57–61. https://doi.org/10.1016/j.htct.2018.05.010
64. Liu Y, Xu J (2019) High-resolution microscopy for imaging cancer pathobiology. Curr Pathobiol Rep 7(3):85–96. https://doi.org/10.1007/s40139-019-00201-w
65. Kumar R, Srivastava R, Srivastava S (2015) Detection and classification of cancer from microscopic biopsy images using clinically significant and biologically interpretable features. J Med Eng 2015:457906. https://doi.org/10.1155/2015/457906
66. Gurina TS, Simms L (2022) Histology, Staining. StatPearls. Treasure Island (FL): StatPearls Publishing Copyright © 2022, StatPearls Publishing LLC
67. Amedee RG, Dhurandhar NR (2001) Fine-needle aspiration biopsy. Laryngoscope 111(9):1551–1557. https://doi.org/10.1097/00005537-200109000-00011
68. Bilous M (2010) Breast core needle biopsy: issues and controversies. Mod Pathol 23 (Suppl 2):S36–S45. https://doi.org/10.1038/modpathol.2010.34
69. Beard CJ, Ponnarasu S, Schmieder GJ (2022) Excisional Biopsy. StatPearls Publishing Copyright © 2022, StatPearls Publishing LLC
70. Pflugfelder A, Weide B, Eigentler TK, Forschner A, Leiter U, Held L et al (2010) Incisional biopsy and melanoma prognosis: facts and controversies. Clin Dermatol 28(3):316–318. https://doi.org/10.1016/j.clindermatol.2009.06.013
71. Lin X, Lin R, Lu F, Yang Y, Wang C, Fang H et al (2021) Laparoscopic biopsy and staging for locally advanced pancreatic cancer: experiences of 76 consecutive patients in a single institution. Langenbeck's Arch Surg 406(7): 2315–2323. https://doi.org/10.1007/s00423-021-02199-5
72. Reid BJ, Weinstein WM, Lewin KJ, Haggitt RC, VanDeventer G, DenBesten L et al (1988) Endoscopic biopsy can detect high-grade dysplasia or early adenocarcinoma in Barrett's esophagus without grossly recognizable neoplastic lesions. Gastroenterology 94(1):81–90. https://doi.org/10.1016/0016-5085(88)90613-0
73. Nischal U, Nischal K, Khopkar U (2008) Techniques of skin biopsy and practical considerations. J Cutan Aesthet Surg 1(2): 107–111. https://doi.org/10.4103/0974-2077.44174
74. Ayre JE (1947) Selective cytology smear for diagnosis of cancer. Am J Obstet Gynecol 53(4):609–617. https://doi.org/10.1016/0002-9378(47)90278-0
75. Bal MS, Goyal R, Suri AK, Mohi MK (2012) Detection of abnormal cervical cytology in Papanicolaou smears. J Cytol 29(1):45–47. https://doi.org/10.4103/0970-9371.93222
76. Karam R, Conner B, LaDuca H, McGoldrick K, Krempely K, Richardson ME et al (2019) Assessment of diagnostic outcomes of RNA genetic testing for hereditary cancer. JAMA Netw Open 2(10):e1913900. https://doi.org/10.1001/jamanetworkopen.2019.13900
77. Carethers JM (2014) DNA testing and molecular screening for colon cancer. Clin Gastroenterol Hepatol 12(3):377–381. https://doi.org/10.1016/j.cgh.2013.12.007
78. Greene MH (1997) Genetics of breast cancer. Mayo Clin Proc 72(1):54–65. https://doi.org/10.4065/72.1.54
79. Bouchard L, Blancquaert I, Eisinger F, Foulkes WD, Evans G, Sobol H et al (2004) Prevention and genetic testing for breast cancer: variations in medical decisions. Soc Sci Med 58(6):1085–1096. https://doi.org/10.1016/s0277-9536(03)00263-6
80. Fernandez-Rozadilla C, Simões AR, Lleonart ME, Carnero A, Carracedo Á (2021) Tumor profiling at the service of cancer therapy. Front Oncol 10:595613. https://doi.org/10.3389/fonc.2020.595613
81. Chakravarty D, Solit DB (2021) Clinical cancer genomic profiling. Nat Rev Genet 22(8): 483–501. https://doi.org/10.1038/s41576-021-00338-8
82. Harnan S, Tappenden P, Cooper K, Stevens J, Bessey A, Rafia R et al (2019) Tumour profiling tests to guide adjuvant chemotherapy decisions in early breast cancer: a systematic review and economic analysis. Health Technol Assess 23(30):1–328. https://doi.org/10.3310/hta23300
83. Kitaeva KV, Rutland CS, Rizvanov AA, Solovyeva VV (2020) Cell culture based in vitro test systems for anticancer drug screening. Front Bioeng Biotechnol 8:322. https://doi.org/10.3389/fbioe.2020.00322

84. Mingaleeva RN, Solovieva VV, Blatt NL, Rizvanov AA (2013) Application of cell and tissue cultures for potential anti-cancer/oncology drugs screening in vitro. Cell Transplant Tissue Eng 8(2):20–28

85. Borenfreund E, Babich H, Martin-Alguacil N (1988) Comparisons of two in vitro cytotoxicity assays—the neutral red (NR) and tetrazolium MTT tests. Toxicol In Vitro 2(1):1–6. https://doi.org/10.1016/0887-2333(88)90030-6

86. Franken NA, Rodermond HM, Stap J, Haveman J, van Bree C (2006) Clonogenic assay of cells in vitro. Nat Protoc 1(5): 2315–2319. https://doi.org/10.1038/nprot.2006.339

87. Gutiérrez L, Stepien G, Gutiérrez L, Pérez-Hernández M, Pardo J, Pardo J et al (2017) 1.09 – Nanotechnology in drug discovery and development. In: Chackalamannil S, Rotella D, Ward SE (eds) Comprehensive medicinal chemistry III. Elsevier, Oxford, pp 264–295

88. Kamiloglu S, Sari G, Ozdal T, Capanoglu E (2020) Guidelines for cell viability assays. Food Front 1(3):332–349. https://doi.org/10.1002/fft2.44

89. Kumar P, Nagarajan A, Uchil PD. Analysis of cell viability by the MTT assay. Cold Spring Harb Protoc. 2018;2018(6): https://doi.org/10.1101/pdb.prot095505

90. Mueller H, Kassack MU, Wiese M (2004) Comparison of the usefulness of the MTT, ATP, and Calcein assays to predict the potency of cytotoxic agents in various human cancer cell lines. J Biomol Screen 9(6):506–515. https://doi.org/10.1177/1087057104265386

91. Kumar P, Nagarajan A, Uchil PD. Analysis of cell viability by the lactate dehydrogenase assay. Cold Spring Harb Protoc. 2018;2018 (6). doi: https://doi.org/10.1101/pdb.prot095497

92. Al-Nasiry S, Geusens N, Hanssens M, Luyten C, Pijnenborg R (2007) The use of Alamar Blue assay for quantitative analysis of viability, migration and invasion of choriocarcinoma cells. Hum Reprod 22(5): 1304–1309. https://doi.org/10.1093/humrep/dem011

93. Gong X, Liang Z, Yang Y, Liu H, Ji J, Fan Y (2020) A resazurin-based, nondestructive assay for monitoring cell proliferation during a scaffold-based 3D culture process. Regen Biomater 7(3):271–281. https://doi.org/10.1093/rb/rbaa002

94. Nowak E, Kammerer S, Küpper JH (2018) ATP-based cell viability assay is superior to trypan blue exclusion and XTT assay in measuring cytotoxicity of anticancer drugs Taxol and Imatinib, and proteasome inhibitor MG-132 on human hepatoma cell line HepG2. Clin Hemorheol Microcirc 69(1–2): 327–336. https://doi.org/10.3233/ch-189120

95. Zanoni M, Piccinini F, Arienti C, Zamagni A, Santi S, Polico R et al (2016) 3D tumor spheroid models for in vitro therapeutic screening: a systematic approach to enhance the biological relevance of data obtained. Sci Rep 6(1):19103. https://doi.org/10.1038/srep19103

96. Ahmann FR, Garewal HS, Schifman R, Celniker A, Rodney S (1987) Intracellular adenosine triphosphate as a measure of human tumor cell viability and drug modulated growth. In Vitro Cell Dev Biol 23(7): 474–480. https://doi.org/10.1007/bf02628417

97. Sharma S, Neale MH, Di Nicolantonio F, Knight LA, Whitehouse PA, Mercer SJ et al (2003) Outcome of ATP-based tumor chemosensitivity assay directed chemotherapy in heavily pre-treated recurrent ovarian carcinoma. BMC Cancer 3:19. https://doi.org/10.1186/1471-2407-3-19

98. Bancone G, Kalnoky M, Chu CS, Chowwiwat N, Kahn M, Malleret B et al (2017) The G6PD flow-cytometric assay is a reliable tool for diagnosis of G6PD deficiency in women and anaemic subjects. Sci Rep 7(1): 9822. https://doi.org/10.1038/s41598-017-10045-2

99. Strober W (2015) Trypan blue exclusion test of cell viability. Curr Protoc Immunol 111: A3.B.1-a3.B. https://doi.org/10.1002/0471142735.ima03bs111

100. Justus CR, Leffler N, Ruiz-Echevarria M, Yang LV (2014) In vitro cell migration and invasion assays. J Vis Exp 88:51046. https://doi.org/10.3791/51046

101. Pijuan J, Barceló C, Moreno DF, Maiques O, Sisó P, Marti RM et al (2019) In vitro cell migration, invasion, and adhesion assays: from cell imaging to data analysis. Front Cell Dev Biol 7:107. https://doi.org/10.3389/fcell.2019.00107

102. Weitz-Schmidt G, Chreng S (2012) Cell adhesion assays. Methods Mol Biol 757:15–30. https://doi.org/10.1007/978-1-61779-166-6_2

103. Kucik DF, Wu C (2005) Cell-adhesion assays. Methods Mol Biol 294:43–54. https://doi.org/10.1385/1-59259-860-9:043

104. Chen HC (2005) Boyden chamber assay. Methods Mol Biol 294:15–22. https://doi.org/10.1385/1-59259-860-9:015

105. Humphries MJ (2009) Cell adhesion assays. Methods Mol Biol 522:203–210. https://doi.org/10.1007/978-1-59745-413-1_14
106. Liang C-C, Park AY, Guan J-L (2007) In vitro scratch assay: a convenient and inexpensive method for analysis of cell migration in vitro. Nat Protoc 2(2):329–333. https://doi.org/10.1038/nprot.2007.30
107. Martinotti S, Ranzato E (2020) Scratch wound healing assay. Methods Mol Biol 2109:225–229. https://doi.org/10.1007/7651_2019_259
108. Yoo SY, Kwon SM (2013) Angiogenesis and its therapeutic opportunities. Mediat Inflamm 2013:127170. https://doi.org/10.1155/2013/127170
109. Aguilar-Cazares D, Chavez-Dominguez R, Carlos-Reyes A, Lopez-Camarillo C, Hernadez de la Cruz ON, Lopez-Gonzalez JS (2019) Contribution of angiogenesis to inflammation and cancer. Front Oncol 9: 1399. https://doi.org/10.3389/fonc.2019.01399
110. PhD SR (2015) What is precision medicine. Accessed 25 Feb 2015
111. Waldman AD, Fritz JM, Lenardo MJ (2020) A guide to cancer immunotherapy: from T cell basic science to clinical practice. Nat Rev Immunol 20(11):651–668. https://doi.org/10.1038/s41577-020-0306-5
112. Themeli M, Kloss CC, Ciriello G, Fedorov VD, Perna F, Gonen M et al (2013) Generation of tumor-targeted human T lymphocytes from induced pluripotent stem cells for cancer therapy. Nat Biotechnol 31(10):928–933. https://doi.org/10.1038/nbt.2678
113. Copier J, Dalgleish A (2006) Overview of tumor cell-based vaccines. Int Rev Immunol 25(5–6):297–319. https://doi.org/10.1080/08830180600992472
114. Motz GT, Coukos G (2011) The parallel lives of angiogenesis and immunosuppression: cancer and other tales. Nat Rev Immunol 11(10):702–711. https://doi.org/10.1038/nri3064
115. Togashi Y, Shitara K, Nishikawa H (2019) Regulatory T cells in cancer immunosuppression – implications for anticancer therapy. Nat Rev Clin Oncol 16(6): 356–371. https://doi.org/10.1038/s41571-019-0175-7
116. Kalos M, June CH (2013) Adoptive T cell transfer for cancer immunotherapy in the era of synthetic biology. Immunity 39(1):49–60. https://doi.org/10.1016/j.immuni.2013.07.002
117. Rosenberg SA, Restifo NP (2015) Adoptive cell transfer as personalized immunotherapy for human cancer. Science 348(6230): 62–68. https://doi.org/10.1126/science.aaa4967
118. Dudley ME, Wunderlich JR, Shelton TE, Even J, Rosenberg SA (2003) Generation of tumor-infiltrating lymphocyte cultures for use in adoptive transfer therapy for melanoma patients. J Immunother 26(4):332–342. https://doi.org/10.1097/00002371-200307000-00005
119. Rosenberg SA, Restifo NP, Yang JC, Morgan RA, Dudley ME (2008) Adoptive cell transfer: a clinical path to effective cancer immunotherapy. Nat Rev Cancer 8(4):299–308. https://doi.org/10.1038/nrc2355
120. June CH, O'Connor RS, Kawalekar OU, Ghassemi S, Milone MC (2018) CAR T cell immunotherapy for human cancer. Science 359(6382):1361–1365. https://doi.org/10.1126/science.aar6711
121. Kakarla S, Gottschalk S (2014) CAR T cells for solid tumors: armed and ready to go? Cancer J 20(2):151–155. https://doi.org/10.1097/PPO.0000000000000032
122. Sterner RC, Sterner RM (2021) CAR-T cell therapy: current limitations and potential strategies. Blood Cancer J 11(4):69. https://doi.org/10.1038/s41408-021-00459-7
123. Jan M, Scarfò I, Larson RC, Walker A, Schmidts A, Guirguis AA et al (2021) Reversible ON- and OFF-switch chimeric antigen receptors controlled by lenalidomide. Sci Transl Med 13(575). https://doi.org/10.1126/scitranslmed.abb6295
124. ClinicalTrials.gov. CAR-Ts for cancer therapy, recruiting and not recruiting 2022 [updated 06 June 2022. 01 June:[Available from: https://clinicaltrials.gov/ct2/results?term=CAR-T&cond=cancer&Search=Apply&recrs=a&recrs=d&age_v=&gndr=&type=&rslt=
125. Yee C, Thompson JA, Byrd D, Riddell SR, Roche P, Celis E et al (2002) Adoptive T cell therapy using antigen-specific CD8+ T cell clones for the treatment of patients with metastatic melanoma: in vivo persistence, migration, and antitumor effect of transferred T cells. Proc Natl Acad Sci U S A 99(25): 16168–16173. https://doi.org/10.1073/pnas.242600099
126. Kim JJ (2015) Applications of iPSCs in cancer research. Biomark Insights 10(Suppl 1): 125–131. https://doi.org/10.4137/bmi.S20065

127. Zamai L, Ponti C, Mirandola P, Gobbi G, Papa S, Galeotti L et al (2007) NK cells and cancer. J Immunol 178(7):4011–4016. https://doi.org/10.4049/jimmunol.178.7.4011
128. ClinicalTrials.gov. CYNK-001, Phase II, GBM: ClinicalTrials.gov; 2022 [updated 02 June 2022. 02/06/2022:[Available from: https://clinicaltrials.gov/ct2/show/NCT05218408?term=cynk-001&phase=1&draw=2&rank=1
129. ClinicalTrials.gov (2022) OnKord, Phase II clinical trials, AML. ClinicalTrials.gov
130. Liu S, Galat V, Galat Y, Lee YKA, Wainwright D, Wu J (2021) NK cell-based cancer immunotherapy: from basic biology to clinical development. J Hematol Oncol 14(1):7. https://doi.org/10.1186/s13045-020-01014-w
131. Bald T, Krummel MF, Smyth MJ, Barry KC (2020) The NK cell-cancer cycle: advances and new challenges in NK cell-based immunotherapies. Nat Immunol 21(8):835–847. https://doi.org/10.1038/s41590-020-0728-z
132. Srivatsan S, Patel JM, Bozeman EN, Imasuen IE, He S, Daniels D et al (2014) Allogeneic tumor cell vaccines: the promise and limitations in clinical trials. Hum Vaccin Immunother 10(1):52–63. https://doi.org/10.4161/hv.26568
133. Andersen BM, Ohlfest JR (2012) Increasing the efficacy of tumor cell vaccines by enhancing cross priming. Cancer Lett 325(2):155–164. https://doi.org/10.1016/j.canlet.2012.07.012
134. Simons JW, Sacks N (2006) Granulocyte-macrophage colony-stimulating factor-transduced allogeneic cancer cellular immunotherapy: the GVAX vaccine for prostate cancer. Urol Oncol 24(5):419–424. https://doi.org/10.1016/j.urolonc.2005.08.021
135. Santos PM, Butterfield LH (2018) Dendritic cell-based cancer vaccines. J Immunol 200(2):443–449. https://doi.org/10.4049/jimmunol.1701024
136. Saxena M, van der Burg SH, Melief CJM, Bhardwaj N (2021) Therapeutic cancer vaccines. Nat Rev Cancer 21(6):360–378. https://doi.org/10.1038/s41568-021-00346-0
137. Blass E, Ott PA (2021) Advances in the development of personalized neoantigen-based therapeutic cancer vaccines. Nat Rev Clin Oncol 18(4):215–229. https://doi.org/10.1038/s41571-020-00460-2
138. Ganzinger M, He S, Breuhahn K, Knaup P (2012) On the ontology based representation of cell lines. PLoS One 7:e48584. https://doi.org/10.1371/journal.pone.0048584
139. Inc. CB (2014) Cell adhesion, migration and invasion assays. Accessed 09 June 2022
140. Lin SL, Chang DC, Chang-Lin S, Lin CH, Wu DT, Chen DT et al (2008) Mir-302 reprograms human skin cancer cells into a pluripotent ES-cell-like state. RNA 14(10):2115–2124. https://doi.org/10.1261/rna.1162708
141. Miyoshi N, Ishii H, Nagai K, Hoshino H, Mimori K, Tanaka F et al (2010) Defined factors induce reprogramming of gastrointestinal cancer cells. Proc Natl Acad Sci U S A 107(1):40–45. https://doi.org/10.1073/pnas.0912407107
142. Mathieu J, Zhang Z, Zhou W, Wang AJ, Heddleston JM, Pinna CM et al (2011) HIF induces human embryonic stem cell markers in cancer cells. Cancer Res 71(13):4640–4652. https://doi.org/10.1158/0008-5472.Can-10-3320
143. Corominas-Faja B, Cufí S, Oliveras-Ferraros-C, Cuyàs E, López-Bonet E, Lupu R et al (2013) Nuclear reprogramming of luminal-like breast cancer cells generates Sox2-overexpressing cancer stem-like cellular states harboring transcriptional activation of the mTOR pathway. Cell Cycle 12(18):3109–3124. https://doi.org/10.4161/cc.26173
144. Kim J, Hoffman JP, Alpaugh RK, Rhim AD, Reichert M, Stanger BZ et al (2013) An iPSC line from human pancreatic ductal adenocarcinoma undergoes early to invasive stages of pancreatic cancer progression. Cell Rep 3(6):2088–2099. https://doi.org/10.1016/j.celrep.2013.05.036

Chapter 2

Cell Cultures at the Air–Liquid Interface and Their Application in Cancer Research

Luisana Di Cristo and Stefania Sabella

Abstract

Air-liquid interface (ALI) cell cultures are considered a valid tool for the replacement of animals in biomedical research. By mimicking crucial features of the human in vivo epithelial barriers (e.g., lung, intestine, and skin), ALI cell cultures enable proper structural architectures and differentiated functions of normal and diseased tissue barriers. Thereby, ALI models realistically resemble tissue conditions and provide in vivo-like responses. Since their implementation, they are routinely used in several applications, from toxicity testing to cancer research, receiving an appreciable level of acceptance (in some cases a regulatory acceptance) as attractive testing alternatives to animals. In this chapter, an overview of the ALI cell cultures will be presented together with their application in cancer cell culture, highlighting the potential advantages and disadvantages of the model.

Key words Alternative in vitro methods, 3R principle, ALI airway model, ALI intestinal model, ALI skin model, toxicity testing, cancer research

1 Introduction

With the term "air-liquid interface (ALI) cell culture," we refer to a cell culture system where cells seeded on a porous filter membrane within an insert (generally a Transwell® permeable support) are exposed to air in their apical side, while cell culture medium is added into the basolateral compartment providing nutrients to cells via diffusion through the porous membrane [1–5]. The cells at the "interface" with the air differentiate and generate a polarized tissue microenvironment.

The ALI culturing approach has been developed to overcome the limitation posed by the submerged cultures, in which the presence of the liquid at the apical side of the cells does not resemble the real conditions of the mimicked epithelial tissue. Indeed, the relevance of the ALI culture system lies in their potential to mimic more closely in vivo situation of epithelial cells

Dania Movia and Adriele Prina-Mello (eds.), *Cancer Cell Culture: Methods and Protocols*,
Methods in Molecular Biology, vol. 2645, https://doi.org/10.1007/978-1-0716-3056-3_2,

© The Author(s), under exclusive license to Springer Science+Business Media, LLC, part of Springer Nature 2023

constantly exposed to gas, such as the respiratory [6–8] and gastric tracts [9, 10], the skin [11–13], and the cornea [14, 15]. Interestingly, ALI cell cultures also prove to be useful to support the functionality of epithelial cells that in vivo do not necessarily have apical access to ambient air, such as epithelia originating from the intestinal [16–18] and female reproductive tract (FRT) mucosa [19, 20]. Moreover, the ALI cell cultures are considered a tissue-mimetic model as they allow long-term culturing by reducing oxygen stress, allowing for a healthy and differentiated cell monolayer [21].

The development of the ALI cell cultures starts with a submerged culture in a Transwell® insert, as the simple culture of cells in a plastic dish does not generate enough cells for establishing such type of cell culture [5]. Indeed, the cells expand submerged and, after reaching the confluence, the medium at the apical compartment is removed to establish the ALI conditions. Cell differentiation then occurs. This process takes approximately 3 weeks regardless of the cell type. In general, the differentiation consists of several changes: morphological (from irregular and squamous cells to polarized cells) and proliferation (from rapidly to limited proliferation activity) changes, ciliation (from cells that lack cilia to ciliated cells), mucus secretion (from cells that do not secrete mucus to mucus-secreting cells), and barrier formation (tight barrier carrying out basal-apical transport, secretion, and defense functions) [4].

The ALI system was developed for the first time in 1980 to model epidermal and respiratory tract tissue in vitro [4]. Since then, the research interest in such culture systems has grown a lot, and nowadays, they are routinely used and considered a valid model of the in vivo situation in the light of reducing the numbers of animals used in experimental toxicity assessment (e.g., xenobiotics, nanoparticles), cancer, and biomedical (e.g., wound healing) fields [4, 5]. Indeed, the ALI system is considered a feasible approach to implement the "3R Principle," which refers to a replacement, reduction, and refinement of animal use in research.

In this regard, there are many regulatory efforts to reduce, refine, and replace the use of animals and therefore advance the alternative approaches using in vitro and in silico models [5, 22]. The United States, Europe, and other groups of countries have formally established the Interagency Coordinating Committee on the Validation of Alternative Methods (ICCVAM) in 2000, the European Partnership for Alternative Approaches to Animal Testing (EPAA) in 2005, and the International Cooperation on Alternative Test Methods (ICATM) in 2009, respectively, to promote a full replacement of the current in vivo tests [22]. Furthermore, in Europe, the Directive 2010/63/EU (EU, 2010) explicitly incorporates the 3R concept in European legislation and establishes the European Union Reference Laboratory-European

Centre for the Validation of Alternative Methods (EURL-ECVAM) at the Joint Research Centre (JRC) as support to the development, validation, and acceptance of alternative methods [22]. In their last available report of 2020, the EURL-ECVAM shows how the development of non-animal methods and approaches has gained ground under the EU Framework Programme for Research and Innovation, Horizon 2020, and in collaborative partnerships [23]. For instance, new approaches have been evaluated and validated in the area of skin and respiratory sensitization, genotoxicity, fish toxicity, and lastly for the quality control of vaccines. The EURL-ECVAM recently opens to EU-funded research projects in the biomedical area and completed the reviews on non-animal models in different disease areas (e.g., respiratory diseases and breast cancer). However, the statistics says that the use of animals is still very high that it is becoming important to invest efforts in promoting the uptake of non-animal methods and models in the research area by establish education and training activity to increase the awareness of the 3R concept. Notably, in September 2021, the European Parliament votes for an EU-wide plan to phase out animal testing by 2035. This resolution is a strong statement as, in European laboratories, nearly ten million animals are sacrificed every year. Although such incredible effort, we are still far from the mission as in vitro regulator-approved methods are poor. The validation process of such alternatives is extremely demanding in terms of efforts and costs as it involves the use of in vivo animals to demonstrate their ability to predict the in vivo environment. However, at the same time, the use of animal data for validating human data is also questionable as interspecies differences exist [22].

This chapter aims to give an overview on the current utilization of the ALI approaches as an alternative in vitro method to animal testing, exploring cell culture models (e.g., lung, intestine, and skin), and several areas of applications (e.g., toxicity testing, cancer research, drug testing, host-pathogen interactions, and wound healing) (Fig. 1). Finally, the challenges that need to be overcome for ALI cultures will be presented. The overview on the available ALI approaches was produced analyzing the public scientific literature using online resources such as PubMed and Google scholar databases. Dates of research cover the period from September to December 2021.

2 ALI Cell Culture Models: State of the Art

ALI cell culture models are widespread across the published literature. In this paragraph, ALI models of lung, intestine, and skin epithelial barriers will be presented in more detail, as they have found valid applications in basic research, industrial application, and regulatory toxicology as relevant alternative models to animals.

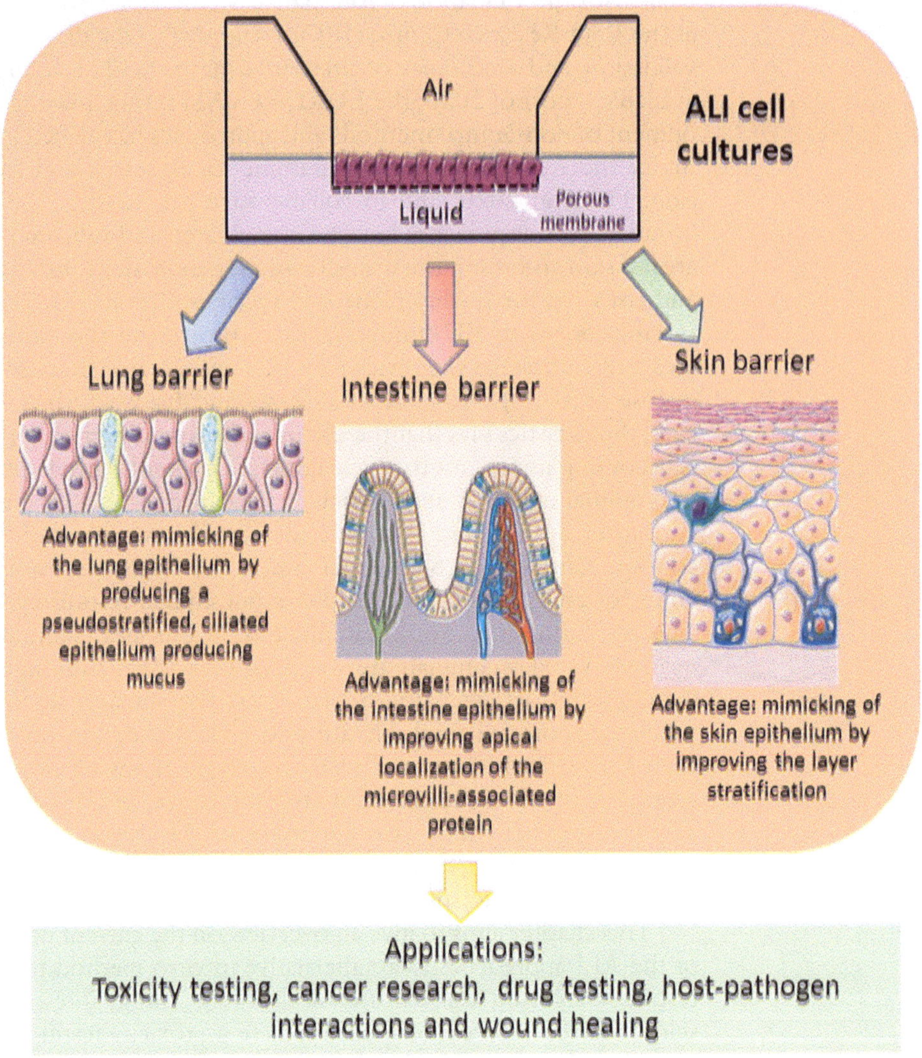

Fig. 1 Schematic representation of the air-liquid interface (ALI) tissue models of lung, intestine, and skin barriers. Examples of ALI cell cultures developed during the last years. The ALI condition enables growing cells in direct contact with the gas phase on the apical side, while the basolateral side is supplied with cell culture medium. Upon differentiation at ALI conditions, epithelial cells resemble some in vivo features of lung, intestine, and skin barriers that make them suitable for several biomedical applications, including cancer research

ALI cell culture models are available also for mimicking the physiological environment of other human tissues, such as the cornea [15, 24–26], the female reproductive [4, 20], the spleen [27, 28], and brain tissues [29, 30]. Notable, among these models, there is the commercialized EpiOcular™ model from MatTek, a reconstituted human corneal epithelium routinely used in regulatory settings. Indeed, the EpiOcular™ Eye Irritation Test (EIT) was validated by EURL ECVAM (EC, 2014a, EC, 2014b) between

2010 and 2013 and then adopted in 2015 by the OECD as Test Guideline (TG) 492 to become a partial replacement of the regulatory accepted in vivo rabbit eye test.

2.1 Lung Barrier

Lungs belong to the respiratory system (from the nasal passages to the alveoli), which provide oxygen to the tissues for energy production as the main function but also act as a barrier to protect the body from airborne pathogens, air pollution, and exposure to toxic chemical substances derived from occupational exposure, household chemicals, and consumer products [31, 32]. The lung epithelium covers the largest body surface (ca. 80 m^2) that includes basal, goblet, and ciliated epithelial cells. From the trachea down to the bronchioles, the lower respiratory tract consists of ciliated epithelium. Here, mucus on the surface, produced by the goblet cells, forms a protective coating to trap inhaled foreign particles facilitating the clearance by the mucociliary apparatus [33]. The defense system of the airway barrier presents also the functionality of innate immune regulation by the production of cytokines together with the adaptive immune system by resident leukocytes [31, 32, 34]. Moreover, the apical side of the epithelial cells is characterized by the presence of tight junctions (TJs) that form a tight barrier between the apical and the basolateral cell compartments that selectively regulate the ion's passage through the paracellular pathway [35, 36].

Gas exchange takes place in the alveoli region. Here the interalveolar septum separates the air compartment from the blood (capillary lumen). This tissue barrier, called the air-blood barrier, is composed of two types of cell layers, the epithelium facing the alveolar lumen and an endothelium facing the capillary lumen. The alveolar epithelium, in contrast to the upper epithelium, is not ciliated, but composed of many squamous type I alveolar epithelial cells with lining functions interspersed with less more cuboidal type II alveolar epithelial cells, which are the surfactant secreting cells, that also provide renewal and repair [37].

The first ALI cell culture model ever developed replicated the airway tissue. Indeed, over the years, the ALI models have gained increasing visibility in respiratory research, as the most relevant in vitro system to use for resembling the lung environment. ALI cell cultures can mimic the main features of the lung epithelium, by producing a peculiar environment where the epithelial cells can proliferate and generate a pseudostratified, ciliated epithelium-producing mucus [38].

To date, numerous methods using different cell cultures (single cultures or co-cultures), different media, and different experimental setups have been developed for establishing the ALI models for the lung tissue, and they are reviewed in different works [5, 32, 39–41]. As by now, many carcinogenic and normal cell types have been investigated in combination with ALI conditions. These include the bronchial-like BEAS-2B, 16HBE14o- and Calu-3 cells, and the alveolar-like A549 and NCI-H441 cells [39]. The Calu-3 cell line (derived by human lung adenocarcinoma) is mostly used for adsorption studies, as this cell type offers the advantage of mucus production and TJs formation [42–47]. On the other side, the BEAS-2B cells have the advantage of deriving from normal human epithelium, and they are therefore used to study airway functionality [7, 39, 48, 49]. However, BEAS-2B cells lack TJs [50]. As opposed to BEAS-2B cells, the 16HBE14o- cell line (also derived from normal human epithelium) forms TJs [39, 51]. The A549 and NCI-H441 cell lines are instead used to develop the alveolar airway barrier. While the most used A549 cells (mimicking the alveolar type II cells) do not form TJs but produce surfactant proteins and they are regularly used in toxicity testing [1–3, 8, 52–55], the NCI-H441 cells form the TJs and are used for transport studies as they express organic cation transporters [39, 56].

To increase the model complexity, ALI multilayered co/triple/tetra-cultures can be used. The advantages of using multiple cells line together lie in creating an intercellular communication that enhances the physiological relevance of the models. The cell types commonly used to create such ALI multilayers are endothelial cells (e.g., EA.hy 926) [5, 39, 57–59], fibroblasts (e.g., MRC-5 cells) [2, 3, 60], dendritic cells (derived from human peripheral blood monocytes) [5, 39, 61, 62], and mast cells (e.g., HMC-1) [5, 39, 57, 59]. These cells are usually implemented to mimic the air-blood barrier together with the alveolar A549 cell lines. Macrophage cells, mostly the human leukemia monocytic cell line differentiated in macrophages (THP-1 cells) or macrophages derived from human peripheral blood monocytes, can also be added to the multilayer cultures to reproduce the residence immune cells for foreign particles clearance function [5, 39, 57–60, 63, 64].

In the last years, the use of primary human cells to implement the airway ALI cell cultures is increased to improve the prediction of the model. These cells represent a more heterogeneous population of different cell types; however, they imply a high availability of human pulmonary tissue and their (donor) variation and several procedures for isolation and differentiation, not to mention the costs [39]. Accordingly, due to the increasing demand for relevant models, a set of ready-to-use human three-dimensional (3D) ALI airway systems has been produced and is nowadays used for basic and regulatory research [6–8, 22, 39, 65–69]. These models include the EpiAirway™ (tracheal/bronchial epithelial cell cultures), the EpiAirway-FT™ (tracheal/bronchial epithelial cells cultured with fibroblasts), and the EpiAlveolar™ (alveolar epithelial cells cultured with pulmonary endothelial cells and fibroblasts) from MatTek, as well as the MucilAir™ (nasal/tracheal or bronchial cell cultures), MucilAir™-HF (nasal/tracheal or bronchial cell cultured with fibroblasts), SmallAir™ (small airway epithelia cells), and SmallAir™-HF (small airway epithelia cells cultured with fibroblasts) from Epithelix Sarl. The main features of such ALI systems are the capability of mucus production, the presence of an active cilia-beating, an active ion transport system, the presence of TJs, intrinsic metabolic activity, and the potential to release cytokines, chemokines, and metalloproteinases in response to chemicals/ stressors. In addition, these ALI cultures are available from a single donor or a pool of (healthy and diseased) donors, and they are characterized by a long lifetime (3–12 months) allowing long-term experiments. Unfortunately, these models, although advanced, have not reached yet an appreciable level of regulatory acceptance, but the high degree of standardization they provide has allowed them to enter into the ECVAM validation process. Indeed, a new test using the EpiAirway™ model for the determination of acute inhalation toxicity has been brought forward for validation and OECD adoption [70].

2.2 Intestinal Barrier

The gut barrier is a functional tissue that regulates the digestion and absorption of nutrients while maintaining a peaceful coexistence with commensal microorganisms and protecting the body from infection [71–73]. The intestinal mucosa consists of three distinct layers: the muscularis mucosa (the deepest layer), the lamina propria (bounded by the epithelium above and the muscularis mucosa below), and the uppermost epithelial layer (the small intestine) [74]. The lamina propria is the structural support for the epithelial cells of the small intestine, and it is composed of lymph vessels, smooth muscle cells, lymphocytes, and macrophages. The small intestine is made of a single-cell layer organized into crypts and villi where crypts are involved in the cell renewal while villi in the nutrient absorption and secretion of mucus [74]. Adherent junctions (AJs) and tight junctions (TJs) ensure the integrity of such intestinal barrier. While TJs act as a paracellular barrier that separates the apical and basolateral compartments, AJs act as a cellular belt maintaining the intercellular adhesion [71, 75]. Although the enterocytes represent the majority of cells in the small intestine as they are involved in metabolic and digestive functions, there are other types of cells, such as the secretory cells, including the goblet and Paneth cells, and the microfold (M) cells [72, 74, 76]. The goblet cells are the cells secreting the mucus that act as not only the first physical barrier against harmful microorganisms and antigens but also as a lubricating agent for intestinal motility. On the other hand, the Paneth cells help to maintain the stem cell niche through paracrine signaling, and they regulate the proliferation and differentiation programs of the other cells. M cells are the innate and adaptive immune cells. They are specialized epithelial cells responsible for the antigen sampling from the gastrointestinal tract lumen and for the immune responses. Lastly, the intestinal epithelium is continually renewed every 3–5 days by the proliferation of the stem cells from the crypts to replace cells extruded from the tips of the villi [77].

The greater progress done to mimic the intestinal barrier functionality has been made by using two-dimensional (2D) culture models. Caco-2 cells, epithelial cells derived from a colorectal adenocarcinoma, expressing several properties of the human enterocytes [76, 78–81] have been extensively used for absorption studies [40]. Indeed, if these cell type is differentiated using ALI culturing conditions, they can form a polarized monolayer representing a reliable system for absorption studies. The polarized monolayer acquires a typical apical brush border with microvilli and TJs. It expresses most receptors, transporters, and drug-metabolizing enzymes found in normal epithelium. Despite all these qualities, the great disadvantage of the Caco-2-based ALI cultures is that they lack the mucus layer. To overcome this, Caco-2 cells are often coupled to a clone of human intestinal epithelial cells HT-29 (HT-29-MTX) capable of differentiating into goblet-like mucus-producing cells [76, 79]. Such co-culture of Caco-2 and HT-29-MTX can be further improved by adding M cells, involved in particle uptake, and therefore useful to study particle absorption [76, 82]. The Raji-B cell line is commonly used for this purpose. They are lymphoblast-like cells able to transform a subpopulation of Caco-2 epithelial cells into phagocytic and transcytotic immune-supporting M-cells [82, 83]. Currently, it is becoming more and more common the use of primary intestinal cells, stem cell-derived intestinal cells, or cell lines (e.g., Caco-2 cells) grown within an extracellular matrix (ECM) [79].

The ALI approach revolution is quite recent in the intestinal research field, but it has shown many advantages. The intestinal epithelial cells are involved in changing oxygen supplementation via the bloodstream as oxygen acts as a regulator of the barrier and the absorptive function of the epithelium [18]. It has been demonstrated that the culturing of intestinal epithelial cells at ALI condition resulted in a significant increase in cell number, epithelial cell layer thickness and expression as well as apical localization of the microvilli-associated protein villin [18]. Moreover, similarities in cell morphology were registered between ALI cell cultures and intestinal epithelial cells in vivo. The culturing of intestinal tissues by ALI method using a collagen gel recapitulates organ structures, multilineage differentiation, and physiology [16]. Furthermore, the ALI cell culture in addition to improving oxygen availability at the medium-cell interface enhances oxidative phosphorylation and suppresses glycolysis [17]. ALI intestinal models have been implemented also to preserve mesenchymal-epithelial interaction when using 3D intestinal models [89]. Here, the ALI culture uses minced whole intestinal tissue embedded in a collagen gel exposed to air. In this way, isolated intestinal cells first form round, cystic organoids that progress to differentiated, crypt-like structures.

With the advancements made in producing cell culture models closer to reality, the use of 3D cultures took hold over the 2D models that suffer from a reduction of cell-cell interaction, lack of cell-matrix interaction, and tissue architecture. This has happened also for intestinal ALI cell culture models. A 3D ALI human-based tissue model grown from cells harvested from a healthy adult donor (EpiIntestinal™ from MatTek) is currently available on the market. The EpiIntestinal™ microtissues incorporate enterocytes, Paneth cells, M cells, and intestinal stem cells into a highly differentiated, polarized epithelium, with functional TJs and brush borders at the apical tissue surface. The EpiIntestinal™ microtissues recapitulate many features of the normal intestinal function (e.g., barrier functionalities and metabolism) [84–87] and have a lifetime of 1 month, allowing sub-chronic exposure. They are found to be a highly reproducible model, and they are widely used by industry and academic institutions. Indeed, they are used to study drug permeation, metabolism, drug-induced gastrointestinal toxicity, pathogen infections, inflammation, and wound healing [87, 88].

2.3 Skin Barrier

The skin is the largest organ of the body (15–17% of the body weight and with a surface area of approximately 1.5–1.7 m^2) and with his stratified squamous epithelium represents the first line of defense against environmental dangers such as pathogens, toxins, and radiations. Besides the defense role, the human skin performs several essential functions, such as heat regulation, sensation, and excretion [11]. The skin is a self-renewing (via repair and desquamation), metabolically active organ composed of three layers, the epidermis (the outermost tissue), the dermis, and the subcutaneous tissue. The dermis is the vascularized layer, which also contains the lymphatics, sweat, and sebaceous glands, while the epidermis has the protective role [40]. The epidermis is a keratinized stratified squamous epithelium composed of four cell layers expressing structural proteins, tight junction proteins, antimicrobial peptides, cytokines, and chemokines [90]. The four layers are the stratum basale (SB), the stratum spinosum (SS), the stratum granulosum (SG), and the stratum corneum (SC) [91, 92]. The latter consists of dead corneocytes that are the end-products of keratinocyte differentiation, and it represents the main penetration barrier of the skin as they are enriched cells embedded in lipid lamellae that form a functional barrier for diffusion of substances through the skin. These lipid lamellae are constituted by ceramides, cholesterol, and free fatty acids [93]. The SC is the result of an evolutionary adaptation to the air to form a dead but functional air-liquid interface barrier [91].

In the last decade, several in vitro skin cell models have been developed [11–13, 94, 95], starting from the conventional 2D co-culture models where cell lines mimicking the keratinocytes present in the epidermis layer are seeded onto fibroblast cells, which represent the most predominant cell population of the dermal layer [95, 96]. Within this culture system, the presence of fibroblasts is essential for the growth of keratinocyte cultures as they can able to secrete substances of the extracellular matrix such as collagen [97]. The most used cell lines to recreate a simple skin barrier are the NIH 3 T3 and HaCaT cells, acting for mouse fibroblast and human keratinocyte cell lines, respectively [95]. However, the culturing of cells on plastic prevents both long-term cultures and stratification of layers. In addition, the skin is constantly exposed to air; thus, the ALI method can improve the ability of these culture systems to mimic the human skin microenvironment [98]. Therefore, the development of relevant skin models, able to recapitulate the in vivo-like architecture by adopting the ALI culturing method, has become the predominant choice for studying skin toxicity, skin wound healing, irritation, sensitization, and corrosion. For instance, the advantage of using the ALI system over the submerged condition was demonstrated in a study where the capability of ovine tendon collagen type I (OTC-I) sponge (a suitable platform for a 3D in vitro skin model) was tested to reproduce an in vivo like epidermal- and dermal-like layers [13].

Nowadays, 3D skin models are mostly organotypic culture systems made by the reconstructed human epidermis (RhE) that consists of normal, human-derived epidermal keratinocytes, which have been cultured at ALI to form a highly differentiated epidermal tissue. These ALI models have high reproducibility and are widely accepted [11, 99]. At the present, there are open-source skin models, the open-source reconstructed epidermis (OS-REp) introduced in the early 1980s [99], as well as commercially available models of only human epidermis or epidermis and dermis (called full-thickness models), such as the EpiDerm™ (epidermal keratinocytes cultures) from MatTek, EpiSkin™ (epidermal keratinocytes cultures), and T-Skin™ (full-thickness model) from EpiSkin, epiCS® (epidermal keratinocytes cultures) from Cell Systems GmbH, and LabCyte_EPI-model (epidermal keratinocytes cultures) from Japan Tissue Engineering Co., Ltd. Interestingly, from a regulatory perspective, such 3D ALI models have received approval as an acceptable alternative in vitro models for skin irritation (OECD TG 439, 2021), corrosion (OECD TG 431, 2019) testings, and for the in vitro phototoxicity test (OECD TG 498, 2021) as results of validation studies [40, 100]. A validation study, coordinated by Cosmetics Europe, is currently ongoing to address genotoxicity (micronucleo and comet assays) and skin metabolism by using RhE models [40]. Considering the aforementioned progress, it has become especially pressing for the cosmetic industry the need for non-animal alternatives for testing cosmetic ingredients, so that from 2013 only animal-free methods may be used for safety assessment of cosmetics in the EU [101].

3 Application of ALI Cultures

Due to their physiological similarity to different in vivo epithelia, ALI tissue cultures are extensively used as in vitro models for several applications, such as toxicity testing (e.g., chemicals or nanoparticles, NPs) and cancer research, as well as clinical research (e.g., drug screening, infection, and wound healing) [32, 40, 102]. As ALI airway cultures are the most popular, this paragraph will be predominantly focused on the applications related to respiratory research with few focused insights on applications related to the other two types of ALI cultures described in the previous section (the intestinal and skin ALI models).

3.1 Toxicity Testing

Nowadays, airway ALI cell cultures are in vitro models adopted and accepted by the research community for toxicity testing. This model has recently coupled to aerosol-based delivery systems to mimic inhalation exposure as for an in vivo model [5, 32, 39]. Indeed, aerosol generation is capable to reproduce the physiological process of deposition of particles/chemicals that meet the lung epithelium. An aerosol can be generated from a dry powder or by the nebulization of a solution, depending on the application. Commercialized and widely used systems developed to generate aerosol are available from Vitrocell® and Cultex® companies. Such systems are characterized by a direct flow where the aerosol hits directly the cells cultured at ALI allowing a homogeneous deposition of the compound of interest (cloud settling) [32, 103]. Moreover, by using these exposure systems, it is also possible to quantify the deposition of the compound by a balance incorporated in the module that allows the determination of the administrative dose to correlate better with the biological effects. However, the deposition rate can be difficult to extrapolate as this process is strongly dependent on the particles' physicochemical properties and on the aerosolization technique that can interfere with the particles deposition (e.g., particles can remain trapped in the nebulizer unit). To help with this issue, some mathematical models have been implemented to calculate deposition using theoretically derived efficiencies for deposition by diffusion, sedimentation, and impaction within the airway or airway bifurcation. These models are the International Commission on Radiological Protection (IRCP) specifically developed to deal with radioactive particles [104] and the Multi-Path Particle Dosimetry (MPPD) model [105].

The coupling of toxicity testing and airway ALI cell cultures includes research on different topics, such as the evaluation of nanoparticles (NPs), combustion-derived products, and chemical toxicity. In general, there is some evidence that a correlation exists between in vivo observations and the data produced by in vitro testing employing ALI airway models over airway submerged cultures for NPs toxicity [5–7, 66, 103, 106, 107]. For instance, it has been reported that repeated exposure (4 and 12 weeks) to aerosolized graphene oxide and silica NPs of the airway ALI model shows results in accordance with the available in vivo inhalation toxicity studies in terms of viability and integrity of the lung tissue [6, 7]. Within this context, the use of ALI cell cultures is precious as they allow to perform long-term studies using repeated exposure regimes at human-relevant exposure doses and therefore can be considered a valuable tool to study the exposure to NPs in occupational settings that normally occurs repeatedly over a long-time frame.

The toxicity related to the combustion-derived products is considered relevant as some of these products are known to be carcinogens and pulmonary irritants [32]. Chronic obstructive pulmonary disease (COPD) is one of the respiratory diseases most diffused [38], and it is highly associated with combustion-derived products exposure [108]. This disease is characterized by a progressive obstruction of the airways that leads to lung failure. Airway ALI models are successfully used to reproduce the adverse tissue responses of COPD reported in vivo. They are widely used to assess cigarette smoke [109–111], combustion-derived fine and ultrafine particles [112, 113], and wood combustion emissions [114, 115] toxicity associated with COPD. Indeed, it has been demonstrated that exposure of airway ALI cultures to cigarette smoke leads to mucus hypersecretion, one of the main characteristics of COPD [111]. In addition, specific disease ALI culture models (lung cells isolated from COPD patients) has been developed and tested to study the toxicity of combustion-derived products in relation to COPD finding a correlation with in vivo evidence of barrier dysfunction and hyperplasia of goblet cells [32, 108, 116, 117].

Airway ALI culture systems are widely implemented also for chemical toxicity testing, particularly for testing respiratory irritants and sensitizers [32, 118]. Indeed, such chemicals are considered substances of high risk at the same level as carcinogens and mutagens, and their exposure is often associated with rhinitis and asthma [118]. The use of airway ALI models, although not approved from a regulatory perspective for testing respiratory chemical sensitizers and irritants, is increased also in this research field as it has been demonstrated that they are suitable in vitro models for identification and characterization of the respiratory sensitizers and irritants and discrimination of chemical respiratory sensitizers from irritants [32, 118–121].

Although ALI models are mostly used for respiratory research, the toxicity of chemicals or hazardous particles needs to be tested also on other different epithelial barriers, such as the skin, which could represent the first barrier of contact with the compound of interest (i.e., cosmetics), or the intestine, which could represent the absorption barrier of a direct (food components) and/or indirect (systemic absorption of an inhaled agent) compound exposure [40].

Information on skin toxicity and related (trans)dermal penetration of a specific compound is of course extremely important for testing specific consumer products, such as chemicals, plant products, a drug with topical application, and cosmetics [40]. However, as mentioned above (paragraph 2.3), progress has been made in this field by using commercialized skin ALI models that are already approved as an acceptable alternative in vitro models for skin irritation and corrosion testing [101]. Therefore, such an application is not included in this discussion.

Information on intestinal toxicity is mandatory to have toxicokinetic information (absorption, distribution, metabolism/biotransformation, ADME) of a compound such a drug in terms of its oral bioavailability that defines the drug fraction that reaches the systemic circulation. To date, only animal models are approved for toxicokinetic experiments according to OECD guidelines, but it is evident that animal-derived toxicokinetic data are not always reliable for extrapolation to humans due to inter-species differences. Therefore, the use of alternative models, such as the ALI intestinal cell cultures, plays an important role in this field but not only for studying drug toxicokinetics but also to investigate the toxicity of some food components (e.g., NPs ingested as additives, nutrient supplements or antimicrobial agents in food packaging). Recently, some progress has been made by using the EpiIntestinal™ ALI models. Indeed, it has been demonstrated that such an ALI model displays a good correlation between in vitro permeability data and the fractions absorbed of drugs in humans gaining insights on drug toxicokinetic and bioavailability [84, 87, 88]. NP toxicity tests on such ALI model also show good results. Here, it has been reported that, conversely to submerged intestinal cell culture systems, the EpiIntestinal ™ tissues require higher concentrations of copper NPs to induce toxicity effects [122]. The same trend of low toxicity was shown also for silver NPs [123], which usually compromises the integrity of the intestinal epithelium in submerged cell cultures [124, 125]. All together, these results highlight the important of using advanced in vitro model for toxicity testing.

3.2 Cancer Research

Several ALI models have been developed to study the mechanism of malignant transformations in lung cancer. For instance, the use of fibroblast cells on epithelial lung cells cultivated at ALI condition can enhance the tumor-like behavior of lung cells by the production of instructive signals including growth factors and chemokines [128] and by the activation of oncogenes driving bronchial dysplasia [129]. Airway spheroids (3D cell cultures that arrange themselves during proliferation into sphere-like formations) cultured at ALI condition over submerged condition have demonstrated the importance of air-grown 3D models to gain information on resistance mechanisms in lung cancer, and they have proven to be a valid method to test chemotherapeutics [128–131]. Multilayered cell cultures of epithelial lung cells cultivated at ALI conditions have been used to study anticancer drugs delivered by a clinical nebulizer [2]. Such ALI model is proven to be an interesting human tool allowing for the efficacy screening of inhaled anticancer drugs. In addition, when fibroblast cells are incorporated in such ALI model, increased resistance to drugs is demonstrated compared to ALI multilayered mono-cultures [3]. This study highlighted the importance of using multiple types of cells under ALI conditions to reflect better the in vivo conditions. In this regard, a commercialized ALI model of lung cancer has been developed, the OncoCilAir™ from OncoTheis, that combines a functional reconstituted human airway epithelium, human lung fibroblasts, and proliferating tumor nodules from mutated non-small-cell lung cancer cell line. Such a model is considered a promising tool in lung cancer research as all of its components are of human origin and specific to individual patients. Moreover, the OncoCilAir™ tissue is able to reproduce health tissue-dependent autofluorescence response typically observed in vivo [134].

ALI culture models are employed also for studying colorectal cancer (CRC), the third most common cancer and the fourth cause of cancer-related fatalities worldwide after lung, liver, and stomach cancers [135, 136]. Depending on the mutation, CRC can be classified as sporadic (the most common), inherited (polyposis and non-polyposis forms), and familiar. The current methods used to treat CRC are surgery and chemotherapy (used mostly as adjuvant before surgery). Recently, the use of monoclonal antibodies has been associated with chemotherapy as an effective tool to target specific tumor receptors. However, drug resistance remains the main emergence of this type of cancer representing one of the reasons for treatment failure [135, 136]. Therefore, the study of the molecular mechanism that induces the resistance to anticancer drugs is critical for CRC treatment, and the use of alternative in vitro methods in this contest can be a useful tool to perform focused experiments that are not possible with animals. To mimic the tumor microenvironment, consisting of epithelial and mesenchymal cells and extracellular substrates, the use of 3D ALI cell cultures has been implemented [16, 137]. It has been reported that 3D tumor ALI cultures (established from CRC tissues) are more resistant to the toxicity induced by 5-fluorouracil (a chemotherapy drug) leading to further insights into the chemoresistance and therefore to the demonstration that the ALI cell cultures may become an application to study anticancer drug resistance [138]. Notably, fragments of patient-derived CRC tissue that are cultured at ALI conditions can preserve the original epithelial and mesenchymal structures [139].

ALI culture models have been implemented also for studying new treatments and pathogenesis of other types of cancer. For instance, for studying the clear cell renal cell carcinoma (ccRCC), ALI 3D cell cultures, isolated from tissue of patients who underwent surgical resection for renal carcinoma, were used to monitor tumor responses to different types of therapies in a controlled manner [28].

For investigating the cutaneous squamous cell carcinoma (SCC), ALI skin cultures were produced by SCC biopsies demonstrating for the first time they are feasible to develop curative treatment and gain insights into the SCC pathogenesis [140]. In addition, it has been demonstrated that in human skin reconstructs, the presence of fibroblasts in a keratinocyte-like ALI cell culture is essential to mimic the tumorigenicity in vivo of the human melanoma [141].

Most of the ALI cell cultures derived from the female reproductive tract (FRT) mucosa are used to produce relevant cultures of cervical epithelial cells that are implemented for dealing with virus-induced neoplasia development, such as the human papillomavirus (HPV) associated with the cervical squamous cell carcinoma [4, 142]. Within this contest, ALI cultures are used to analyze HPV infection, the role of viral proteins throughout the viral life cycle, and the oncogenesis defining low- or high-risk HPV type. Currently, new ALI models of FRT have been employed to study cancer development in oviduct epithelial cells [4].

Lastly, ALI cell cultures have been implemented also for assessing the glioblastoma pathogenesis. Here, a tissue engineering method was used to develop 3D ALI cultures of human pluripotent stem cells differentiated in nervous tissue together with glioblastoma stem-like cells to establish a human tissue sharing similarities with the glioblastoma development in the brain environment. Such culture type allows the identification of clinically relevant genes induced by the tumor/host tissue interaction providing comprehension into the pathogenesis of the glioblastoma [143].

3.3 Translational Research

ALI airway models coupled to an aerosol-generating system are commonly used to assess the clinical efficacy of inhaled drugs by assessing their deposition, absorption, and diffusion through the barrier. Several examples of such application are currently summarized in the review of Cao et al. [32]. In this review, the authors show that the ALI airway models represent a reliable tool to differentiate between substances with low and high absorption, to perform drug screening, to investigate drug absorption kinetics mostly thanks to their barrier features that resemble the in vivo condition [142–146].

Polarized ALI airway cell cultures are considered a valuable system also for studying the mechanism of the pathogen-host interaction, such as bacteria and virus as they are considered a natural infection target. Examples of such utilization can be found in the reviews of Cao et al. and Baldassi et al. [32, 108].

Recently, such a model has been implemented for studying the severe acute respiratory syndrome coronavirus 2 (SARS-CoV-2) that has caused the current COVID-19 pandemic. Here, the screening of different therapeutical agents and their effectiveness on virus inactivation has been tested using the ALI airway cell cultures as the ALI models facilitate virus isolation and characterization over the conventional submerged cultures that lack expression of several entry factors that allow virus proliferation [108]. In this regard, an ALI airway model was used to study the biological characteristics of the virus after its isolation for bronchial lavage fluid of patients suffering from SARS-CoV-2-mediated Covid-19 disease [149].

One of the applications of the ALI intestinal model lies in the intrinsic features that characterize intestinal development, physiology, and diseases. Indeed, many aspects of gastrointestinal (GI) health are impacted by the gut microbiome, so a defective interaction between the host and a specific gut microbe can lead to a bunch of infectious diseases. Studying the microbe–intestinal epithelial cell interactions in vivo is complex; thus, the use of alternatives is highly demanding [21]. In this contest, 3D ALI intestinal cell cultures formed by induced pluripotent stem cells (iPSCs) or adult crypt from intestinal stem cells (ISC) are the preferred cell culture system as they could enhance host-microbe interaction studies [21].

Moreover, it has been demonstrated that the ALI cultivation method is a well-grounded instrument also for studying the interaction with intestinal pathogens, such as for studying the interaction of *Helicobacter pylori* with the GI epithelial surface over several weeks [150]. Here, the generation of columnar epithelium of the human GI mucosa resembling the in vivo conditions was obtained thanks to the ALI cultivation method.

ALI skin models are implemented in many aspects of skin research such as the study of wound healing, bacterial infection, and screening of antimicrobial agents [90]. For instance, human skin equivalents (HSE), consisting of both human keratinocytes and fibroblasts, are in vitro tissues cultured at ALI used to study the re-epithelialization and epithelial-mesenchymal cross talk following wounding as they can mimic the healing of human skin [151, 152]. Epidermal skin equivalents (only keratinocytes) or HSE cultures are used at ALI conditions also for studying bacterial-skin interaction (e.g., *Acinetobacter baumannii and Pseudomonas aeruginosa*) [153, 154]. In this context, they represent a valid model to study skin colonization, evaluate the effects of novel antimicrobial agents, and develop approaches to reduce the bacterial burden in skin wounds.

4 Conclusions and Outlook

Currently, efforts are made to develop alternatives to animal models as part of the 3R principle. Apart from the ethical issues derived from the use of animals in the research laboratories, the species-specific differences between humans and the most used animal models (rats or mice) cannot be underestimated, and it remains the main scientific issue. Non-animal models of different human epithelial barriers cultivated under the ALI condition can be considered a useful tool to recapitulate the physiology of human barriers of several tissues, such as the lung, the skin, and the intestine, and thus can replace the use of animals in research testing. This review has described how human ALI systems resemble in vivo exposure conditions in a much more relevant way, as compared to the standard submerged approaches and how, during the last years, such models have reached a considerable level of quality in different research fields.

In the cosmetic industry, the rush in producing alternative in vitro skin models has been triggered by the new Cosmetics Regulation (EU, 2009a), which has banned animal experiments and has forced the research community in finding new instruments of research. Such progress has found the summit in the use of human ALI cell cultures of reconstructing skin equivalents, leading to a full replacement of animals in the safety testing of cosmetics and chemicals and the introduction of new in vitro OECD guidelines (OECD TGs 439, 431, and 498). However, such standardized skin tests do not consider the different exposure scenarios as well as prolonged exposure. In this context, the use of mathematical models to extrapolate data from skin testing to a specific exposure scenario can enhance the quality of the already validated skin testing [40]. On the other hand, when looking at the existing validated ALI models of the lung and intestinal epithelium, sparse progress has been made in comparison to the skin due to the complexity of such organ tissues. Nevertheless, both in industry and academia, many research activities are currently ongoing toward the implementation of complex in vitro human models more realistic of the lung and intestinal mucosa. In this context, the ALI EpiAirway™ model is under a validation program as in vitro alternative for acute inhalation toxicity studies, and the ALI intestinal culture EpiIntestinal™ has proven to reach a good level of in vivo similarity in terms of cell morphology [40, 87]. However, there is still a gap to fill. These ALI models, however, lack critical inputs from the microbiota that have a specific role in nutrient and drug metabolism, maintenance of structural integrity of the epithelial barrier, immunomodulation, and protection against pathogens [155].

Besides the advancements reached in skin, lung, and intestine models, a big limitation of the conventional ALI models remains: they are static and not dynamic culture systems, so processes such as nutrient exchange and cell-cell interaction cannot be investigated [38]. Therefore, the next step for achieving the 3R is the routine integration of ALI models in a fluidic platform able to mimic the physiological flows that recreate tissue perfusion and shear stresses stimuli [156]. Here, an "organ-on-a-chip" is defined as a microfluidic cell culture system that contains continuously perfused chambers with living cells arranged inside achieving the multicellular architectures and tissue-tissue interfaces. Interestingly, with this technology, it is also possible to combine different cellular models to recreate the so-called "body or multi-organ-on-a-chip" aiming to replace systemic animal testing.

To conclude, although improvement and standardization of in vitro alternative models to animals are still needed, the advantages associated with the use of ALI models are evident, and future efforts should be aiming at improving such systems for regulatory acceptance.

References

1. Di Cristo L, Maguire CM, Mc Quillan K et al (2018) Towards the identification of an in vitro tool for assessing the biological behavior of aerosol supplied nanomaterials. Int J Environ Res Public Health 15:563
2. Movia D, Bazou D, Volkov Y, Prina-Mello A (2018) Multilayered Cultures of NSCLC cells grown at the Air-Liquid Interface allow the efficacy testing of inhaled anti-cancer drugs. Sci Rep 8:1–19
3. Movia D, Bazou D, Prina-Mello A (2019) ALI multilayered co-cultures mimic biochemical mechanisms of the cancer cell-fibroblast cross-talk involved in NSCLC MultiDrug Resistance. BMC Cancer 19:1–21
4. Chen S, Schoen J (2019) Air-liquid interface cell culture: from airway epithelium to the female reproductive tract. Reprod Domest Anim 54:38–45
5. Lacroix G, Koch W, Ritter D et al (2018) Air–liquid Interface in vitro models for respiratory toxicology research: consensus workshop and recommendations. Appl Vitro Toxicol 4:91–106
6. Di Cristo L, Boccuni F, Iavicoli S, Sabella S (2020) A human-relevant 3D in vitro platform for an effective and rapid simulation of workplace exposure to nanoparticles: silica nanoparticles as case study. Nano 10:1761
7. Di Cristo L, Grimaldi B, Catelani T et al (2020) Repeated exposure to aerosolized graphene oxide mediates autophagy inhibition and inflammation in a three-dimensional human airway model. Materials Today Bio 6: 100050
8. Movia D, Di Cristo L, Alnemari R et al (2017) The curious case of how mimicking physiological complexity in in vitro models of the human respiratory system influences the inflammatory responses. A preliminary study focused on gold nanoparticles. J Interdiscip Nanomed 2:110–130
9. Ootani A, Toda S, Fujimoto K, Sugihara H (2000) An air–liquid interface promotes the differentiation of gastric surface mucous cells (GSM06) in culture. Biochem Biophys Res Commun 271:741–746
10. Li X, Ootani A, Kuo C (2016) An air–liquid interface culture system for 3D organoid culture of diverse primary gastrointestinal tissues. In: Gastrointestinal physiology and diseases. Springer, pp 33–40
11. Alexander FA, Eggert S, Wiest J (2018) Skin-on-a-chip: transepithelial electrical resistance and extracellular acidification measurements through an automated air-liquid interface. Genes 9:114
12. Larouche D, Cantin-Warren L, Desgagné M et al (2016) Improved methods to produce tissue-engineered skin substitutes suitable for the permanent closure of full-thickness skin injuries. BioResearch open access 5:320–329

13. Fauzi MB, Rashidbenam Z, Bin Saim A, Binti Hj Idrus R (2020) Preliminary study of in vitro three-dimensional skin model using an ovine collagen type i sponge seeded with co-culture skin cells: submerged versus air-liquid interface conditions. Polymers 12: 2784
14. Greco D, Vellonen K-S, Turner HC et al (2010) Gene expression analysis in SV-40 immortalized human corneal epithelial cells cultured with an air-liquid interface. Mol Vis 16:2109
15. Kawata K, Aoki S, Futamata M et al (2019) Mesenchymal cells and fluid flow stimulation synergistically regulate the kinetics of corneal epithelial cells at the air–liquid interface. Graefes Arch Clin Exp Ophthalmol 257: 1915–1924
16. Usui T, Sakurai M, Umata K et al (2018) Preparation of human primary colon tissue-derived organoid using air liquid Interface culture. Curr Protoc Toxicol 75:22–26
17. Klasvogt S, Zuschratter W, Schmidt A et al (2017) Air–liquid interface enhances oxidative phosphorylation in intestinal epithelial cell line IPEC-J2. Cell Death Discov 3:1–7
18. Nossol C, Diesing A-K, Walk N et al (2011) Air–liquid interface cultures enhance the oxygen supply and trigger the structural and functional differentiation of intestinal porcine epithelial cells (IPEC). Histochem Cell Biol 136:103–115
19. Zhu Y, Yang Y, Guo J et al (2017) Ex vivo 2D and 3D HSV-2 infection model using human normal vaginal epithelial cells. Oncotarget 8: 15267
20. Lee Y, Dizzell SE, Leung V et al (2016) Effects of female sex hormones on susceptibility to HSV-2 in vaginal cells grown in air-liquid interface. Viruses 8:241
21. Han X, Mslati M, Davies E et al (2021) Creating a more perfect union: Modeling intestinal bacteria-epithelial interactions using organoids. Cell Mol Gastroenterol Hepatol
22. Movia D, Bruni-Favier S, Prina-Mello A (2020) In vitro alternatives to acute inhalation toxicity studies in animal models—a perspective. Front Bioeng Biotechnol 8:549
23. Zuang V, Dura A, et al (2021) Non-animal methods in science and regulation – EURL ECVAM Status Report 2020. Publications Office of the European Union, Luxembourg. https://doi.org/10.2760/719755
24. Kandarova H, Letasiova S, Adriaens E et al (2018) CON4EI: EpiOcular™ Eye Irritation Test (EpiOcular™ EIT) for hazard identification and labelling of eye irritating chemicals. Toxicol In Vitro 49:21–33
25. Kandarova H, Letasiova S, Adriaens E et al (2018) CON4EI: CONsortium for in vitro Eye Irritation testing strategy-EpiOcular™ time-to-toxicity (EpiOcular ET-50) protocols for hazard identification and labelling of eye irritating chemicals. Toxicol In Vitro 49:34–52
26. Wang S, Ghezzi CE, Gomes R et al (2017) In vitro 3D corneal tissue model with epithelium, stroma, and innervation. Biomaterials 112:1–9
27. Sekiya S, Kikuchi T, Shimizu T (2019) Perfusion culture maintained with an air-liquid interface to stimulate epithelial cell organization in renal organoids in vitro. BMC Biomed Eng 1:1–12
28. Esser LK, Branchi V, Leonardelli S et al (2020) Cultivation of clear cell renal cell carcinoma patient-derived organoids in an air-liquid interface system as a tool for studying individualized therapy. Front Oncol 10: 1775
29. Giandomenico SL, Mierau SB, Gibbons GM et al (2019) Cerebral organoids at the air–liquid interface generate diverse nerve tracts with functional output. Nat Neurosci 22: 669–679
30. Ao Z, Cai H, Havert DJ et al (2020) One-stop microfluidic assembly of human brain organoids to model prenatal cannabis exposure. Anal Chem 92:4630–4638
31. Alper S, Janssen WJ (2018) Lung innate immunity and inflammation. Springer
32. Cao X, Coyle JP, Xiong R et al (2020) Invited review: human air-liquid-interface organotypic airway tissue models derived from primary tracheobronchial epithelial cells—overview and perspectives. In Vitro Cell Dev Biol Anim:1–29
33. Wanner A, Salathé M, O'Riordan TG (1996) Mucociliary clearance in the airways. Am J Respir Crit Care Med 154:1868–1902
34. Wang H, He L, Liu B et al (2018) Establishment and comparison of air-liquid interface culture systems for primary and immortalized swine tracheal epithelial cells. BMC Cell Biol 19:1–10
35. Flynn AN, Itani OA, Moninger TO, Welsh MJ (2009) Acute regulation of tight junction ion selectivity in human airway epithelia. Proc Natl Acad Sci 106:3591–3596
36. Godfrey RWA (1997) Human airway epithelial tight junctions. Microsc Res Tech 38:488–499

37. Knudsen L, Ochs M (2018) The micromechanics of lung alveoli: structure and function of surfactant and tissue components. Histochem Cell Biol 150:661–676
38. Movia D, Prina-Mello A (2020) Preclinical development of orally inhaled drugs (OIDs)—are animal models predictive or shall we move towards in vitro non-animal models? Animals 10:1259
39. Upadhyay S, Palmberg L (2018) Air-liquid interface: relevant in vitro models for investigating air pollutant-induced pulmonary toxicity. Toxicol Sci 164:21–30
40. Gordon S, Daneshian M, Bouwstra J et al (2015) Non-animal models of epithelial barriers (skin, intestine and lung) in research, industrial applications and regulatory toxicology. ALTEX 32:327–378
41. Clippinger AJ, Allen D, Jarabek AM et al (2018) Alternative approaches for acute inhalation toxicity testing to address global regulatory and non-regulatory data requirements: an international workshop report. Toxicol In Vitro 48:53–70
42. Rotoli BM, Gatti R, Movia D et al (2015) Identifying contact-mediated, localized toxic effects of MWCNT aggregates on epithelial monolayers: a single-cell monitoring toxicity assay. Nanotoxicology 9:230–241
43. Rotoli BM, Bussolati O, Costa AL et al (2012) Comparative effects of metal oxide nanoparticles on human airway epithelial cells and macrophages. J Nanopart Res 14: 1–14
44. Di Cristo L, Bianchi MG, Chiu M et al (2019) Comparative in vitro cytotoxicity of realistic doses of benchmark multi-walled carbon nanotubes towards macrophages and airway epithelial cells. Nano 9:982
45. Inoue D, Furubayashi T, Tanaka A et al (2020) Quantitative estimation of drug permeation through nasal mucosa using in vitro membrane permeability across Calu-3 cell layers for predicting in vivo bioavailability after intranasal administration to rats. Eur J Pharm Biopharm 149:145–153
46. Sibinovska N, Žakelj S, Roškar R, Kristan K (2020) Suitability and functional characterization of two Calu-3 cell models for prediction of drug permeability across the airway epithelial barrier. Int J Pharm 585:119484
47. He R-W, Gerlofs-Nijland ME, Boere J et al (2020) Comparative toxicity of ultrafine particles around a major airport in human bronchial epithelial (Calu-3) cell model at the air–liquid interface. Toxicol In Vitro 68:104950
48. Facio AJ, Yon J, Corbière C et al (2022) Toxicological impact of organic ultrafine particles (UFPs) in human bronchial epithelial BEAS-2B cells at air-liquid interface. Toxicol In Vitro 78:105258
49. Pinkston R, Zaman H, Hossain E et al (2020) Cell-specific toxicity of short-term JUUL aerosol exposure to human bronchial epithelial cells and murine macrophages exposed at the air–liquid interface. Respir Res 21:1–15
50. Stewart CE, Torr EE, Mohd Jamili NH et al (2012) Evaluation of differentiated human bronchial epithelial cell culture systems for asthma research. J Allergy:2012
51. Callaghan PJ, Ferrick B, Rybakovsky E et al (2020) Epithelial barrier function properties of the 16HBE14o-human bronchial epithelial cell culture model. Biosci Rep 40: BSR20201532
52. Rothen-Rutishauser B, Blank F, Mühlfeld C, Gehr P (2008) In vitro models of the human epithelial airway barrier to study the toxic potential of particulate matter. Expert Opin Drug Metab Toxicol 4:1075–1089
53. Hutchinson D, Müller J, McCarthy JE et al (2018) Cadmium nanoparticles citrullinate cytokeratins within lung epithelial cells: cadmium as a potential cause of citrullination in chronic obstructive pulmonary disease. Int J Chron Obstruct Pulmon Dis 13:441
54. Bitterle E, Karg E, Schroeppel A et al (2006) Dose-controlled exposure of A549 epithelial cells at the air–liquid interface to airborne ultrafine carbonaceous particles. Chemosphere 65:1784–1790
55. Öhlinger K, Kolesnik T, Meindl C et al (2019) Air-liquid interface culture changes surface properties of A549 cells. Toxicol In Vitro 60:369–382
56. Salomon JJ, Muchitsch VE, Gausterer JC et al (2014) The cell line NCl-H441 is a useful in vitro model for transport studies of human distal lung epithelial barrier. Mol Pharm 11:995–1006
57. Klein SG, Serchi T, Hoffmann L et al (2013) An improved 3D tetraculture system mimicking the cellular organisation at the alveolar barrier to study the potential toxic effects of particles on the lung. Part Fibre Toxicol 10:1–18
58. Wang Y, Adamcakova-Dodd A, Steines BR et al (2020) Comparison of in vitro toxicity of aerosolized engineered nanomaterials using air-liquid interface mono-culture and co-culture models. NanoImpact 18:100215
59. Fizean I, Chary A, Cambier S et al (2018) Responsiveness assessment of a 3D tetraculture alveolar model exposed to diesel exhaust particulate matter. Toxicol In Vitro 53:67–79

60. Barosova H, Karakocak BB, Septiadi D et al (2020) An in vitro lung system to assess the proinflammatory hazard of carbon nanotube aerosols. Int J Mol Sci 21:5335
61. Rothen-Rutishauser BM, Kiama SG, Gehr P (2005) A three-dimensional cellular model of the human respiratory tract to study the interaction with particles. Am J Respir Cell Mol Biol 32:281–289
62. Fytianos K, Chortarea S, Rodriguez-Lorenzo L et al (2017) Aerosol delivery of functionalized gold nanoparticles target and activate dendritic cells in a 3D lung cellular model. ACS Nano 11:375–383
63. He R-W, Braakhuis HM, Vandebriel RJ et al (2021) Optimization of an air-liquid interface in vitro cell co-culture model to estimate the hazard of aerosol exposures. J Aerosol Sci 153:105703
64. Meindl C, Öhlinger K, Zrim V et al (2021) Screening for effects of inhaled nanoparticles in cell culture models for prolonged exposure. Nano 11:606
65. Jackson GR Jr, Maione AG, Klausner M, Hayden PJ (2018) Prevalidation of an acute inhalation toxicity test using the EpiAirway in vitro human airway model. Appl in Vitro Toxicol 4:149–158
66. Barosova H, Maione AG, Septiadi D et al (2020) Use of EpiAlveolar lung model to predict fibrotic potential of multiwalled carbon nanotubes. ACS Nano 14:3941–3956
67. Barilli A, Visigalli R, Ferrari F et al (2020) Organic cation transporters (OCTs) in EpiAirway™, a cellular model of Normal human bronchial epithelium. Biomedicine 8: 127
68. McGee Hargrove M, Parr-Dobrzanski B, Li L et al (2021) Use of the MucilAir airway assay, a new approach methodology, for evaluating the safety and inhalation risk of agrochemicals. Applied In Vitro Toxicology 7:50–60
69. Cervena T, Vrbova K, Rossnerova A et al (2019) Short-term and long-term exposure of the MucilAir™ model to polycyclic aromatic hydrocarbons. Altern Lab Anim 47:9–18
70. Zuang V, Dura A, Asturiol Bofill D, et al (2020) EURL ECVAM status report on the development, validation and regulatory acceptance of alternative methods and approaches (2019). EUR 30100 EN, Publications Office of the European Union
71. Jung K, Eyerly B, Annamalai T et al (2015) Structural alteration of tight and adherens junctions in villous and crypt epithelium of the small and large intestine of conventional nursing piglets infected with porcine epidemic diarrhea virus. Vet Microbiol 177:373–378
72. Viggiano D, Ianiro G, Vanella G et al (2015) Gut barrier in health and disease: focus on childhood. Eur Rev Med Pharmacol Sci 19: 1077–1085
73. Wells JM, Brummer RJ, Derrien M et al (2017) Homeostasis of the gut barrier and potential biomarkers. Am J Physiol-Gastrointestinal Liver Physiol 312:G171–G193
74. Doherty MM, Charman WN (2002) The mucosa of the small intestine. Clin Pharmacokinet 41:235–253
75. Jeon MK, Klaus C, Kaemmerer E, Gassler N (2013) Intestinal barrier: molecular pathways and modifiers. World J Gastrointest Pathophysiol 4:94
76. Lefebvre DE, Venema K, Gombau L et al (2015) Utility of models of the gastrointestinal tract for assessment of the digestion and absorption of engineered nanomaterials released from food matrices. Nanotoxicology 9:523–542
77. Peterson LW, Artis D (2014) Intestinal epithelial cells: regulators of barrier function and immune homeostasis. Nat Rev Immunol 14: 141–153
78. Lea T (2015) Caco-2 cell line. The impact of food bioactives on health. pp 103–111
79. Ponce de León-Rodríguez M del C, Guyot J-P, Laurent-Babot C (2019) Intestinal in vitro cell culture models and their potential to study the effect of food components on intestinal inflammation. Crit Rev Food Sci Nutr 59:3648–3666
80. Carnovale C, Guarnieri D, Di Cristo L et al (2021) Biotransformation of silver nanoparticles into oro-gastrointestinal tract by integrated in vitro testing assay: generation of exposure-dependent physical descriptors for nanomaterial grouping. Nano 11:1587
81. Guarnieri D, Sánchez-Moreno P, Del Rio Castillo AE et al (2018) Biotransformation and biological interaction of graphene and graphene oxide during simulated oral ingestion. Small 14:1800227
82. Araújo F, Sarmento B (2013) Towards the characterization of an in vitro triple co-culture intestine cell model for permeability studies. Int J Pharm 458:128–134
83. Gibb M, Pradhan SH, Mulenos MR et al (2021) Characterization of a human in vitro intestinal model for the Hazard assessment of nanomaterials used in cancer immunotherapy. Appl Sci 11:2113

84. Cui Y, Claus S, Schnell D et al (2020) In-depth characterization of EpiIntestinal microtissue as a model for intestinal drug absorption and metabolism in human. Pharmaceutics 12:405
85. Ayehunie S, Stevens Z, Landry T, et al (2014) Novel 3D human small intestinal tissue model (EpiIntestinal™) to assess drug permeation & inflammation. In: AAPS NERDG annual meeting, Farmington, CT, USA
86. Simon F, Garcia J, Guyot L et al (2020) Impact of interleukin-6 on drug-metabolizing enzymes and transporters in intestinal cells. AAPS J 22:1–10
87. Fedi A, Vitale C, Ponschin G et al (2021) In vitro models replicating the human intestinal epithelium for absorption and metabolism studies: a systematic review. J Control Release
88. Markus J, Landry T, Stevens Z et al (2021) Human small intestinal organotypic culture model for drug permeation, inflammation, and toxicity assays. In Vitro Cell Dev Biol Anim 57:160–173
89. Short SP, Costacurta PW, Williams CS (2017) Using 3D organoid cultures to model intestinal physiology and colorectal cancer. Curr Colorectal Cancer Rep 13:183–191
90. Choi K-YG, Wu BC, Lee AH-Y et al (2020) Utilizing organoid and air-liquid interface models as a screening method in the development of new host defense peptides. Front Cell Infect Microbiol 10:228
91. Matsui T, Amagai M (2015) Dissecting the formation, structure and barrier function of the stratum corneum. Int Immunol 27:269–280
92. Brettmann EA, de Guzman SC (2018) Recent evolution of the human skin barrier. Exp Dermatol 27:859–866
93. Bouwstra JA, Dubbelaar FER, Gooris GS, Ponec M (2000) The lipid organisation in the skin barrier. Acta Dermatol Venereol, pp 23–30
94. Pourchet LJ, Thepot A, Albouy M et al (2017) Human skin 3D bioprinting using scaffold-free approach. Adv Healthc Mater 6: 1601101
95. Teimouri A, Yeung P, Agu R (2018) 2D vs. 3D cell culture models for in vitro topical (dermatological) medication testing. IntechOpen
96. Gangatirkar P, Paquet-Fifield S, Li A et al (2007) Establishment of 3D organotypic cultures using human neonatal epidermal cells. Nat Protoc 2:178–186
97. Rheinwald JG, Green H (1975) Serial cultivation of strains of human epidermal keratinocytes: the formation of keratinizing colonies from single cells. Cell 6:331–343
98. Aoki S, Takezawa T, Sugihara H, Toda S (2016) Progress in cell culture systems for pathological research. Pathol Int 66:554–562
99. Mewes KR, Fischer A, Zöller NN et al (2016) Catch-up validation study of an in vitro skin irritation test method based on an open source reconstructed epidermis (phase I). Toxicol In Vitro 36:238–253
100. Spielmann H, Hoffmann S, Liebsch M et al (2007) The ECVAM international validation study on in vitro tests for acute skin irritation: report on the validity of the EPISKIN and EpiDerm assays and on the Skin Integrity Function Test. Altern Lab Anim 35:559–601
101. Rogiers V, Benfenati E, Bernauer U et al (2020) The way forward for assessing the human health safety of cosmetics in the EU – workshop proceedings. Toxicology 436:152421
102. Lo Y-H, Karlsson K, Kuo CJ (2020) Applications of organoids for cancer biology and precision medicine. Nat Cancer 1:761–773
103. Secondo LE, Liu NJ, Lewinski NA (2017) Methodological considerations when conducting in vitro, air–liquid interface exposures to engineered nanoparticle aerosols. Crit Rev Toxicol 47:225–262
104. Klumpp J, Bertelli L (2017) KDEP: a resource for calculating particle deposition in the respiratory tract. Health Phys 113:110–121
105. Manojkumar N, Srimuruganandam B, Nagendra SS (2019) Application of multiple-path particle dosimetry model for quantifying age specified deposition of particulate matter in human airway. Ecotoxicol Environ Saf 168:241–248
106. Chortarea S, Barosova H, Clift MJD et al (2017) Human asthmatic bronchial cells are more susceptible to subchronic repeated exposures of aerosolized carbon nanotubes at occupationally relevant doses than healthy cells. ACS Nano 11:7615–7625
107. Fröhlich E (2018) Comparison of conventional and advanced in vitro models in the toxicity testing of nanoparticles. Artif Cells Nanomed Biotechnol 46:1091–1107
108. Baldassi D, Gabold B, Merkel OM (2021) Air- liquid Interface cultures of the healthy and diseased human respiratory tract: promises, challenges, and future directions. Adv NanoBiomed Res 1:2000111
109. Gindele JA, Kiechle T, Benediktus K et al (2020) Intermittent exposure to whole cigarette smoke alters the differentiation of

primary small airway epithelial cells in the air-liquid interface culture. Sci Rep 10:1–17
110. Bishop E, Haswell L, Adamson J et al (2019) An approach to testing undiluted e-cigarette aerosol in vitro using 3D reconstituted human airway epithelium. Toxicol In Vitro 54:391–401
111. Schamberger AC, Staab-Weijnitz CA, Mise-Racek N, Eickelberg O (2015) Cigarette smoke alters primary human bronchial epithelial cell differentiation at the air-liquid interface. Sci Rep 5:1–9
112. Juarez-Facio AT, Castilla C, Corbière C et al (2022) Development of a standardized in vitro approach to evaluate microphysical, chemical, and toxicological properties of combustion-derived fine and ultrafine particles. J Environ Sci 113:104–117
113. Vaughan A, Stevanovic S, Banks AP et al (2019) The cytotoxic, inflammatory and oxidative potential of coconut oil-substituted diesel emissions on bronchial epithelial cells at an air-liquid interface. Environ Sci Pollut Res 26:27783–27791
114. Kanashova T, Sippula O, Oeder S et al (2018) Emissions from a modern log wood masonry heater and wood pellet boiler: composition and biological impact on air-liquid interface exposed human lung cancer cells. J Mol Clin Med 1:23–35
115. Ihantola T, Di Bucchianico S, Happo M et al (2020) Influence of wood species on toxicity of log-wood stove combustion aerosols: a parallel animal and air-liquid interface cell exposure study on spruce and pine smoke. Part Fibre Toxicol 17:1–26
116. Wang Q, Ahmad Khan N, Muthumalage T et al (2019) Dysregulated repair and inflammatory responses by e-cigarette-derived inhaled nicotine and humectant propylene glycol in a sex-dependent manner in mouse lung. FASEB Bio Adv 1:609–623
117. Amatngalim GD, Schrumpf JA, Henic A et al (2017) Antibacterial defense of human airway epithelial cells from chronic obstructive pulmonary disease patients induced by acute exposure to nontypeable Haemophilus influenzae: modulation by cigarette smoke. J Innate Immun 9:359–374
118. Huang S, Wiszniewski L, Constant S, Roggen E (2013) Potential of in vitro reconstituted 3D human airway epithelia (MucilAir™) to assess respiratory sensitizers. Toxicol In Vitro 27:1151–1156
119. Gohlsch K, Mückter H, Steinritz D et al (2019) Exposure of 19 substances to lung A549 cells at the air liquid interface or under submerged conditions reveals high correlation between cytotoxicity in vitro and CLP classifications for acute lung toxicity. Toxicol Lett 316:119–126
120. Dwivedi AM, Upadhyay S, Johanson G et al (2018) Inflammatory effects of acrolein, crotonaldehyde and hexanal vapors on human primary bronchial epithelial cells cultured at air-liquid interface. Toxicol In Vitro 46:219–228
121. Chary A, Serchi T, Moschini E et al (2019) An in vitro coculture system for the detection of sensitization following aerosol exposure. ALTEX-Altern Anim Ex 36:403–418
122. Henson TE, Navratilova J, Tennant AH et al (2019) In vitro intestinal toxicity of copper oxide nanoparticles in rat and human cell models. Nanotoxicology 13:795–811
123. Pindáková L, Kašpárková V, Kejlová K et al (2017) Behaviour of silver nanoparticles in simulated saliva and gastrointestinal fluids. Int J Pharm 527:12–20
124. Miethling-Graff R, Rumpker R, Richter M et al (2014) Exposure to silver nanoparticles induces size-and dose-dependent oxidative stress and cytotoxicity in human colon carcinoma cells. Toxicol In Vitro 28:1280–1289
125. Williams KM, Gokulan K, Cerniglia CE, Khare S (2016) Size and dose dependent effects of silver nanoparticle exposure on intestinal permeability in an in vitro model of the human gut epithelium. J Nanobiotechnol 14:1–13
126. Horie M, Saito A, Mikami Y et al (2012) Characterization of human lung cancer-associated fibroblasts in three-dimensional in vitro co-culture model. Biochem Biophys Res Commun 423:158–163
127. Correia LL, Johnson J-A, McErlean P et al (2017) SOX2 drives bronchial dysplasia in a novel organotypic model of early human squamous lung cancer. Am J Respir Crit Care Med 195:1494–1508
128. Han K, Pierce SE, Li A et al (2020) CRISPR screens in cancer spheroids identify 3D growth-specific vulnerabilities. Nature 580: 136–141
129. Di Liello R, Ciaramella V, Barra G et al (2019) Ex vivo lung cancer spheroids resemble treatment response of a patient with NSCLC to chemotherapy and immunotherapy: case report and translational study. ESMO open 4:e000536
130. Meenach SA, Tsoras AN, McGARRY RC et al (2016) Development of three-dimensional lung multicellular spheroids in air-and liquid-interface culture for the evaluation of

anticancer therapeutics. Int J Oncol 48: 1701–1709
131. Gupta SK, Torrico Guzmán EA, Meenach SA (2017) Coadministration of a tumor-penetrating peptide improves the therapeutic efficacy of paclitaxel in a novel air-grown lung cancer 3D spheroid model. Int J Cancer 141: 2143–2153
132. Kilin V, Mas C, Constant S et al (2017) Health state dependent multiphoton induced autofluorescence in human 3D in vitro lung cancer model. Sci Rep 7:1–10
133. Mármol I, Sánchez-de-Diego C, Pradilla Dieste A et al (2017) Colorectal carcinoma: a general overview and future perspectives in colorectal cancer. Int J Mol Sci 18:197
134. Elbadawy M, Usui T, Yamawaki H, Sasaki K (2018) Development of an experimental model for analyzing drug resistance in colorectal cancer. Cancers 10:164
135. Ootani A, Li X, Sangiorgi E et al (2009) Sustained in vitro intestinal epithelial culture within a Wnt-dependent stem cell niche. Nat Med 15:701–706
136. Usui T, Sakurai M, Enjoji S, et al (2016) Establishment of a novel model for anticancer drug resistance in three-dimensional primary culture of tumor microenvironment. Stem Cells Int 2016
137. Idris M, Alves MM, Hofstra RM et al (1876) Intestinal multicellular organoids to study colorectal cancer. Biochimica et Biophysica Acta (BBA)-reviews on. Cancer 2021:188586
138. Commandeur S, De Gruijl FR, Willemze R et al (2009) An in vitro three-dimensional model of primary human cutaneous squamous cell carcinoma. Exp Dermatol 18:849–856
139. Meier F, Nesbit M, Hsu M-Y et al (2000) Human melanoma progression in skin reconstructs: biological significance of bFGF. Am J Pathol 156:193–200
140. Bienkowska-Haba M, Luszczek W, Myers JE et al (2018) A new cell culture model to genetically dissect the complete human papillomavirus life cycle. PLoS Pathog 14: e1006846
141. Nayernia Z, Turchi L, Cosset E et al (2013) The relationship between brain tumor cell invasion of engineered neural tissues and in vivo features of glioblastoma. Biomaterials 34:8279–8290
142. Reus AA, Maas WJ, Jansen HT et al (2014) Feasibility of a 3D human airway epithelial model to study respiratory absorption. Toxicol In Vitro 28:258–264
143. Acosta MF, Muralidharan P, Meenach SA et al (2016) In vitro pulmonary cell culture in pharmaceutical inhalation aerosol delivery: 2-D, 3-D, and in situ bioimpactor models. Curr Pharm Des 22:2522–2531
144. Lenz A-G, Stoeger T, Cei D et al (2014) Efficient bioactive delivery of aerosolized drugs to human pulmonary epithelial cells cultured in air–liquid interface conditions. Am J Respir Cell Mol Biol 51:526–535
145. Lin H, Li H, Cho H-J et al (2007) Air-liquid interface (ALI) culture of human bronchial epithelial cell monolayers as an in vitro model for airway drug transport studies. J Pharm Sci 96:341–350
146. Hoffmann W, Gradinaru J, Farcal L et al (2018) Establishment of a human 3D tissue-based assay for upper respiratory tract absorption. Appl In Vitro Toxicol 4:139–148
147. Lu R, Zhao X, Li J et al (2020) Genomic characterisation and epidemiology of 2019 novel coronavirus: implications for virus origins and receptor binding. Lancet 395:565–574
148. Boccellato F, Woelffling S, Imai-Matsushima A et al (2019) Polarised epithelial monolayers of the gastric mucosa reveal insights into mucosal homeostasis and defence against infection. Gut 68:400–413
149. Egles C, Garlick JA, Shamis Y (2010) Three-dimensional human tissue models of wounded skin. In: Epidermal cells. Springer, pp 345–359
150. El Ghalbzouri A, Hensbergen P, Gibbs S et al (2004) Fibroblasts facilitate re-epithelialization in wounded human skin equivalents. Lab Investig 84:102–112
151. de Breij A, Haisma EM, Rietveld M et al (2012) Three-dimensional human skin equivalent as a tool to study Acinetobacter baumannii colonization. Antimicrob Agents Chemother 56:2459–2464
152. Shepherd J, Douglas I, Rimmer S et al (2009) Development of three-dimensional tissue-engineered models of bacterial infected human skin wounds. Tissue Eng Part C Methods 15:475–484
153. Jandhyala SM, Talukdar R, Subramanyam C et al (2015) Role of the normal gut microbiota. World J Gastroenterol: WJG 21:8787
154. Bhatia SN, Ingber DE (2014) Microfluidic organs-on-chips. Nat Biotechnol 32:760–772

Chapter 3

Three-Dimensional Spheroids for Cancer Research

Melissa Anne Tutty and Adriele Prina-Mello

Abstract

In vitro cell culture is one of the most widely used tools used today for increasing our understanding of various things such as protein production, mechanisms of drug action, tissue engineering, and overall cellular biology. For the past decades, however, cancer researchers have relied heavily on conventional two-dimensional (2D) monolayer culture techniques to test a variety of aspects of cancer research ranging from the cytotoxic effects of antitumor drugs to the toxicity of diagnostic dyes and contact tracers. However, many promising cancer therapies have either weak or no efficacy in real-life conditions, therefore delaying or stopping altogether their translating to the clinic. This is, in part, due to the reductionist 2D cultures used to test these materials, which lack appropriate cell-cell contacts, have altered signaling, do not represent the natural tumor microenvironment, and have different drug responses, due to their reduced malignant phenotype when compared to real in vivo tumors. With the most recent advances, cancer research has moved into 3D biological investigation. Three-dimensional (3D) cultures of cancer cells not only recapitulate the in vivo environment better than their 2D counterparts, but they have, in recent years, emerged as a relatively low-cost and scientifically accurate methodology for studying cancer. In this chapter, we highlight the importance of 3D culture, specifically 3D spheroid culture, reviewing some key methodologies for forming 3D spheroids, discussing the experimental tools that can be used in conjunction with 3D spheroids and finally their applications in cancer research.

Key words Spheroid, Cancer research, 3D models

1 Introduction

Today, cell culture is one of the most widely used in vitro tools for increasing our understanding of cell biology, protein production, mechanism of drug action, and tissue engineering [1]. It is also hugely important in the understanding of disease, making it an integral part of cancer research. Most experimentation undertaken within cancer biology heavily relies on in vitro cell experiments undertaken in conventional two-dimensional (2D) environments, where anticancer drugs are tested prior to in vivo testing, and for decades, in vitro 2D cultures have been the gold standard for determining acute toxicity of drugs and nanomedicines. Currently, 2D in vitro assays are based on either primary cell cultures derived

Dania Movia and Adriele Prina-Mello (eds.), *Cancer Cell Culture: Methods and Protocols*,
Methods in Molecular Biology, vol. 2645, https://doi.org/10.1007/978-1-0716-3056-3_3,

© The Author(s), under exclusive license to Springer Science+Business Media, LLC, part of Springer Nature 2023

directly from harvested tissue, or immortalized cell lines, that is, primary cells that have been genetically transformed to generate rapidly proliferating, easily cultured cells with artificial phenotypes that can be grown for prolonged periods. 2D in vitro models grown in two dimensions have access to consistent amounts of nutrients and growth factors from culture medium, thus allowing for homogenous cell growth and proliferation [2]. This coupled with the low cost, efficiency, and ease of use of the 2D in vitro culturing methods, making 2D models a very attractive preclinical tool. Despite these advantages, however, major flaws can be seen with conventional 2D cultures, and they have many limitations. While useful as early predictors of toxicity, immortalized cell lines have reduced metabolic competencies in comparison to their primary cell counterparts. The large majority of these methods also provide little control over the morphology of cells, which determines the biophysical cues, which affect activity of cells in vivo [2]. Furthermore, conventional 2D cell culture relies on the adherence of dissociated cells to flat surfaces, typically sterile polystyrene or glass, which in turn further forces cells into an artificial stretched morphology. Cell polarity is also affected, and there are disturbances in the interactions between the cellular and extracellular environments [1]. 2D cell culture lacks appropriate cell/cell contacts and a natural tumor microenvironment, as well as proper tumor signaling and drug response, which leads to a reduced malignant phenotype when trying to mimic the tumor microenvironment (TME). The TME is a heterogenous and complex structure made of many different cellular (e.g., endothelial cells, mesenchymal stem cells, fibroblasts, lymphocytes, etc.) and noncellular components (e.g., extracellular matrix (ECM), chemokines and cytokines, and growth factors), all of which play critical roles in the development and progression of cancer [3, 4]. Because of these fundamental issues, the vast majority of promising preclinical drugs either have weak efficacy, or none at all, in real patients with cancer, delaying the discovery of new therapies and causing poor prognoses [5]. While significant investments are made today for cancer drug discovery and development, the approval rate for drugs is staggeringly slow, with <5% of drugs reaching the clinic. Most drugs deemed promising show little to no efficacy in patients with actual tumors, which in turn slows down delay of chemotherapeutic drug development. As a means of overcoming these issues, in recent years, more advanced in vitro culture methodologies have emerged as a means of providing a more in vivo–like microenvironment for testing drugs and nanomedicines and for studying cancer, with the ultimate goal of accelerating the passage of medicines through the drug discovery pipeline, while also reducing the number of animals used in experimentation. Due to their more in vivo-like spatial arrangement, cell signaling, cell/cell and cell/matrix interactions, and drug resistance, three-dimensional (3D) cultures of cancer cells

Table 1
Key advantages and disadvantages of 2D and 3D culture methodologies

Type of cell culture	Advantages	Disadvantages
2D	Low cost Fast reproduction Easy to manipulate	Reduced cell/cell and cell/matrix interactions Limited perfusion of oxygen and nutrients More susceptible to pharmacological action from stimuli. Reduced cellular differentiation Different genetic profile to in vivo tissue Increased proliferation compared to in vivo tumors
3D	Heterogeneity in perfusion of oxygen and nutrients Enhanced cell/cell and cell/matrix interactions Physiologically relevant drug penetration More in vivo relevant genetic profile In vivo-like zonation observed, that is, proliferative, quiescent, and necrotic zones	More expensive Certain techniques more complicated and laborious; require specialized equipment

have become more common for better recapitulating in vivo-like cellular environments, and they are now praised as being more scientifically accurate in the preclinical assessment of new drug or nanomedicine candidates, prior to moving into expensive and time-consuming in vivo animal models, as they have the potential to remove underperforming therapies from the drug-discovery pipeline early on, thus saving both time and money. Key advantages and disadvantages of both culture types can be seen in Table 1, and an overview of the key differences between them can be seen in Table 2. The first uses of 3D cell culture date back to the 1970s, where Hamburg and Salmon developed a culture method involving soft agar solutions [6]. Since then, interest in 3D culture has increased exponentially, with many different 3D culture methods widely used today (this is discussed in further detail below).

2 The Tumor Microenvironment (TME) and Its Implications in Cancer

The success of cancer chemotherapy and immunotherapy relies heavily on a deep knowledge of the tumor microenvironment, or TME, and the mechanisms of immune invasion by which stroma, tumor, and immune cells interact and function in complex network [7]. One of the largest challenges today with regard to the development of cancer therapies and overall clinical outcome is

Table 2
Brief overview of the differences between 2D and 3D culture. Morphology, response to compounds, in vivo likeness, and other key parameters determining the predictive value of the in vitro model have been taken into consideration

	Type of culture	
Property	**2D**	**3D**
Morphology	Sheet-like, flat, stretched cells grown in monolayers	In vivo like shape. Cells self-assemble/aggregate to natural state
Proliferation	Can proliferate at a higher rate than in vivo	Proliferate at faster or slower rate depending on cell type/3D system
Protein/gene expression	Often display different expression levels to in vivo models	Protein/gene expression levels similar to those in in vivo models
Access to essential materials	Unlimited access to O_2, metabolites, nutrients and signaling molecules	Varied access due to 3D morphology
Cell-cell interactions	Cannot recapitulate complex cell-cell/cell-matrix interactions due to flat morphology	Appropriate interactions between cell-cell and cell-matrix environments
Multi-cellular systems	Good for studying immune response	Good for co-cultures but complicated when more than two cell types introduced
Sensitivity to drugs/nanomaterials	More sensitive; often succumb to treatment more readily due to reductionist morphology	May be more resistant to treatment, better predictors of in vivo drug and nanomaterial response
Exposure to stimuli	Cells grown in 2D equally exposed to nutrient/growth factors/drugs distributed in culture medium	Depending on spheroid size, materials may not penetrate core and reach all cells
Quality of culture	Reproducible high-performance and simple but highly reductionist	Quality depends on method, but more accurate and have longer lifespan
In vivo likeness	Do not mimic natural presentation of tissues	Great likeness to in vivo as organs and tissues are in 3D form
Cost of maintaining culture	Cheap, all reagents/materials commercially available	More expensive (depends on type), more time-consuming, batch-to-batch variation if using culture substrates, less availability
Time needed for cultures	Minutes to few hours	3–4 h to days

understanding the TME. While cancer immunotherapy is becoming more common in the last decade, we are still far from fully understanding how to harness the immune environment to optimize benefit clinically. The TME consists of a highly varied heterogenous population of malignant cells, rich in tumor stroma; proliferating tumor cells; blood vessels; infiltrating inflammatory cells, for example, tumor-associated macrophages (TAMs) and cancer-associated fibroblasts (CAFs); and other associated tissue cells, as well as the molecules produced and released by these cell types, and it has many critical roles in tumor differentiation, dissemination, epigenetics, and immune invasion.

To date, a variety of different treatment options against cancer exist, including chemotherapy, radiotherapy, surgery, hormonal therapy, and gene targeting [8]. Each of these therapies, with the exception of surgical removal, targets cancerous cells either directly, via cytotoxic drugs, which damage DNA, or radiation, which causes apoptosis, or indirectly, via breakdown of the TME, which in turn deprives tumor cells of the essential materials they need to grow and advance. Despite the importance of these therapies in the treatment of cancer, however, they can induce new biological responses in tumors, via angiogenic or immune modulation. This is what causes resistance to drugs, a primary cause of the failure of cancer treatment in many patients and poor prognosis [5]. There are many TME-associated barriers, which reduce the delivery of drugs to tumors [9]. These include hypoxia (mentioned in more detail below), altered enzyme levels, abnormal vasculature, an immunosuppressive environment, an acidic pH, rigid ECM, and changed metabolic pathways. Abnormal vasculature makes distribution of chemotherapies in cancer cells difficult, which in turn negatively impacts the effectiveness of therapies [10]. Acidic pH also can contribute to resistance to drugs, with increased anaerobic metabolism causing increased production of lactic acid, a factor that causes a drop in pH to around 6.2–7.2 [11]. Cellular hypoxia, a hallmark of cancer that occurs when cancer cells in inner areas of a tumor have an insufficient blood supply, plays critical roles in the development and malignancy of tumors [12]. While in non-tumorous cells, hypoxia does not cause harm, in cancer cells, they switch from oxidative phosphorylation to aerobic glycolysis, in a disorder called Warburg effect. Various immune cells including neutrophils, mast cells, natural killer cells, and macrophages also secrete many soluble factors that promote, among other issues like immunosuppression and chronic inflammation, drug resistance [13–15]. Resistance to anticancer drugs can be further promoted by various mechanisms associated with immune escape, something that occurs during tumor progression [10]. Tumor cells also have the ability to change cellular organization within ECM, which further causes a physical barrier for anticancer drugs to penetrate [16].

With all of these issues, it is critical to develop new therapies that target the tumor-promoting microenvironmental factors, with an overall goal of blocking crosstalk between the TME and cancerous cells. A variety of different strategies exist for this, including co-targeting the TME and cell, altering immune response, or inhibiting ligand-receptor interactions and downstream pathways [5]. However, to date, the study of tumors and the TME is normally done in reductionist environments, which do not consider the many physiological barriers that can cause drug resistance, including accumulation and penetration. Therefore, it is critically important to develop newer, more physiologically relevant models that have the capacity to recapitulate the barriers of the TME in order to accurately and sensitively assess new drug candidates and nanomedicines for cancer. Cell spheroids act as a perfect substitute to existing methodologies as they can mimic the physiology of solid tumors, in terms of not only heterogenous cell population but also complex tumor architecture, ECM deposition, and mass transport barriers, allowing better prediction of anticancer drug activity.

3 Characteristics of Tumors and Spheroids and Types of 3D Spheroid Cultures

Varying cancer cells have the ability to self-assemble in spheroids, depending on their culture environment and whether it favors cell/cell or cell/ECM interactions over cell/substrate interactions. Through close cell/cell or cell/ECM interactions, 3D spheroids can be formed, which very closely mimic the natural spatial arrangement of cells we see in vivo and the environment of avascular tumors, where cells proliferate, aggregate, and differentiate [5]. Briefly, the general concept surrounding 3D spheroids is the formation of 3D structures in which cells form layers, mimicking both the physical and biological structure and function of cells in the in vivo environment. Once a spheroid has grown to a diameter of around 200 μm or larger, they will in general have a round phenotype and will consist of three concentric zones comprised of varying populations of cells. In 3D conformation, cell receives stimuli from local environment, in a similar manner to what we observe in vivo [17]. In comparison to 2D cultures, which are mainly composed of proliferative cells due to necrotic cells being washed away during culture medium replenishment, 3D spheroids are comprised of cells at varying stages, that is, proliferating, quiescent, apoptotic, hypoxic, and necrotic (Fig. 1). In a manner that closely mimics in vivo, the outer cell layer is strongly exposed to culture medium and essential nutrients; therefore, cells remain viable and highly proliferative. Deeper into the spheroid, in the middle layer, there is limited diffusion, and cells receive fewer essential nutrients, oxygen, and growth factors and therefore tend to be in quiescent or hypoxic states, mimicking in vivo tissues, in

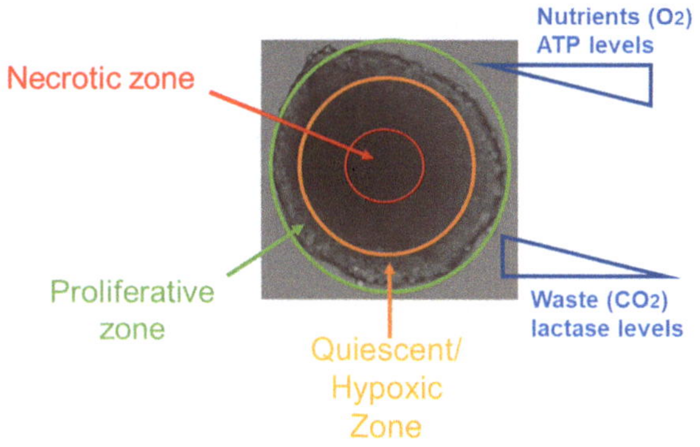

Fig. 1 Structure of a spheroid. 3D spheroids are formed from cells at varying stages to form different zones, that is, proliferating, quiescent, apoptotic, hypoxic, and necrotic zones, mimicking the cellular heterogenicity seen in solid tissues

particular tumors, and deep into the spheroid core, all cells receive little to no nutrients or oxygen, a necrotic core form [18, 19]. In spheroids over 500 μm, hypoxia in inner zones cause altered gene expression, due to stabilization of HIF-1α, a transcription factor, which in turn triggers the Warburg effect and promotes aerobic glycolysis and lactic acid production. This lowers the pH of the inner layer of spheroids [5]. It is this oxygen and lack of essential nutrients, along with the buildup of metabolic waste, that induced the large amount of cell death in the innermost portion of the spheroids.

In 3D environments, in vivo-like cell polarity and morphology are also maintained and can be restored to cells, which were earlier cultured in 2D [17, 20]. An example of this can be seen with liver cells, which show zonation similar to that found in vivo when grown in 3D hepatic cells cluster to form cellular spheroids/micro-tissues. Because of the characteristic zonation in spheroids and the fact they recapitulate several key in vivo tumor characteristics, as well as the gene expression and signaling pathways that are closely related to real tumors, spheroids are an extremely promising model for may applications including oncology, cancer initiation, metastatic processes, and invasion, as well as biodistribution and toxicity studies.

In recent years, many of these 3D culture systems have been developed, and they have quickly become vital experimental tools for many research purposes. Undoubtedly, these advanced systems show great promise in drug and nanomedicine discovery, by acting as a badly needed bridge between conventional cell culture and in vivo animal models. However, much effort and work still need to be done to not only assure reproducibility but to also move this technology further, with the establishment of standardized and validated protocols for 3D culture models [21]. There are a variety of different types of 3D spheroid culture which exist and include monoculture spheroids, co-culture spheroids, and multicellular tumor spheroids.

Monoculture spheroids consist of spheroids formed from one primary or immortalized cell type, and to date, almost all cell types have been cultured in this manner, including liver [22–26], lung [27–29], ovarian [30], prostate [31], and kidney [32]. Most commonly used for drug screening and toxicity studies, monoculture spheroids are normally cultured in ultra-low attachment (ULA) plates, hanging-drop setups, using magnetic levitation, or in rotary cultures. Each of these methods is discussed in more detail below. For spheroid models to be suited to drug discovery, it is critical that they be formed from relevant cell types. Generation of co-culture spheroids, that is, spheroids formed from two complimentary cell types, for example, hepatocytes and Kupffer cells [23], is also highly relevant to drug discovery and toxicity screening, as they provide further information and can accurately model the complex interactions that happen between different cells types, such as endothelial cells, immune cells, and fibroblasts. These interactions are crucial for predicting the therapeutic effect of a compound. Other assays like tumor invasion are severely hampered by only using single cell spheroids embedded in ECM, due to the fact they cannot account for cell/cell adhesion in tumors, making co-culture spheroids another highly relevant 3D model [33].

Multicellular tumor spheroids, or MCTS, are possibly the most well-characterized and studied organotypic model of cancer today. These spheroid types are formed either from isolated tumor cells alone or in combination with other cell types. They can be cultured with or without scaffolds, and their morphologies are varied depending on the nature of the cell type or culture conditions used [34]. This model is extremely attractive as they recapitulate in vivo characteristics in many different areas, including with regard to in vivo tumor gene expression, signaling, growth kinetics, and cell heterogeneity [35]. MCTS size plays a large role in their applicability to studying cancer. Larger MCTS, that is, over 500 μm, have been shown to display similar physicochemical gradients to avascular tumors or micro metastases that have a size of between 0.5 and 1 mm^3. This is due to the lack of diffusion of essential nutrients, waste, O_2, and other soluble factors such as chemokines, cytokines, and growth factors. It is these actors that make MCTS perfect models for studying physicochemical gradients and the effect they have on the characteristics of tumor cells [36]. Lack of O_2 in MCTS induces hypoxia, which in turn triggers changes in gene expression. This causes promotion of lactic acid due to aerobic glycolysis, which further reduces pH levels in the core of MCTS [37]. Furthermore, the buildup of waste materials that cannot escape due to the tightly packed and dense cell layers induces cell death at the core. When a MCTS is sectioned, concentric rings of a varied cell population can be observed, made up of an inner necrotic pocket of cells, with apoptotic cells in a peri necrotic zone, enclosed in a quiescent middle layer of viable cells, followed by an outer layer of migratory and proliferative cells (as schematically shown in Fig. 1) [38]. It is this characteristic structure that makes MCTS an extremely useful model for studying cancer, specifically cancers that can be characterized by densely packed tumor cell clusters, or tumor embolus. An example of this is inflammatory breast cancer, or IBC, a lethal breast cancer subtype that is characterized by incidence of dermal and stromal tumor emboli in breast tissue. When the drainage of lymph fluid is prevented by tumor emboli, following obstruction of lymphatic vessels, swelling, and skin reddening occurs in breast tissue [39, 40].

Many different types of cancer cells have been cultured as MCTS. Examples include breast cancer MCTS cultured from BT-20 and MCF-7 cells in polydimethylsiloxane (PDMS), where it was found that 3D spheroids were more resistant to TRAIL-mediated apoptosis and stem cell-like characteristics than their 2D counterparts (due to COX-2/PGE-2 signaling pathway activation) [41] and ovarian MCTS cultures from HEYA8, SKOV3, and HEY cells using ultra-low attachment plates [42]. Multicellular spheroids of the liver, formed from immortalized hepatic cell lines (or PHH) and other notable nonparenchymal liver cells including liver sinusoidal endothelial cells (LSECs), Kupffer cells (KCs), or hepatic stellate cells (HSCs), provide useful in vitro models with structure and functions, which very closely mimic the human liver [43]. For example, hepatocyte-HSC co-culture spheroids exhibit the same liver-specific morphological and structural properties as it is observed in monocultures. Additionally, they exhibit a 30% increase in albumin secretion, maintained up to 2 months in culture, and enhanced CYP enzyme expression and activity [44, 45]. Co-culture models of KCs and LSECs remain viable for up to 35 days and exhibit enhanced expression of transporters and sensitivity to hepatotoxins [46], making them very useful models for liver toxicity screening. Hepatic spheroids have proved useful in the preclinical assessment of materials like NanoBioMaterials (NBMs) [47–49] and are often used to study conditions such as drug-induced liver injury (DILI) [50], which cannot be accurately predicted using animal models. Hepatic multicellular tumor spheroids (MCTS), that is, 3D spheroids formed from hepatocarcinoma cell lines, have successfully used for assessing NBM cytotoxicity and for predicting in vivo antitumor activity, as detailed by Mikhail and co-workers. Here, a MCTS model was treated with docetaxel (DTX)-loaded micelles, showing that spheroids were significantly more resistant to treatment in comparison to the corresponding 2D monocultures and responded in an in vivo–like manner, demonstrating that MCTS are viable and in vivo-like platforms for evaluating the impact of NBMs, in conditions that closely mimic in vivo tumor microenvironment [35, 51, 52]. MCTS have also been shown to be possess characteristics that make them more amenable to high-throughput screening assays [53].

4 Techniques for Generating 3D Spheroids

To date, there are a wide variety of methods for cultivation of cellular spheroids. Cells are cultured in low-binding surfaces or materials without adhesives, and cell self-assemble into these structures by aggregating to form loose spherical structures after 24 h in culture, which further compacting into tighter more compact structures after 3 days (depending on cell type). Spheroids tend to reach full compactness by day 7, with varying morphologies depending on the cell type cultures and the method used. In terms of in vivo likeness, spheroids approximately 300–500 μm in size have been found to best mimic in vivo tumors, specifically regarding the hypoxic zones and proliferative gradients. There are many different methods used for culturing spheroids, both scaffold-free and scaffold-based, and these will be discussed in more detail below.

4.1 Scaffold-Free Methods

Scaffold-free methods of spheroid formation are possibly to most commonly used methods of cancer spheroid culture today, and they include hanging drop methods, ultra-low attachment/adherent (ULA) culture plates, rotary and rocked cultures, and magnetic levitation. Advantageous as they not only allow cells to self-aggregate and form microtissues as they would in an in vivo situation but by also using a substrate free method, it is possible to avoid complex interactions and phenotypes imparted from culture substrates or matrices, and they form direct cell-cell contacts and maintain viability longer, overcoming some the most commonly seen negative effects of conventional 2D and more complex 3D models. In terms of spheroid formation with specific cell types, using scaffold-free methodologies impart much stronger in vivo–like structure and function, including like cuboidal cell morphology, polarization of hepatocytes, and the formation of secondary structures in the form of canalicular networks, which is observed with liver spheroids [43].

4.1.1 Ultra-Low Attachment (ULA) Plates

Using ULA 96-well plates is one of the most common methods of spheroid cultivation due to their low cost and ease of use. Here, plate surfaces are either coated with a non-adherent substrate to prevent the adhesion of cells, in turn promoting aggregation and spheroid formation, or plates have wells that are shaped specifically to promote natural aggregation of cells to one single, centrally located spheroid, that is, round-bottomed, conical, or V-shaped [54]. In this setting, the formation of spheroids can also be enhanced by centrifugation, which promoted cellular proximity. Agarose or Poly-HEMA (poly(2-hydroxyethyl methacrylate)) is most commonly used to coat plates, by adding approximately 50 μL of solution into each well of 96-well plates, at varying concentrations [55]. By utilizing this technique, many spheroids can simultaneously generate on the same plate, making reproducibility and scale-up easy. Endpoint analysis, including microscopy, is also easier in ULA environments than other culture methods. Spheroids formed using this method are suitable for many areas of cancer research, in particular for drug screening as there is no need to transfer spheroids from one plate to another for cytotoxicity assays. Despite their advantages, however, this method is cell line dependent, with some not forming compact spheroids with this method [56].

4.1.2 Hanging Drop

Another popular scaffold-free method of producing spheroids for cancer research is via the hanging drop method. Commercialized by InSphero™, this method can also routinely be undertaken simply using petri dishes. Here, approximately 25 μL of cell suspension is dropped carefully on the inside of a petri dish lid, which has been pre-treated with phosphate-buffered saline (PBS) to avoid dehydration of cellular suspension. The petri dish lid is then inverted, and as a consequence of surface tension, the drops of cell solution remain suspended in a "hanging drop." Due to the force of gravity, the cells are able to spontaneously self-aggregate, causing the formation of a single spheroid [34, 57]. With this method, large numbers of spheroids can be created on a single petri dish, and the size of spheroids can be carefully controlled by changing starting cell density. On the other hand, however, this method is more labor intensive than others as there are many steps involved, and there is a risk of contamination and cell damage if media evaporates [69–72]. While forming one single spheroid per drop, issues can also arise with heterogenicity in terms of spheroid morphology and size, which in turn impacts standardization and scale-up if large numbers of spheroids are required, specifically for cancer drug screening assays where standardization is key. Recently, various tools have been developed to reduce these limitations [58, 59], including the PANDA system, or pressure-assisted network for droplet accumulation, which is made up of a pressure chip that can create compact and homogenous hanging drop arrays, which in turn facilitates quick and cost-effective spheroid production [58]. Another method, 3D printed hanging-drop dripping arrays, exists, which allows the in situ anticancer compound screening, analysis of tumor metastasis, and migration assays, while also promoting heterotypic spheroid interaction [60].

4.1.3 Rotary Cultures

Rotary cell culture systems were designed in the early 1990s and grow 3D cell spheroids in microgravity. Rotary cell culture systems (RCCS), or cell culture bioreactors, consist of rotating walls that create microgravity environments that maintain cells in suspension and allow the formation of spheroids. Here, cells in suspension rotate slowly about an x-axis, maintaining a continuous free fall [57]. Initial rotation is slow, at approximately 15 rpm, but as spheroids form and become larger, rotation is increased to approximately 25 rpm to keep them in suspension. Spheroids from both primary cells and immortalized cells can be formed using rotary cultures, as well as co-culture spheroids of multiple different cancer cell types. Rotary cultures are useful for long-term-spheroid cultures, and the culture conditions can be highly controlled. Spheroid yield using this method is also high; however, large variability can be observed with this method with regard to spheroid size, and there can be difficulty monitoring spheroid assembly in real time.

4.1.4 Magnetic Levitation

Magnetic levitation is a 3D culture technique that forms spheroids via the treatment of cells with magnetic nanoparticles followed by incubation with magnetic forces. These forces overcome natural gravitational force, allowing cells to levitate and in turn form aggregates [61]. The magnetic nanoparticle assemblies can consist of gold or iron oxide nanoparticles or polymer polylysine. After cells absorb the nanoparticles, a magnet is placed above them (i.e., magnetic levitation) or below them (i.e., 3D magnetic printing) encouraging aggregation and spheroid formation. This is a fast method for spheroid production, with spheroids forming using this method in less than 16 h [62]. As prior preparation of the nanoparticle solutions is necessary however and only a small number of spheroids can be generated, this method is somewhat more labor intensive compared to others.

4.2 Scaffold Formation Methods

While the most common types of 3D models employ cell spheroids used for studying angiogenesis, mimicking the tumor microenvironment and for screening anticancer compounds for toxicity, their fundamental lack of ECM and somewhat simplified biological significance on their own can make them fall short as appropriate models of cancer [63]. In contrast to this, 3D scaffold-based spheroid models have the ability to influence the key mechanical and biochemical signals, which are required for accurate cell-cell and cell-ECM interactions, as well as mimicking the nutrient deficient and indeed hypoxic conditions of the TME [64, 65]. In recent years, the use of 3D scaffold models for studying the complex interactions that occur between cells and the TME has gained increasing attention. Today, 3D scaffold spheroid models are becoming more increasingly used in the culture of patient-derived primary cells, assessing toxicity of drugs and testing new anticancer drugs and therapies for their efficacy on patient samples. It is also possible using scaffolds to co-culture both healthy and malignant cells, therefore recapitulating the heterogeneity observed in the body [63]. With scaffold techniques for spheroid formation, systems that anchor cells in some manner are utilized to mimic the natural structure of the ECM, which allows enhance cell/cell and cell/matrix interactions. Scaffolds can be produced from various materials and include porous microcarriers made up of natural, synthetic, and semisynthetic hydrogels that are made up of cross-linked polymers [5]. Porous scaffolds have been used in bioengineering for many years and have more recently become more popular in cell culture [66, 67] due to their ability to allow increased diffusion of oxygen, nutrients, and metabolic material, while also mimicking the natural architecture of the ECM, allowing good attachment, proliferation, and cellular support [68, 69]. While polymers are the most common materials for scaffolds, natural substances such as hyaluronic acid, collagen, chitosan, and gelatin are also common, as are synthetic materials like poly (lactide-co-glycolide) [83–88]. Scaffold methods are useful in situations where spheroid size needs to be carefully controlled and where larger spheroids need to be produced. They also have been shown to enhance therapeutic resistance and tumor cell invasion [70]; however, collecting spheroids and separating them once they have reached their target side is difficult compared to non-scaffold methods [44]. The most common scaffold methodologies include matrices and hydrogels, micropatterned plates, spinner flasks, microcarriers beads, and microfluidic devices, all discussed in more detail below.

4.2.1 Matrices and Hydrogels

Another popular method employed for culturing spheroids is the use of culture matrices or hydrogels as scaffolds for cell growth [71]. Once such matrix, Matrigel™, has proven to be very useful in the culture of spheroids and has been used to model both the liver [72, 73] and lung [27, 74], among others, with many varying cancer cell lines used. Matrigel™ is a solubilized basement membrane preparation that is extracted from the Engelbreth-Holm-Swarm (EHS) mouse sarcoma, a tumor that is rich in ECM proteins including laminin, its major component, collagen IV, and growth factors [75]. Today, it is one of the most common ECM-based hydrogels used for 3D cancer cell cultures, allowing improved attachment and differentiation of normal and transformed cells and better mimicking the in vivo environment for 2D and 3D cell applications. Culturing primary cells as spheroids in Matrigel™ has also provided more functionally stable and more sensitive cultures than 2D counterparts [23].

In recent years, hydrogels have been also extensively used in ECM-based 3D culture system as a means of mimicking the natural cellular microenvironment [76]. Hydrogels consist of cross-linked polymers with high water content, and they are widely used in 3D cell culture to easily stimulate the growth characteristics of cells. Hydrogels with high porosity provide high fluxes of water content, nutrients, oxygen, and other soluble biomolecules. Natural hydrogels, which are strongly biocompatible, mainly include constituents of ECM proteins such as collagen fibers and hyaluronic acid, as well as chitosan, alginates, or silk, with both synthetic and natural hydrogels acting as suitable scaffolds for the formation of 3D spheroids. Hydrogels have been proven to have many advantages over other methods, with the drug response profiles of prostate, lung, and breast cancer cell lines having shown large differences in dose-response curves of both docetaxel and fulvestrant when cells are grown in collagen when compared to 2D cultures or even 3D culture systems such as ULA plates, or other natural scaffolds like Matrigel [77].

4.2.2 Microcarrier Beads

Microcarrier beads or micropatterned bead systems have been used for decades to form 3D cultures. In this method, cells can adhere to either synthetic (i.e., poly(d,l-lactide-co-glycolide, dextran, etc.) or natural (i.e., cellulose or collagen) coated beads, which in turn forms sphere-like structures [78, 79]. The beads act as a solid support for cellular attachment and allow aggregation, which is useful for cell lines, which do not aggregate spontaneously. Microcarrier beads are an easy, reproducible, and fast method of generating spheroids and allow the formation of heterogenous spheroids from various different cell types; however, one disadvantage is that in vivo-like tumor physiological conditions can't be mimicked.

4.2.3 Spinner Flasks

Spinner flasks create cell spheroids by using constant stirring to promote cell/cell collisions and to prevent cells in suspension from settling. To produce spheroids in spinner flasks, well mixed single suspensions of cells are seeded with mass transfer and the fluid environment of the flask carefully controlled by convective forces generated from magnetic stir bars or an impeller [57]. For spheroid formation using this method, a constant speed of rotation is necessary—too slow and spheroids will settle in the flask, and too high will cause cell damage due to shear stress. Spinner cultures are useful as they can be easily scaled up, producing large numbers of spheroids at one time. Another advantage is that fluid movement will aid mass transport both in and out of spheroids. Culture conditions can be easily controlled by reducing agitation speed and allowing the cells to settle and by media changes or by addition of various growth factors or drugs. Spinner culture has been used with a variety of primary cells, cell lines, and mixtures of two different cell types to form heterotypic spheroids [80]. By decreasing the agitation speed and allowing the cells to settle, culture conditions can be controlled through media changes and addition of drugs and growth factors. Due to the movement in the flask, however, this method may not be useful for cells that have a low cohesiveness or ones that are sensitive to shear stress. Spinner flasks are also not useful for cells that undergo apoptosis while in suspension. Microscopy and imaging of spheroids is also difficult using this method, as cells aggregate and cannot be visualized easily.

4.2.4 Microfluidic Devices

A primary disadvantage of conventional in vitro models is the issue of providing a constant flow of culture medium/nutrients to the cells, and the artificial way medium is renewed sequentially, to either keep cells alive and healthy, or to introduce drugs or materials for toxicity screening [81]. Some of the most recent 3D culture advances have been achieved in an attempt to overcome these limitations, with organ-on-a-chip microsystems developed and capable to reproduce many key structural, functional, biochemical, and mechanical features of living organs, in a single, small device [82]. Microfluidics has also proven to be extremely useful in modelling cancer [83–85].

When using microfluidics devices, cells flow through small networks of channels into micro-chambers. Here, they are portioned and are exposed to rotational flow, which causes aggregation of cells [86]. Microfluidics devices work with immortalized cell lines, primary cells, and multiple cell-type co-cultures [87, 88]. Microfluidic devices are useful for the high-throughput production of spheroids that are size-controlled and for high-throughput analysis due to the fact many of these platforms have biosensors that facilitate real-time imaging and system monitoring [89, 90]. Furthermore, the perfusion system in microfluidics chips allows for careful control of fluid shear stress and the concentration of soluble factors around spheroids [57]. However, despite the many benefits associated with microfluidic models, they are not without fault due to their complex nature. Issues like disaggregation and loss of cells can occur to the constant perfusion of cell culture medium, particularly as cells proliferate and reach confluence. Furthermore, the effect of sheer stress on cultured cells has yet to be fully elucidated in some cases [91]. Finally, the large setup and maintenance costs associated with these models and systems, as well as the large numbers of cells needed to build them and the associated issues with their handling and data analysis, are also limiting factors for the uptake of these complex platform technologies for high-throughput screening in cancer research.

5 Experimental Tools Used with Tumor Spheroids

As a general rule, many of the methods applied to traditional 2D cultures can be applied to 3D spheroids; however, there are limitation associated with some of them. Many different experimental techniques can be applied to characterize and study tumor spheroids. These can be used on spheroids in various states, that is, viable, fixed, or dissociated, and can also be used on spheroids treated with nanomedicines or other anticancer drugs. Common methods used to assess parameters such as morphology or phenotype of tumor spheroids, evaluation of ant-cancer effect drugs/nanomedicines, protein or gene expression, metabolic status, and migration and invasion include biochemical assays (e.g., LDH or CellTiter-Glo™), various microscopy techniques (e.g., confocal microscopy, SEM, TEM), or flow cytometry (e.g., apoptosis, proliferation, and live/dead assays). Some of the most common methodologies used are discussed below, with a detailed description of these techniques and many more found in reviews from Costa et al. [66] and Friedrich et al. [35].

5.1 Optical and Electron Microscopy

The easiest and most commonly used technique for observing and recording changes in cancer spheroid morphology and phenotype is via immunostaining and subsequent microscopy techniques. This can either be done in a simple manner using Brightfield, epifluorescence, or laser scanning confocal microscopy (LSCM) and appropriate software such as ImageJ or Leica LasX [25, 92, 93], or using more advanced and complex methodologies such scanning electron microscopy (SEM) and transmission electron microscopy (TEM) [94, 95]. Fluorescence microscopy and immunostaining can provide critical information about spheroids, ranging from confirming the presence of important secondary structures, for example, bile canaliculi in 3D liver spheroids [25, 96], to ECM deposition in spheroids, which have been stained with laminin, collage IV, or fibronectin antibodies [97, 98]. Key structural information like cytoskeleton arrangement and other parameters such as apoptosis or proliferation can be also examined using fluorescence or confocal microscopy, using widely available fluorescence antibodies like rhodamine phalloidin, Ki-67, Hoechst/DAPI, Annexin V, various caspases, propidium iodide (PI), or TUNEL staining [99, 100]. These dyes can be used to visualize drug or nanoparticle presence in spheroids [25, 51, 93]. LSCM provides more detail of spheroid architecture than previously mentioned methods, but it is limited to spheroids that are relatively small in size due to the limitations of both light and dye penetration [101]. One potential way of overcoming this is by treating spheroids as microtissues and processing them for histological using hematoxylin and eosin (H&E) staining, which identifies varying cell types while also providing information about shape, structures, and patterns of cells and providing an overall picture of spheroid health. Here, hematoxylin stains the nuclei a purple blue color, and eosin stains the ECM and cytoplasm pink, with various other structures staining as different shades or combinations of these colors. By using these dyes to differentiate between both the nuclear and cytoplasmic portions, a general overview of internal spheroid structure can be obtained [25, 102]. Staining for other specific events like proliferation (Ki-67) and cell death (caspases), among others, can also be undertaken [103, 104].

For more detailed microscopic analysis of spheroids, electron microscopy maybe used. Despite being more complex and time-consuming in terms of both sample preparation and imaging, methods such as SEM and TEM provide much higher-resolution images, with integral structures clearly visible [105]. SEM produces very high-resolution images of the superficial topography of tumor spheroids, in comparison to TEM (or multiphoton microscopy), which produces 3D-rendered images of their internal structures [106]. These methods not only provide information on cell/cell interactions inside spheroids, for example, the formation of junctions between adjoining cells and ECM, but also other information including organelle swelling, apoptosis, and the interaction of cells with drugs and nanoparticles [105, 107].

5.2 Molecular Biology Methods and Biological Assays

Many conventional methods for molecular biology can also be applied to spheroids, with some having greater success than others. Standard methods like qRT-PCR and western blotting are useful to evaluate expression of genes and proteins, in both 2D and 3D cultures, with or without treatment of drugs. Both of these methods are based upon dissociation of spheroids and cell lysis to extract both genetic material and proteins from spheroids; however, due to the densely packed cells in spheroids, dissociating spheroids can often be tedious or difficult.

As the delivery of both oxygen and nutrients to spheroids is hindered by their dense cell layers and due to the fact waste products accumulate readily within them, cell viability within spheroids is massively affected. A variety of different assays can be used to assess both cell viability and cytotoxicity. The resazurin or acid phosphatase (AP) assays are cost-effective, sensitive, and high-throughput assays, which measure viability in 2D cultures and 3D spheroids [55, 108]. Other assays like the 2,5-diphenyl-2H-tetrazolium bromide, or MTT assay [109], and the Alamar Blue assay [110], are often used with spheroids; however, supernatant-based assays or ones that do not require spheroid dissociation are often preferred. As stated, full dissociation can be difficult to achieve with spheroids, and thus, the production of reagents or kits that allow assessment of spheroids without dissociation has been prioritized. One such method is the CellTiter-Glo™ reagent for spheroid viability. This commercially available assay is widely used to determine the presence of viable cells in spheroids, a parameter that is determined by the presence of ATP, indicative of metabolically active cells, and is appropriate for even large spheroids as it has better penetration capacity than many other reagents [111–113]. The lactate dehydrogenase or LDH assay, which determines the presence of the cytosolic enzyme LDH released into cell culture supernatants following damage to plasma membrane, is also preferred for use with spheroids due to the fact it is supernatant based and does not require spheroid dissociation [113].

5.3 Flow Cytometry

Another well-known methodology applied to tumor spheroids is flow cytometry, which provides many different types of information on cells including cell cycle, phenotype, viability, proliferation, apoptosis and necrosis, and uptake of materials like drugs and nanomedicines. Due to their large number of single cells from which they are formed, spheroids can be dissociated into single cells for easy analysis via flow cytometry. This method provides statistical information in single cell resolution, making it possible to more accurately investigate drug function in cells, which are packed in a more in vivo-like 3D orientation [114, 115]. To date, flow cytometry has been used to study many different events in tumor spheroids. Beaumont et al. have used flow cytometry to characterize the cell cycle status of cells with respect to their positioning within melanoma spheroids [115], also tracking cell cycle status and viability (discussed in more detail below) in different subsections of tumor spheroids under varying conditions over time. Patra et al. have also utilized flow cytometry for analyzing HepG2 liver spheroids. Here, HepG2 cells formed spheroids of two different sizes within microfluidic chambers, before they were treated with varying combinations of the anticancer compounds cisplatin, resveratrol, and tirapazamine. Then, flow cytometric analysis was undertaken on large number of these uniform spheroids, not only demonstrating that spheroid size plays a critical role in spheroid response to drugs but also that there are great advantages in combining the two technologies for ant-cancer drug screening applications [114].

Briefly mentioned above, one of the most common assays coupled with flow cytometry is the live/dead assay. This type of single cell analysis works via staining live cells with Calcein AM and dead and membrane-damaged ones with ethidium homodimer. This method is particularly useful for spheroids as it allows identification of the various subtypes within tumor spheroids, including proliferating, quiescent, and necrotic cells [106]. However, there are limitations associated with the use of flow cytometry with cell spheroids. To analyze spheroids using this method, spheroids must be enzymatically (using reagents such as trypsin or Accutase®) or mechanically (via pipetting or physical agitation) disassociated in order to get single cell suspensions suitable to be measured via flow cytometry [116]. Dissociation of spheroids is not only laborious and time-consuming (relative to other spheroid assays) but also damages spheroid integrity, which in turn affects viability. Furthermore, due to cells lost during the process of dissociation, large numbers of spheroids are needed for this type of analysis. Another limitation is the inability to evaluate drug penetration in spheroids using flow cytometry due to the need for dissociation. In the last decades, a newer flow methodology has been designed to overcome this issue, the COPAS™ flow cytometer, or Complex Object Parametric Analyzer and Sorter [117]. This instrument has been specifically designed to enable particle/cell measurement, from sizes in the range 20–1500 μm, making it highly useful or assessing viability of tumor spheroids and also in the study of cancer.

Using flow cytometry coupled with spheroids is often also applied to other ever-advancing areas of cancer therapy, such as immunotherapy. The discovery of new immunotherapies that have the ability to specifically target and enhance the responses of T-cells, for example, CAR-T, to cancer is rapidly expanding, and spheroids have proven useful for the evaluation of immunotherapies due to their reproducibility, robustness, and greater in vivo relevance. Previously, reductionist models such as suspension cells or 2D cell monolayers would have been used here; however, these methods are not optimal due to their lack of key features of the tumor microenvironment. Traditional methods for assessing T cell response in 3D models involved the use of imaging-based techniques; however, when spheroids are applied, these methods are often limited as multiple workflows and often multiple instruments are required for quantifying different parameters, and protocols may be time-consuming and laborious, requiring much optimization. There is also an increased risk of data variability as data from different assays need to be correlated. To overcome these issues, various advanced flow cytometry-based assays have been put forward as solutions to address these issues, one being the TIL assay, which measures the degree of T-cell infiltration in spheroids and determines phenotype of non-infiltrated and infiltrated cells, and the second being an ICK spheroid method, whereby cytokine quantification and subset analysis are used to examine activation, killing, memory profiles, and exhaustion of T-cells as they are being co-cultured with spheroids. The iQUE® Advanced High-Throughput Flow Cytometry and a selection of T-cell kits are utilized for both these methods [118].

6 Applications of 3D Cultures in Cancer Research

6.1 Chemoresistance Assessment

Chemoresistance, or the ability of cancer cells to either evade or survive the presence of drugs, is a huge concern and key challenge in oncology and a major reason why current anticancer therapies fail. There are various molecular mechanisms why cancer cells can promote survival and avoid death when responding to common chemotherapies, and they are made up of complex signaling pathways that can be activated by many stimuli [119, 120]. The progression of cancer and chemoresistance are very closely linked to the extracellular environment, which gain specific properties during the development of tumors.

Studying chemoresistance in 2D cultures is not optimal, due to the differences in the ECM between traditional 2D monocultures and what is observed in spheroids. It is now widely accepted that cancer cells, when grown in 3D environments such as spheroids, more accurately mimic drug behavior, in terms of resistance than cells that are grown in 2D reductionist environments [121]. It is thought that this difference is due to the tumor microenvironment and the 3D spatial organization of spheroids, with increased cell/cell and cell/matrix interactions leading to changes in gene expression [5, 122], and examples of this have been seen in various publications. The upregulation of various cell-adhesion molecules, genes, and proteins, including miR-146a, SNED1, and DARP32, has been shown to increase chemoresistance in tumor spheroids formed from pancreatic cells when compared to 2D cultures [123]. Spheroids cultured from DU145 cells are protected against the apoptotic effects of docetaxel by fibronectin, through interactions with insulin-like growth factor 1 receptor (IGF-1) [124]. Further into the molecular mechanism, when grown in 3D multicellular spheroids, as opposed to 2D monolayers, many apoptotic stimuli, including histone deacetylase, proteasome inhibitors, tumor necrosis factor-related apoptosis-inducing ligand (TRAIL), and ribotoxic stressors, were not effective against mesothelioma cells [121]. The enhanced resistance to chemotherapy drugs observed in spheroids is thought to be due to many factors that are associated with composition and organization, including acidic microenvironment, hypoxia, altered cell metabolism, cell heterogeneity, and cell/cell and cell/matrix interactions [122, 125–127]. The way these factors illicit chemoresistance is well-reviewed in one such publication from Nunes et al. [128]. While the majority of studies show that cells in spheroids are in fact exhibit more chemoresistance than cells cultures in 2D, some studies have reported that cells in multicellular tumor spheroids are either equally as sensitive, or more sensitive to anticancer drugs than their 2D counterparts, for example, PS-341, a proteasome inhibitor, has been shown to be as effective in killing prostate and ovarian cells grown as 3D spheroids and tumor cells grown in 2D monolayers [129]. Other studies have also shown that spheroids are more radioresistant than 2D monocultures [130].

6.2 Spheroids in the Study of Nanomedicines

Today, nanomaterials are used in several fields, including chemistry, bioengineering, biotechnology, electronics, and medicine. One of the most common therapeutic uses for nanomaterials involves their use in the treatment and diagnosis of cancer. Here, nanomaterials or NBMs, that is, nanomaterials designed for biomedical applications, are used either as carriers of chemotherapy drugs or, have their physical properties, are harnessed for other uses such as magnetic hyperthermia or deep tissue imaging [131]. These materials are formulated from a wide array of materials, for example, gold, iron oxide, and various polymers, with various coatings or sizes, depending on their potential uses. The key primary advantage associated with NBMs, and the reason they are preferred for oncology, is their preferential accumulation in tumor sites via the enhanced permeability and retention (EPR) effect, that is, the mechanism by which NBMs accumulate in tissues that have a greater vascular permeability, such as cancer sites or sites of inflammation [132].

In recent years, the use of spheroids in the study of nanomedicines and NBMs has become more common, and to date, there is great emphasis currently being placed on more advanced in vitro methodologies for assessing interactions and possible effects of NBMs [133]. This is due to the fact traditional in vitro cell models lack not only the phenotypic details of the in vivo environment but also because physiological function and cross talk between cells also is highly reductionist in traditional models. As detailed in this chapter, spheroids have been described as a potential methodology to overcome these issues, bridging the gap between traditional 2D models and in vivo animal work and acting as better predictors of NBM toxicity [134].

In recent years, various new mono- and co-culture 3D models have been developed, combining different relevant cell types into one organotypic model, which can be used to determine not only NBM toxicity but also their interactions with various cell types in a 3D environment [133]. While the majority of 3D models are still not standardized or validated methods for assessing the cellular interactions and toxicities associated with NBMs, in recent years, many studies have used them in preclinical investigation, for applications in both the optimization of NBM physicochemical properties and screening of their therapeutic effects [135]. In general, spheroid cultures tend to be more resistant to the impact of NBMs when compared to 2D cultures. A variety of factors contribute to the decreased toxicity of NBMs in 3D models. Firstly, as compact spheroids form tight and gap junctions, only cells that are on the spheroid periphery have access to the NBMs, whereas the cells deep within spheroid core often cannot be reached by NBMs. Spheroids lack in fact capillary veins, which facilitate easy transport of NBMs to spheroid core. In contrast, 2D monolayers easily absorb certain types of NBMs due to all cells being in contact with treatment. Secondly, highly differentiated cells have a relatively low endocytic ability as compared to cells cultured in 2D [136]. 3D spheroid models can be useful in the prediction of the tumoral penetration capacity of NBMs of varying sizes and coatings, therefore assisting in the optimization of their physicochemical characteristics [135, 137, 138]. Liver (HepG2) spheroids have been used, for example, by Jiang et al. to underpin how a NBMs charge effects its uptake [139]. Here, positively and negatively charged polystyrene NBMs, often used to carry diagnostic dyes, were loaded into multicellular spheroids under both static and dynamic conditions, in the absence and indeed presence of serum proteins, before their penetrative abilities were investigated using laser scanning confocal microscopy. In this article, it was shown that negatively charged NBMs were more attached to the spheroids surface and could penetrate spheroids more readily. Spheroids have also been used to fine-tune the shape of NBMs, with Zhao et al. demonstrating that spherical poly(1-O-MAFru)-b-PMMA NBMs 30 nm in size exhibit greater uptake than their rod-shaped counterparts, even though both have the same diameter [140]. Fledderman et al. have also used cells cultured as 3D spheroids to assess penetration and uptake of negatively charged SiO_2 nanoparticles, 99 nm in size, in order to obtain a greater understanding of nanoparticle uptake in solid tissues [29, 137, 141, 142].

3D spheroids have also been useful for the screening of the potential of NBM therapeutic efficacy [134, 143–153]. Due to the structural similarities between spheroids and in vivo solid tumors and tissues, the efficacy observed in 3D models is normally reduced when compared to 2D cultures [135]. Other studies have used spheroids in the preclinical assessment of NBM cytotoxicity. Tutty et al. have also undertaken a large-scale study using liver spheroids to assess the impact on viability and cytotoxicity induced by a panel of NBMs often used in cancer diagnosis or treatment, namely, AuNP, a liposome, and three formulations of poly(alkylcyanoacrylate), or PACA NBMs [93]. The results presented here suggest that 3D cell culture models can improve on the predictive outcome for nanoparticle preclinical assessment and toxicity screening [49].

A small number of studies to date have also utilized hepatic spheroids to assess the uptake and penetration of NBMs that can be used in cancer diagnosis and treatment, specifically AuNPs and IONPs, which are not only used as contrast agents but have multimodal applications in that they can be used with radiation therapy for hyperthermia treatment [154–156]. One of the most recent developments with regard to the application of spheroids in cancer is their application to magnetic hyperthermia, and to date, a small number of studies have probed their potential in this area. England et al. have comprehensively reviewed the uptake and distribution of various sized AuNPs (50–100 nm) in spheroids, making comparisons to 2D monolayer cultures and in vivo mouse models [107] in an attempt to overcome the main biological and physical barriers to NBM transport. Miyamoto et al. have also utilized spheroids to assess uptake and localization of various magnetic IONPs, with IONPs shown to be incorporated deep into spheroids and observed in both planar and vertical directions, while also being localized to the lysosomes of spheroids, in a similar manner to what is observed in vivo [157]. Peng et al. have adapted spheroids for the assessment of triphenylphosphonium cation-superparamagnetic iron oxide nanoparticles (TPP-SPIONS), with this study revealing that spheroids, in comparison to cell monolayers, better reflected the in vivo tumor environment, with spheroids less responsive to TPP-SPIONs during magnetic hyperthermia treatment [158].

To date, cancer spheroids are rapidly becoming a preferred cell culture format for the in vitro screening of nanomedicines and NBMs, due to the reductionist environment presented in 2D cultures and the overall metabolic differences overserved between animals and humans. As toxicity is one of the main reasons for which NBMs are either rejected or removed from the market, it is essential to use the most sensitive and accurate screening methods during their hazard assessment. Three-dimensional cancer spheroid cultures offer the possibility to study many elements important to the preclinical assessment of NBMs not feasible in conventional 2D cultures, including the uptake, diffusion, and distribution of materials into under-vascularized and/or hypoxic regions, something which is not possible with 2D monolayers [107, 159]. Not only are 3D tumor spheroids a much better representation of the natural tumor environment, but they also exhibit higher invasiveness and resistance to not only anticancer therapies but also radiotherapy when compared to 2D models. In recent years to improve the prediction of a drug or NBMs clinical efficacy, spheroid models have been made more sophisticated, incorporating new relevant cell types to form multicellular spheroids. These models are very attractive for the preclinical assessment of NBMs, as the cell types normally responsible for the uptake of NBMs in the body are largely ignored in single cell models, for example, in the liver whereby spheroid models are ordinarily formed from hepatocytes, which only take up approximately 7% of NBMS, compared to the non-parenchymal cells like Kupffer cells (KCs), liver sinusoidal endothelial cells (LSECs), and hepatic stellate cells, which take up the vast majority of NBMs [160, 161]. These more complex spheroid models are very attractive in cancer research and could vastly improve the selection of tested NBMs prior to clinical trials, thereby avoiding costly in vivo experiments [161].

6.3 Invasion and Migration

Cancer spheroids have also proven useful for assessing invasion and migration. Determining the invasive and migratory capacity of tumor and stromal cells, and underpinning their underlying mechanisms, is highly relevant in cancer research, from diagnosis to drug development to treatment. Gaining motility and migratory potential is an extremely important hallmark of malignant tumors. Hypoxia and the interaction of soluble mediators with stromal cells, both common characteristics of solid tumors, drives cell invasion and migration, via a series of steps involving changes in cell/cell and cell/ECM adhesion, actin skeleton remodeling, and protein degradation of ECM, among others [162, 163]. Therefore, it is inaccurate and highly reductionist environment for studying the multistep process that is metastasis, as this process relies heavily on a three-dimensional environment where tumor cells can passage and integrate freely. Because of these reasons, 3D tumor spheroids are regarded today as highly relevant in vitro models for studying the processes of migration and invasion [34, 55, 164]. Ideally, migration and invasion assays should be undertaken in 3D cultures, specifically co-cultures that also include a variety of nontumor cells, including dendritic and endothelial cells, macrophages, cancer-activated fibroblasts (CAFs), and immune cells, to better simulate the invasion and migration we see in solid tumors. For example, adhesion and ECM molecule expression patterns are similar in 3D spheroids to what is observed in vivo, with 3D spheroids also inducing metastasis-associated protein expression [55, 164, 165]. Nontumor cells in the tumor microenvironment are also circulating and through paracrine signaling continuously interact with cancer cells, for example, fibroblasts promote contact-dependent cancer cell invasion and motility in 3D co-culture spheroids of colorectal cells, a result validated in vivo and simply not possible in a reductionist 2D culture [166]. Heterotypic 3D cultures of a variety of cells including CAFs and stromal cells, via release of growth factors and cytokines, promote an epithelial to mesenchymal transition (EMT), something that leads to development of tumors and metastasis [167, 168].

A variety of techniques exist today for studying cancer cell invasion and migration in 3D spheroids [28, 55, 169–172]. Spheroids can be placed in transwell chamber slides on top of thick layers of ECM-like materials like Matrigel or collagen, and invasion can be measured following incubation with various chemo-attractants by counting the number of cells that move from top chamber to lower [164]. Furthermore, by incorporating other cell types like endothelial cells or fibroblasts on top of the matrix layer, the capacity of cells to cross cellular barriers can also be assessed [34], something which is very relevant in mimicking how cancer cells cross blood vessel barriers. Invasion assays incorporating spheroids can also be used for assessing drug efficacy [173]. More complex and sophisticated methods are comprising of 3D invasion/migration assays coupled to computers for quantification. These complex methods are highly reproducible and perform robust and optimized experiments for calculating the invasive potential of cells, for example, the speed and extent of tumor spheroid invasion can be sensitively determined using a 3D invasion assay coupled to an imaging cytometer [34, 164, 174].

6.4 Current Outlook: The Application of Cancer Spheroids to Pharma, Personalized Medicine, and the 3Rs Concept

Owing to their varied applications and strong in vivo relevance, the current outlook for 3D cancer spheroids is promising, and in recent years, they have been receiving more attention for a variety of reasons. Due to the sheer heterogeneity of malignant tumors, conventional treatment methods, that is, chemotherapy and radiation therapy, often show low success rates or are only successful in certain cohorts of cancer patients. As a consequence of this, it is crucial to consider a patient's individual characteristics in order to find the most appropriate treatment regime [175]. Developing personalized medicine not only involves methodologies for diagnosing and treating pathological conditions but also treating them, based upon an individual patient's characteristics [176]. These characteristics include genetic, transcriptomics, epigenetic, metabolomic, and proteomic markers of an entire patient's cells, tissue, and body. With regard to cancer, it is also possible to assess gene expression profiles and identify changes, which are occurring. By using personalized medicine, information can be obtained about a specific patient's illness, allowing suitable drug selection and dosing, in combination with optimization of treatments down the line [177]. Furthermore, personalized medicine also influences the preclinical assessment of drugs, not only evaluating their safety and efficacy but also reducing costs associated with pharmaceutical development.

To date, 3D tumor models already play significant roles in developing anticancer drugs and greatly assist in the identification of new cancer therapies [178], with spheroids formed from certain cell lines exhibiting resistance to the cytotoxic drugs taxol, doxorubicin, 5-flurouracil, and cisplatin [179, 180], with MDA-MB-157 spheroids being completely resistant to paclitaxel [181]. It can be hypothesized that the main contributing factor to this resistance is, in a similar manner to solid tumors, insufficient drug penetration, and distribution in the spheroid. To date, the realm of personalized medicine is one of the most promising directions for cancer spheroids. Here, cancer spheroids are cultured using tumor cells directly obtained from patient tissue as opposed to immortalized or primary cell lines and, by using high-throughput personalized medicine methods, can provide appropriate therapies for that patient [178]. These spheroids can be formed from most different tumor types, providing great opportunity for the creation of biobanks that have appropriate patient material, which can be used to not only screen anticancer therapies but also develop new therapeutic agents. There are many examples of the use spheroids with regard to personalized medicine. In tumor cells, hypoxia may cause changes, leading to altered gene expression. This has been seen in particular with breast cancer spheroids and hypoxia inducible factor (HIF) 1, whereby HIF α inhibition in spheroids causes accumulation of doxorubicin, in turn reducing resistance to the drug [182]. Furthermore, cancer spheroids have also been used for evaluating drug delivery systems, for example, a drug delivery system for the NCL-240 proapoptotic drug was developed by Pattni et al., who also assessed its efficacy in spheroids formed from the NCI/ADR-RES ovarian cancer cell line [183]. Patents are also being created for new antitumor substances, and various tumor cell inhibitors are being studied using 3D cancer spheroids. Some of these patented materials include curaxin (US20150045406), a transforming growth factor (TGF)-β inhibitor (WO2020104648) and an ATP inhibitor (957.EP3094317). Additionally, cancer organoids are also being used to assist in the development of cancer "vaccines," with inventions being patented (e.g., (US20190343939)) [178]. A final promising direction for cancer spheroids and personalized medicine is in the creation of biobanks. Here, biological materials for the screening of new drug compounds can be stored in large biobanks, for example, the living breast cancer organoid biobank created by Sachs et al. [178, 184]. Using this biobank, many studies have been undertaken, including DNA and RNA sequencing and histological studies. Other spheroid biobanks have also been created, that is, the genetically engineered stomach organoid biobank created by Nanki et al. [185] and the gastrointestinal tumor biobank created by Vlachogiannis et al. [186].

Beyond the use of cancer spheroids in personalized medicine, they are also being investigated, developed, validated, and utilized by larger pharmaceutical companies, such as Genentech and Astra Zeneca. These companies have identified that the incorporation of 3D models into the drug development pipeline has the potential to not only deliver more translatable results to the clinic but also reduce the numbers of animals used in preclinical research, a factor that will be discussed further below [187]. While more costly than the conventional 2D testing methods used for decades, they are still seen as having great importance, especially if used at the target validation phase of drug discovery, increasing target confidence [188]. Recently, 3D spheroid and organoid research has attracted great industry investment on the back of their track record and profound potential in modelling neoplastic disease [189]. Due to their ease of culture compared to other more complex methods, their reproducibility, and amenability to high-throughput (HTP) assays, 3D spheroids (specifically hepatic spheroids) are also utilized routinely by pharmaceutical companies in their testing regimes and for assessing mechanistic hepatotoxicity and drug induced liver injury (DILI), with great success [190–192]. The focus now on cancer diagnosis and treatment has imparted momentum to the expansion of the spheroid and organoid market, and in recent years, due to the many drawbacks of 2D cell culture in cancer research, their increasing role in closely recapitulating the heterogeneity and complexity of human tumors has been recognized by industry.

One final consideration when discussing tumor spheroids is their applications in the 3R concept of replacing, reducing, and refining the use of animals in in vivo experimentation. The pressing need for more ethically acceptable and physiologically relevant and predictive in vitro models has motivated researchers to develop 3D cell culture models, which may be used as alternatives to animals. Historically, the use of animal models in preclinical research has provided a basic indication of the efficacy and fate of anticancer drugs in organs, while also providing vital information on dosing and potential toxicities. Although the use of animals in the prediction of human toxicities dates to the late 1930s [193], and today both the EMA and the FDA require preclinical assessment to be undertaken in animals prior to any new anticancer drug or therapy entering human clinical trials [194], and despite the assumption that animals are good predictors of human toxicity, much information exists to demonstrate the contrary, and there have been serious questions being currently raised as to the fitness of animals to adequately predict human responses [195]. In recent years, it has been shown that only 60% of drugs entering clinical trials are successful in Phase I (safety) trials, with approximately half of these failures occurring due to unanticipated human toxicity, unseen in animal models. Various factors have been cited as the cause of these discrepancies between humans and animals and the reasons why patients are more sensitive to adverse reactions than their animal experimental counterparts, including environmental factors, pharmacogenetic factors, lifestyle, susceptibility, and heterogeneity of patient populations. Due to these reasons, it has been suggested that animals should be less widely used and potentially replaced with other more reliable, in vivo relevant tissue-mimetic cell-based methods like tumor spheroids.

7 Concluding Remarks

It is becoming a more widely accepted consensus among the scientific community, both within research, pharma, and regulatory bodies that 3D spheroid models are a beneficial tool and key enabler of anticancer treatments due to their greater physiological relevance to the pre-existing methods for testing these therapies. As discussed in this chapter, many key features such as 3D spatial organization, cell/cell and cell/ECM interactions, diffusion gradients, complex cell signaling, and drug resistance are missing in conventional 2D cell culture models, making 3D cancer spheroids a much more valuable and accurate tool for assessing the therapeutic potential of anticancer drug candidates and therefore increasing predictability of efficacy in vivo [5]. Spheroids have the ability to do this by not only identifying but also eliminating nonviable therapies early on in the drug discovery pipeline, speeding up the translation of therapies to the clinic, while also reducing the numbers of animals used in experimentation in a complementary manner. Great advancements have also been made in recent years with regard to spheroid heterogeneity, and recently many studies have moved beyond simple cancer cell spheroids by incorporating components of the immune system (e.g., leukocytes) [196, 197], ECM production (i.e., fibroblasts) [198–200], and vasculature (i.e., endothelial cells) [201–203], in order to represent the diversity of real tumors.

Going forward, the key to further enabling spheroid technology is through the standardization of spheroid generation and characterization protocols, which will overcome some of the hurdles faced with spheroids today, including variation in morphology and reproducibility. Suitable protocols for the large-scale and high-throughput production of spheroids are also critical to reduce the workflow involved in culture and analysis. With this being said, considerable progress has been made in recent years with regard in cancer spheroids, and it is anticipated that they will be, in the near future, a feasible model that will not only model cancer more accurately than current models but also significantly impact the transfer of anticancer drugs bench to bedside. To conclude, the use of 3D cancer spheroids in cancer research, for addressing many scientific questions ranging from drug accumulation and antitumor activity to tumor penetration, is fast becoming a reality and in the near future is anticipated to be a key step in preclinical assessment, bridging the gap between reductionist 2D models and costly, time-consuming, and less ethically acceptable in vivo studies.

Acknowledgments

We acknowledged the partial financial support from the European Union under H2020 REFINE (ref 761104), and EXPERT (ref 825828).

The authors would also like to thank Dr. G. McManus, Microscopy and Imaging facility manager; Professor C. O'Farrelly, Chair in Comparative Immunology, Trinity College Dublin; Dr. Dania Movia, Trinity College Dublin; and Professor M. Santin, from University of Brighton, for the technical discussions and insight on 3D spheroids imaging and biology.

References

1. Kapalczynska M, Kolenda T, Przybyla W, Zajaczkowska M, Teresiak A, Filas V et al (2018) 2D and 3D cell cultures – a comparison of different types of cancer cell cultures. Arch Med Sci 14(4):910–919. https://doi.org/10.5114/aoms.2016.63743
2. Duval K, Grover H, Han L-H, Mou Y, Pegoraro AF, Fredberg J et al (2017) Modeling physiological events in 2D vs. 3D cell culture. Physiology (Bethesda) 32(4):266–277. https://doi.org/10.1152/physiol.00036.2016

3. Quail DF, Joyce JA (2013) Microenvironmental regulation of tumor progression and metastasis. Nat Med 19(11):1423–1437. https://doi.org/10.1038/nm.3394
4. Belli C, Trapani D, Viale G, D'Amico P, Duso BA, Della Vigna P et al (2018) Targeting the microenvironment in solid tumors. Cancer Treat Rev 65:22–32. https://doi.org/10.1016/j.ctrv.2018.02.004
5. Pinto B, Henriques AC, Silva PMA, Bousbaa H (2020) Three-dimensional spheroids as in vitro preclinical models for cancer research. Pharmaceutics 12(12):1186. https://doi.org/10.3390/pharmaceutics12121186
6. Hamburger AW, Salmon SE (1977) Primary bioassay of human tumor stem cells. Science (New York, NY) 197(4302):461–463. https://doi.org/10.1126/science.560061
7. Labani-Motlagh A, Ashja-Mahdavi M, Loskog A (2020) The tumor microenvironment: a milieu hindering and obstructing antitumor immune responses. Front Immunol 11:940. https://doi.org/10.3389/fimmu.2020.00940
8. Wang JJ, Lei KF, Han F (2018) Tumor microenvironment: recent advances in various cancer treatments. Eur Rev Med Pharmacol Sci 22(12):3855–3864. https://doi.org/10.26355/eurrev_201806_15270
9. Jing X, Yang F, Shao C, Wei K, Xie M, Shen H et al (2019) Role of hypoxia in cancer therapy by regulating the tumor microenvironment. Mol Cancer 18(1):157. https://doi.org/10.1186/s12943-019-1089-9
10. Wu T, Dai Y (2017) Tumor microenvironment and therapeutic response. Cancer Lett 387:61–68. https://doi.org/10.1016/j.canlet.2016.01.043
11. White KA, Grillo-Hill BK, Barber DL (2017) Cancer cell behaviors mediated by dysregulated pH dynamics at a glance. J Cell Sci 130(4):663–669. https://doi.org/10.1242/jcs.195297
12. Hannon G, Tansi FL, Hilger I, Prina-Mello A (2021) The effects of localized heat on the hallmarks of cancer. Adv Ther 4(7):2000267. https://doi.org/10.1002/adtp.202000267
13. Jarosz-Biej M, Smolarczyk R, Cichoń T, Kułach N (2019) Tumor microenvironment as a "game changer" in cancer radiotherapy. Int J Mol Sci 20(13):3212
14. Riera-Domingo C, Audigé A, Granja S, Cheng W-C, Ho P-C, Baltazar F et al (2020) Immunity, hypoxia, and metabolism–the Ménage à Trois of cancer: implications for immunotherapy. Physiol Rev 100(1):1–102. https://doi.org/10.1152/physrev.00018.2019
15. Pitt JM, Marabelle A, Eggermont A, Soria JC, Kroemer G, Zitvogel L (2016) Targeting the tumor microenvironment: removing obstruction to anticancer immune responses and immunotherapy. Ann Oncol 27(8):1482–1492. https://doi.org/10.1093/annonc/mdw168
16. Najafi M, Farhood B, Mortezaee K (2019) Extracellular matrix (ECM) stiffness and degradation as cancer drivers. J Cell Biochem 120(3):2782–2790. https://doi.org/10.1002/jcb.27681
17. Kapałczyńska M, Kolenda T, Przybyła W, Zajączkowska M, Teresiak A, Filas V et al (2018) 2D and 3D cell cultures – a comparison of different types of cancer cell cultures. Arch Med Sci 14(4):910–919. https://doi.org/10.5114/aoms.2016.63743
18. Kim JB (2005) Three-dimensional tissue culture models in cancer biology. Semin Cancer Biol 15(5):365–377. https://doi.org/10.1016/j.semcancer.2005.05.002
19. Gurski LA, Petrelli NJ, Jia X, Farach-Carson MC (2010) 3D matrices for anti-cancer drug testing and development. Oncol Issues 25(1):20–25. https://doi.org/10.1080/10463356.2010.11883480
20. Benya PD, Shaffer JD (1982) Dedifferentiated chondrocytes reexpress the differentiated collagen phenotype when cultured in agarose gels. Cell 30(1):215–224. https://doi.org/10.1016/0092-8674(82)90027-7
21. Edmondson R, Broglie JJ, Adcock AF, Yang L (2014) Three-dimensional cell culture systems and their applications in drug discovery and cell-based biosensors. Assay Drug Dev Technol 12(4):207–218. https://doi.org/10.1089/adt.2014.573
22. Llewellyn SV, Conway GE, Shah U-K, Evans SJ, Jenkins GJS, Clift MJD et al (2020) Advanced 3D liver models for in vitro genotoxicity testing following long-term nanomaterial exposure. J Vis Exp 160:e61141. https://doi.org/10.3791/61141
23. Bell CC, Dankers ACA, Lauschke VM, Sison-Young R, Jenkins R, Rowe C et al (2018) Comparison of hepatic 2D Sandwich cultures and 3D spheroids for long-term toxicity applications: a multicenter study. Toxicol Sci 162(2):655–666. https://doi.org/10.1093/toxsci/kfx289
24. Chang TT, Hughes-Fulford M (2009) Monolayer and spheroid culture of human liver hepatocellular carcinoma cell line cells demonstrate distinct global gene expression

patterns and functional phenotypes. Tissue Eng Part A 15(3):559–567. https://doi.org/10.1089/ten.tea.2007.0434
25. Gaskell H, Sharma P, Colley HE, Murdoch C, Williams DP, Webb SD (2016) Characterization of a functional C3A liver spheroid model. Toxicol Res 5(4):1053–1065. https://doi.org/10.1039/c6tx00101g
26. Kizawa H, Nagao E, Shimamura M, Zhang G, Torii H (2017) Scaffold-free 3D bio-printed human liver tissue stably maintains metabolic functions useful for drug discovery. Biochem Biophys Rep 10:186–191. https://doi.org/10.1016/j.bbrep.2017.04.004
27. Fessart D, Begueret H, Delom F (2013) Three-dimensional culture model to distinguish normal from malignant human bronchial epithelial cells. Eur Respir J 42(5):1345–1356. https://doi.org/10.1183/09031936.00118812
28. Shabalina EY, Skorova EY, Chudakova DA, Anikin VB, Reshetov IV, Mynbaev OA et al (2021) The matrix-dependent 3D spheroid model of the migration of non-small cell lung cancer: a step towards a rapid automated screening. Front Mol Biosci 8:610407. https://doi.org/10.3389/fmolb.2021.610407
29. Xu WH, Han M, Dong Q, Fu ZX, Diao YY, Liu H et al (2012) Doxorubicin-mediated radiosensitivity in multicellular spheroids from a lung cancer cell line is enhanced by composite micelle encapsulation. Int J Nanomedicine 7:2661–2671. https://doi.org/10.2147/ijn.s30445
30. Boylan KLM, Manion RD, Shah H, Skubitz KM, Skubitz APN (2020) Inhibition of ovarian cancer cell spheroid formation by synthetic peptides derived from Nectin-4. Int J Mol Sci 21(13):4637. https://doi.org/10.3390/ijms21134637
31. Rodríguez-Dorantes M, Cruz-Hernandez CD, Cortés-Ramírez SA, Cruz-Burgos JM, Reyes-Grajeda JP, Peralta-Zaragoza O et al (2021) Prostate cancer spheroids: a three-dimensional model for studying tumor heterogeneity. In: Robles-Flores M (ed) Cancer cell signaling: methods and protocols. Springer US, New York, pp 13–17
32. Omer D, Pleniceanu O, Gnatek Y, Namestnikov M, Cohen-Zontag O, Goldberg S et al (2021) Human kidney spheroids and monolayers provide insights into SARS-CoV-2 renal interactions. J Am Soc Nephrol 32(9):2242–2254. https://doi.org/10.1681/asn.2020111546
33. Huang YL, Shiau C, Wu C, Segall JE, Wu M (2020) The architecture of co-culture spheroids regulates tumor invasion within a 3D extracellular matrix. Biophys Rev Lett 15(3):131–141. https://doi.org/10.1142/s1793048020500034
34. Nath S, Devi GR (2016) Three-dimensional culture systems in cancer research: focus on tumor spheroid model. Pharmacol Ther 163:94–108. https://doi.org/10.1016/j.pharmthera.2016.03.013
35. Friedrich J, Seidel C, Ebner R, Kunz-Schughart LA (2009) Spheroid-based drug screen: considerations and practical approach. Nat Protoc 4(3):309–324. https://doi.org/10.1038/nprot.2008.226
36. Groebe K, Mueller-Klieser W (1991) Distributions of oxygen, nutrient, and metabolic waste concentrations in multicellular spheroids and their dependence on spheroid parameters. Eur Biophys J 19(4):169–181. https://doi.org/10.1007/BF00196343
37. Alvarez-Pérez J, Ballesteros P, Cerdán S (2005) Microscopic images of intraspheroidal pH by 1H magnetic resonance chemical shift imaging of pH sensitive indicators. Magn Reson Mater Phys Biol Med 18(6):293–301. https://doi.org/10.1007/s10334-005-0013-z
38. Bell HS, Whittle IR, Walker M, Leaver HA, Wharton SB (2001) The development of necrosis and apoptosis in glioma: experimental findings using spheroid culture systems. Neuropathol Appl Neurobiol 27(4):291–304. https://doi.org/10.1046/j.0305-1846.2001.00319.x
39. Vermeulen PB, van Golen KL, Dirix LY (2010) Angiogenesis, lymphangiogenesis, growth pattern, and tumor emboli in inflammatory breast cancer. Cancer 116(S11):2748–2754. https://doi.org/10.1002/cncr.25169
40. Lehman HL, Dashner EJ, Lucey M, Vermeulen P, Dirix L, Laere SV et al (2013) Modeling and characterization of inflammatory breast cancer emboli grown in vitro. Int J Cancer 132(10):2283–2294. https://doi.org/10.1002/ijc.27928
41. Chandrasekaran S, Marshall JR, Messing JA, Hsu JW, King MR (2014) TRAIL-mediated apoptosis in breast cancer cells cultured as 3D spheroids. PLoS One 9(10):e111487. https://doi.org/10.1371/journal.pone.0111487
42. Tong JG, Valdes YR, Barrett JW, Bell JC, Stojdl D, McFadden G et al (2015) Evidence for differential viral oncolytic efficacy in an in vitro model of epithelial ovarian cancer metastasis. Mol Ther Oncol 2:15013. https://doi.org/10.1038/mto.2015.13

43. Peshwa MV, Wu FJ, Sharp HL, Cerra FB, Hu WS (1996) Mechanistics of formation and ultrastructural evaluation of hepatocyte spheroids. In Vitro Cell Dev Biol Anim 32(4): 197–203. https://doi.org/10.1007/bf02722946
44. Wong SF, No da Y, Choi YY, Kim DS, Chung BG, Lee SH (2011) Concave microwell based size-controllable hepatosphere as a three-dimensional liver tissue model. Biomaterials 32(32):8087–8096. https://doi.org/10.1016/j.biomaterials.2011.07.028
45. Riccalton-Banks L, Liew C, Bhandari R, Fry J, Shakesheff K (2003) Long-term culture of functional liver tissue: three-dimensional coculture of primary hepatocytes and stellate cells. Tissue Eng 9(3):401–410. https://doi.org/10.1089/107632703322066589
46. Messner S, Agarkova I, Moritz W, Kelm JM (2013) Multi-cell type human liver microtissues for hepatotoxicity testing. Arch Toxicol 87(1):209–213. https://doi.org/10.1007/s00204-012-0968-2
47. Lee J, Lilly GD, Doty RC, Podsiadlo P, Kotov NA (2009) In vitro toxicity testing of nanoparticles in 3D cell culture. Small 5(10): 1213–1221. https://doi.org/10.1002/smll.200801788
48. Elje E, Mariussen E, Moriones OH, Bastus NG, Puntes V, Kohl Y et al (2020) Hepato (Geno)Toxicity assessment of nanoparticles in a HepG2 liver spheroid model. Nanomaterials (Basel) 10(3):545. https://doi.org/10.3390/nano10030545
49. Dubiak-Szepietowska M, Karczmarczyk A, Jonsson-Niedziolka M, Winckler T, Feller KH (2016) Development of complex-shaped liver multicellular spheroids as a human-based model for nanoparticle toxicity assessment in vitro. Toxicol Appl Pharmacol 294:78–85. https://doi.org/10.1016/j.taap.2016.01.016
50. Vorrink SU, Zhou Y, Ingelman-Sundberg M, Lauschke VM (2018) Prediction of drug-induced hepatotoxicity using long-term stable primary hepatic 3D spheroid cultures in chemically defined conditions. Toxicol Sci 163(2):655–665. https://doi.org/10.1093/toxsci/kfy058
51. Mikhail AS, Eetezadi S, Allen C (2013) Multicellular tumor spheroids for evaluation of cytotoxicity and tumor growth inhibitory effects of nanomedicines in vitro: a comparison of docetaxel-loaded block copolymer micelles and Taxotere(R). PLoS One 8(4): e62630. https://doi.org/10.1371/journal.pone.0062630
52. Ramaiahgari SC, den Braver MW, Herpers B, Terpstra V, Commandeur JN, van de Water B et al (2014) A 3D in vitro model of differentiated HepG2 cell spheroids with improved liver-like properties for repeated dose high-throughput toxicity studies. Arch Toxicol 88(5):1083–1095. https://doi.org/10.1007/s00204-014-1215-9
53. Kunz-Schughart LA, Kreutz M, Knuechel R (1998) Multicellular spheroids: a three-dimensional in vitro culture system to study tumour biology. Int J Exp Pathol 79(1):1–23. https://doi.org/10.1046/j.1365-2613.1998.00051.x
54. Fang Y, Eglen RM (2017) Three-dimensional cell cultures in drug discovery and development. SLAS Discov 22(5):456–472. https://doi.org/10.1177/1087057117696795
55. Vinci M, Gowan S, Boxall F, Patterson L, Zimmermann M, Court W et al (2012) Advances in establishment and analysis of three-dimensional tumor spheroid-based functional assays for target validation and drug evaluation. BMC Biol 10:29. https://doi.org/10.1186/1741-7007-10-29
56. Ekert JE, Johnson K, Strake B, Pardinas J, Jarantow S, Perkinson R et al (2014) Three-dimensional lung tumor microenvironment modulates therapeutic compound responsiveness in vitro – implication for drug development. PLoS One 9(3):e92248. https://doi.org/10.1371/journal.pone.0092248
57. Achilli T-M, Meyer J, Morgan JR (2012) Advances in the formation, use and understanding of multi-cellular spheroids. Expert Opin Biol Ther 12(10):1347–1360. https://doi.org/10.1517/14712598.2012.707181
58. Cho CY, Chiang TH, Hsieh LH, Yang WY, Hsu HH, Yeh CK et al (2020) Development of a novel hanging drop platform for engineering controllable 3D microenvironments. Front Cell Dev Biol 8:327. https://doi.org/10.3389/fcell.2020.00327
59. Wu HW, Hsiao YH, Chen CC, Yet SF, Hsu CH (2016) A PDMS-based microfluidic hanging drop Chip for Embryoid body formation. Molecules 21(7):882. https://doi.org/10.3390/molecules21070882
60. Zhao L, Xiu J, Liu Y, Zhang T, Pan W, Zheng X et al (2019) A 3D printed hanging drop dripper for tumor spheroids analysis without recovery. Sci Rep 9(1):19717. https://doi.org/10.1038/s41598-019-56241-0
61. Souza GR, Molina JR, Raphael RM, Ozawa MG, Stark DJ, Levin CS et al (2010) Three-dimensional tissue culture based on magnetic cell levitation. Nat Nanotechnol 5(4):

291–296. https://doi.org/10.1038/nnano.2010.23

62. Verjans ET, Doijen J, Luyten W, Landuyt B, Schoofs L (2018) Three-dimensional cell culture models for anticancer drug screening: worth the effort? J Cell Physiol 233(4): 2993–3003. https://doi.org/10.1002/jcp.26052
63. Unnikrishnan K, Thomas LV, Ram Kumar RM (2021) Advancement of scaffold-based 3D cellular models in cancer tissue engineering: an update. Front Oncol 11:733652. https://doi.org/10.3389/fonc.2021.733652
64. Efraim Y, Schoen B, Zahran S, Davidov T, Vasilyev G, Baruch L et al (2019) 3D structure and processing methods direct the biological attributes of ECM-based cardiac scaffolds. Sci Rep 9(1):5578. https://doi.org/10.1038/s41598-019-41831-9
65. Li W, Hu X, Yang S, Wang S, Zhang C, Wang H et al (2018) A novel tissue-engineered 3D tumor model for anti-cancer drug discovery. Biofabrication 11(1):015004. https://doi.org/10.1088/1758-5090/aae270
66. Costa EC, Moreira AF, de Melo-Diogo D, Gaspar VM, Carvalho MP, Correia IJ (2016) 3D tumor spheroids: an overview on the tools and techniques used for their analysis. Biotechnol Adv 34(8):1427–1441. https://doi.org/10.1016/j.biotechadv.2016.11.002
67. Rijal G, Li W (2017) A versatile 3D tissue matrix scaffold system for tumor modeling and drug screening. Sci Adv 3(9):e1700764. https://doi.org/10.1126/sciadv.1700764
68. Xiao Y, Zhou M, Zhang M, Liu W, Zhou Y, Lang M (2019) Hepatocyte culture on 3D porous scaffolds of PCL/PMCL. Colloids Surf B Biointerfaces 173:185–193. https://doi.org/10.1016/j.colsurfb.2018.09.064
69. Kuriakose AE, Hu W, Nguyen KT, Menon JU (2019) Scaffold-based lung tumor culture on porous PLGA microparticle substrates. PLoS One 14(5):e0217640. https://doi.org/10.1371/journal.pone.0217640
70. Wang X, Dai X, Zhang X, Li X, Xu T, Lan Q (2018) Enrichment of glioma stem cell-like cells on 3D porous scaffolds composed of different extracellular matrix. Biochem Biophys Res Commun 498(4):1052–1057. https://doi.org/10.1016/j.bbrc.2018.03.114
71. Godoy P, Hewitt NJ, Albrecht U, Andersen ME, Ansari N, Bhattacharya S et al (2013) Recent advances in 2D and 3D in vitro systems using primary hepatocytes, alternative hepatocyte sources and non-parenchymal liver cells and their use in investigating mechanisms of hepatotoxicity, cell signaling and ADME. Arch Toxicol 87(8):1315–1530. https://doi.org/10.1007/s00204-013-1078-5
72. Molina-Jimenez F, Benedicto I, Dao Thi VL, Gondar V, Lavillette D, Marin JJ et al (2012) Matrigel-embedded 3D culture of Huh-7 cells as a hepatocyte-like polarized system to study hepatitis C virus cycle. Virology 425(1): 31–39. https://doi.org/10.1016/j.virol.2011.12.021
73. Tao F, Sayo K, Sugimoto K, Aoki S, Kojima N (2020) Development of a tunable method to generate various three-dimensional microstructures by replenishing macromolecules such as extracellular matrix components and polysaccharides. Sci Rep 10(1):6567. https://doi.org/10.1038/s41598-020-63621-4
74. Meenach SA, Tsoras AN, McGarry RC, Mansour HM, Hilt JZ, Anderson KW (2016) Development of three-dimensional lung multicellular spheroids in air- and liquid-interface culture for the evaluation of anticancer therapeutics. Int J Oncol 48(4):1701–1709. https://doi.org/10.3892/ijo.2016.3376
75. Hughes CS, Postovit LM, Lajoie GA (2010) Matrigel: a complex protein mixture required for optimal growth of cell culture. Proteomics 10(9):1886–1890. https://doi.org/10.1002/pmic.200900758
76. Book review (2012) Pharmaceutical biotechnology: drug discovery and clinical applications (2nd edition). Biotechnol J 7(9): 1061–1062. https://doi.org/10.1002/biot.201200115
77. Stock K, Estrada MF, Vidic S, Gjerde K, Rudisch A, Santo VE et al (2016) Capturing tumor complexity in vitro: comparative analysis of 2D and 3D tumor models for drug discovery. Sci Rep 6:28951. https://doi.org/10.1038/srep28951
78. Antoni D, Burckel H, Josset E, Noel G (2015) Three-dimensional cell culture: a breakthrough in vivo. Int J Mol Sci 16(3): 5517–5527
79. Huang L, Abdalla AME, Xiao L, Yang G (2020) Biopolymer-based microcarriers for three-dimensional cell culture and engineered tissue formation. Int J Mol Sci 21(5):1895
80. Kelm JM, Fussenegger M (2004) Microscale tissue engineering using gravity-enforced cell assembly. Trends Biotechnol 22(4):195–202. https://doi.org/10.1016/j.tibtech.2004.02.002
81. Gebhardt R, Hengstler JG, Muller D, Glockner R, Buenning P, Laube B et al

(2003) New hepatocyte in vitro systems for drug metabolism: metabolic capacity and recommendations for application in basic research and drug development, standard operation procedures. Drug Metab Rev 35(2–3):145–213. https://doi.org/10.1081/dmr-120023684
82. Huh D, Hamilton GA, Ingber DE (2011) From 3D cell culture to organs-on-chips. Trends Cell Biol 21(12):745–754. https://doi.org/10.1016/j.tcb.2011.09.005
83. Sontheimer-Phelps A, Hassell BA, Ingber DE (2019) Modelling cancer in microfluidic human organs-on-chips. Nat Rev Cancer 19(2):65–81. https://doi.org/10.1038/s41568-018-0104-6
84. Mondadori C, Crippa M, Moretti M, Candrian C, Lopa S, Arrigoni C (2020) Advanced microfluidic models of cancer and immune cell extravasation: a systematic review of the literature. Front Bioeng Biotechnol 8:907. https://doi.org/10.3389/fbioe.2020.00907
85. Boussommier-Calleja A, Li R, Chen MB, Wong SC, Kamm RD (2016) Microfluidics: a new tool for modeling cancer-immune interactions. Trends Cancer 2(1):6–19. https://doi.org/10.1016/j.trecan.2015.12.003
86. Okuyama T, Yamazoe H, Mochizuki N, Khademhosseini A, Suzuki H, Fukuda J (2010) Preparation of arrays of cell spheroids and spheroid-monolayer cocultures within a microfluidic device. J Biosci Bioeng 110(5):572–576. https://doi.org/10.1016/j.jbiosc.2010.05.013
87. Huang CP, Lu J, Seon H, Lee AP, Flanagan LA, Kim HY et al (2009) Engineering microscale cellular niches for three-dimensional multicellular co-cultures. Lab Chip 9(12):1740–1748. https://doi.org/10.1039/b818401a
88. Hsiao AY, Torisawa YS, Tung YC, Sud S, Taichman RS, Pienta KJ et al (2009) Microfluidic system for formation of PC-3 prostate cancer co-culture spheroids. Biomaterials 30(16):3020–3027. https://doi.org/10.1016/j.biomaterials.2009.02.047
89. Jin HJ, Cho YH, Gu JM, Kim J, Oh YS (2011) A multicellular spheroid formation and extraction chip using removable cell trapping barriers. Lab Chip 11(1):115–119. https://doi.org/10.1039/c0lc00134a
90. Agastin S, Giang UB, Geng Y, Delouise LA, King MR (2011) Continuously perfused microbubble array for 3D tumor spheroid model. Biomicrofluidics 5(2):24110. https://doi.org/10.1063/1.3596530
91. Lee SA, No da Y, Kang E, Ju J, Kim DS, Lee SH (2013) Spheroid-based three-dimensional liver-on-a-chip to investigate hepatocyte-hepatic stellate cell interactions and flow effects. Lab Chip 13(18):3529–3537. https://doi.org/10.1039/c3lc50197c
92. Leary E, Rhee C, Wilks BT, Morgan JR (2018) Quantitative live-cell confocal imaging of 3D spheroids in a high-throughput format. SLAS Technol 23(3):231–242. https://doi.org/10.1177/2472630318756058
93. Tutty MA, Adriele P-M (2021) Pre-clinical 2D and 3D toxicity response to a panel of nanomaterials; Comparative assessment of NBM-induced liver toxicity. Drug Deliv Transl Res 12(9):2157–2217
94. Jaros J, Petrov M, Tesarova M, Hampl A (2017) Revealing 3D ultrastructure and morphology of stem cell spheroids by electron microscopy. Methods Mol Biol 1612:417–431. https://doi.org/10.1007/978-1-4939-7021-6_30
95. Relucenti M, Francescangeli F, De Angelis ML, D'Andrea V, Miglietta S, Pilozzi E et al (2021) The ultrastructural analysis of human colorectal cancer stem cell-derived spheroids and their mouse xenograft shows that the same cells types have different ratios. Biology (Basel) 10(9):929. https://doi.org/10.3390/biology10090929
96. Kyffin JA, Sharma P, Leedale J, Colley HE, Murdoch C, Harding AL et al (2019) Characterisation of a functional rat hepatocyte spheroid model. Toxicol In Vitro 55:160–172. https://doi.org/10.1016/j.tiv.2018.12.014
97. Correa de Sampaio P, Auslaender D, Krubasik D, Failla AV, Skepper JN, Murphy G et al (2012) A heterogeneous in vitro three dimensional model of tumour-stroma interactions regulating sprouting angiogenesis. PLoS One 7(2):e30753. https://doi.org/10.1371/journal.pone.0030753
98. Nederman T, Norling B, Glimelius B, Carlsson J, Brunk U (1984) Demonstration of an extracellular matrix in multicellular tumor spheroids. Cancer Res 44(7):3090–3097
99. Granato G, Ruocco MR, Iaccarino A, Masone S, Calì G, Avagliano A et al (2017) Generation and analysis of spheroids from human primary skin myofibroblasts: an experimental system to study myofibroblasts deactivation. Cell Death Discov 3(1):17038. https://doi.org/10.1038/cddiscovery.2017.38

100. Bergdorf KN, Phifer CJ, Bechard ME, Lee MA, McDonald OG, Lee E et al (2021) Immunofluorescent staining of cancer spheroids and fine-needle aspiration-derived organoids. STAR Protocols 2(2):100578. https://doi.org/10.1016/j.xpro.2021.100578
101. Piccinini F, Tesei A, Arienti C, Bevilacqua A (2015) Cancer multicellular spheroids: volume assessment from a single 2D projection. Comput Methods Prog Biomed 118(2): 95–106. https://doi.org/10.1016/j.cmpb.2014.12.003
102. Gebhard C, Gabriel C, Walter I (2016) Morphological and Immunohistochemical characterization of canine osteosarcoma spheroid cell cultures. Anat Histol Embryol 45(3): 219–230. https://doi.org/10.1111/ahe.12190
103. Laurent J, Frongia C, Cazales M, Mondesert O, Ducommun B, Lobjois V (2013) Multicellular tumor spheroid models to explore cell cycle checkpoints in 3D. BMC Cancer 13:73. https://doi.org/10.1186/1471-2407-13-73
104. Huang C-H, Lei KF, Tsang N-M (2019) Apoptosis and cell cycle arrest of hepatocellular carcinoma spheroids treated by an alternating electric field. Biotechnol Prog 35(3): e2787. https://doi.org/10.1002/btpr.2787
105. Chelobanov B, Poletaeva J, Epanchintseva A, Tupitsyna A, Pyshnaya I, Ryabchikova E (2020) Ultrastructural features of gold nanoparticles interaction with HepG2 and HEK293 cells in monolayer and spheroids. Nanomaterials (Basel) 10(10):2040. https://doi.org/10.3390/nano10102040
106. Ma H-l, Jiang Q, Han S, Wu Y, Tomshine JC, Wang D et al (2012) Multicellular tumor spheroids as an in vivo–like tumor model for three-dimensional imaging of chemotherapeutic and nano material cellular penetration. Mol Imaging 11(6):7290.2012.00012. https://doi.org/10.2310/7290.2012.00012
107. England CG, Gobin AM, Frieboes HB (2015) Evaluation of uptake and distribution of gold nanoparticles in solid tumors. Eur Phys J Plus 130(11). https://doi.org/10.1140/epjp/i2015-15231-1
108. Wen Z, Liao Q, Hu Y, You L, Zhou L, Zhao Y (2013) A spheroid-based 3-D culture model for pancreatic cancer drug testing, using the acid phosphatase assay. Braz J Med Biol Res 46(7):634–642. https://doi.org/10.1590/1414-431x20132647
109. Bresciani G, Hofland LJ, Dogan F, Giamas G, Gagliano T, Zatelli MC (2019) Evaluation of spheroid 3D culture methods to study a pancreatic neuroendocrine neoplasm cell line. Front Endocrinol 10:682. https://doi.org/10.3389/fendo.2019.00682
110. Eilenberger C, Kratz SRA, Rothbauer M, Ehmoser EK, Ertl P, Küpcü S (2018) Optimized alamarBlue assay protocol for drug dose-response determination of 3D tumor spheroids. MethodsX 5:781–787. https://doi.org/10.1016/j.mex.2018.07.011
111. Zanoni M, Piccinini F, Arienti C, Zamagni A, Santi S, Polico R et al (2016) 3D tumor spheroid models for in vitro therapeutic screening: a systematic approach to enhance the biological relevance of data obtained. Sci Rep 6(1):19103. https://doi.org/10.1038/srep19103
112. Murali VS, Chang B-J, Fiolka R, Danuser G, Cobanoglu MC, Welf ES (2019) An image-based assay to quantify changes in proliferation and viability upon drug treatment in 3D microenvironments. BMC Cancer 19(1):502. https://doi.org/10.1186/s12885-019-5694-1
113. Tutty MA, Prina-Mello A, Vella G (2022) Pre-clinical 2D and 3D toxicity response to a panel of nanomaterials; comparative assessment of NBM-induced liver toxicity. Res Square. https://doi.org/10.21203/rs.3.rs-1277778/v1
114. Patra B, Peng C-C, Liao W-H, Lee C-H, Tung Y-C (2016) Drug testing and flow cytometry analysis on a large number of uniform sized tumor spheroids using a microfluidic device. Sci Rep 6(1):21061. https://doi.org/10.1038/srep21061
115. Beaumont KA, Anfosso A, Ahmed F, Weninger W, Haass NK (2015) Imaging- and flow cytometry-based analysis of cell position and the cell cycle in 3D melanoma spheroids. J Vis Exp 106:e53486. https://doi.org/10.3791/53486
116. Grässer U, Bubel M, Sossong D, Oberringer M, Pohlemann T, Metzger W (2018) Dissociation of mono- and co-culture spheroids into single cells for subsequent flow cytometric analysis. Ann Anat – Anat Anz 216:1–8. https://doi.org/10.1016/j.aanat.2017.10.002
117. Pulak R (2006) Techniques for analysis, sorting, and dispensing of C. elegans on the COPAS flow-sorting system. Methods Mol Biol 351:275–286. https://doi.org/10.1385/1-59745-151-7:275
118. K. McBain*, M. Oliver, L. Kelsey, C. Szybut and T. Dale (2021) Quantifying T cell response in 3D tumor spheroids using advanced flow cytometry workflows. https://www.sartorius.com/en/products/flow-

cytometry/flow-cytometryresources/quantifying-t-cell-response-in-3d-tumor-spheroids-using-advanced-flow-cytometry-workflows-applicationnote

119. Senthebane DA, Rowe A, Thomford NE, Shipanga H, Munro D, Mazeedi M et al (2017) The role of tumor microenvironment in Chemoresistance: to survive, keep your enemies closer. Int J Mol Sci 18(7). https://doi.org/10.3390/ijms18071586
120. Zheng HC (2017) The molecular mechanisms of chemoresistance in cancers. Oncotarget 8(35):59950–59964. https://doi.org/10.18632/oncotarget.19048
121. Barbone D, Yang TM, Morgan JR, Gaudino G, Broaddus VC (2008) Mammalian target of rapamycin contributes to the acquired apoptotic resistance of human mesothelioma multicellular spheroids. J Biol Chem 283(19):13021–13030. https://doi.org/10.1074/jbc.M709698200
122. Huanwen W, Zhiyong L, Xiaohua S, Xinyu R, Kai W, Tonghua L (2009) Intrinsic chemoresistance to gemcitabine is associated with constitutive and laminin-induced phosphorylation of FAK in pancreatic cancer cell lines. Mol Cancer 8:125. https://doi.org/10.1186/1476-4598-8-125
123. Longati P, Jia X, Eimer J, Wagman A, Witt M-R, Rehnmark S et al (2013) 3D pancreatic carcinoma spheroids induce a matrix-rich, chemoresistant phenotype offering a better model for drug testing. BMC Cancer 13(1):95. https://doi.org/10.1186/1471-2407-13-95
124. Thomas F, Holly JMP, Persad R, Bahl A, Perks CM (2010) Fibronectin confers survival against chemotherapeutic agents but not against radiotherapy in DU145 prostate cancer cells: involvement of the insulin like growth factor-1 receptor. The Prostate 70(8):856–865. https://doi.org/10.1002/pros.21119
125. Weigelt B, Lo AT, Park CC, Gray JW, Bissell MJ (2010) HER2 signaling pathway activation and response of breast cancer cells to HER2-targeting agents is dependent strongly on the 3D microenvironment. Breast Cancer Res Treat 122(1):35–43. https://doi.org/10.1007/s10549-009-0502-2
126. Sethi T, Rintoul RC, Moore SM, MacKinnon AC, Salter D, Choo C et al (1999) Extracellular matrix proteins protect small cell lung cancer cells against apoptosis: a mechanism for small cell lung cancer growth and drug resistance in vivo. Nat Med 5(6):662–668. https://doi.org/10.1038/9511
127. Aoudjit F, Vuori K (2001) Integrin signaling inhibits paclitaxel-induced apoptosis in breast cancer cells. Oncogene 20(36):4995–5004. https://doi.org/10.1038/sj.onc.1204554
128. Nunes AS, Barros AS, Costa EC, Moreira AF, Correia IJ (2019) 3D tumor spheroids as in vitro models to mimic in vivo human solid tumors resistance to therapeutic drugs. Biotechnol Bioeng 116(1):206–226. https://doi.org/10.1002/bit.26845
129. Frankel A, Man S, Elliott P, Adams J, Kerbel RS (2000) Lack of multicellular drug resistance observed in human ovarian and prostate carcinoma treated with the proteasome inhibitor PS-341. Clin Cancer Res 6(9):3719–3728
130. Ferrante A, Rainaldi G, Indovina P, Indovina PL, Santini MT (2006) Increased cell compaction can augment the resistance of HT-29 human colon adenocarcinoma spheroids to ionizing radiation. Int J Oncol 28(1):111–118
131. Salata OV (2004) Applications of nanoparticles in biology and medicine. J Nanobiotechnol 2(1):3. https://doi.org/10.1186/1477-3155-2-3
132. Shi Y, van der Meel R, Chen X, Lammers T (2020) The EPR effect and beyond: strategies to improve tumor targeting and cancer nanomedicine treatment efficacy. Theranostics 10(17):7921–7924. https://doi.org/10.7150/thno.49577
133. Wick P, Grafmueller S, Petri-Fink A, Rothen-Rutishauser B (2014) Advanced human in vitro models to assess metal oxide nanoparticle-cell interactions. MRS Bull 39(11):984–989. https://doi.org/10.1557/mrs.2014.219
134. Alepee N, Bahinski A, Daneshian M, De Wever B, Fritsche E, Goldberg A et al (2014) State-of-the-art of 3D cultures (organs-on-a-chip) in safety testing and pathophysiology. Altex 31(4):441–477. https://doi.org/10.14573/altex.1406111
135. Mó I, Sabino IJ, Melo-Diogo D, Lima-Sousa-R, Alves CG, Correia IJ (2020) The importance of spheroids in analyzing nanomedicine efficacy. Nanomedicine 15(15):1513–1525. https://doi.org/10.2217/nnm-2020-0054
136. Zhang X, Jiang T, Chen D, Wang Q, Zhang LW (2020) Three-dimensional liver models: state of the art and their application for hepatotoxicity evaluation. Crit Rev Toxicol 50(4):279–309. https://doi.org/10.1080/10408444.2020.1756219
137. Capek I (2006) Chapter 1: Nanotechnology and nanomaterials. In: Capek I (ed) Studies in interface science. Elsevier, Amsterdam, pp 1–69

138. Millard M, Yakavets I, Zorin V, Kulmukhamedova A, Marchal S, Bezdetnaya L (2017) Drug delivery to solid tumors: the predictive value of the multicellular tumor spheroid model for nanomedicine screening. Int J Nanomedicine 12:7993–8007. https://doi.org/10.2147/ijn.s146927
139. Huang K, Boerhan R, Liu C, Jiang G (2017) Nanoparticles penetrate into the multicellular spheroid-on-chip: effect of surface charge, protein Corona, and exterior flow. Mol Pharm 14(12):4618–4627. https://doi.org/10.1021/acs.molpharmaceut.7b00726
140. Zhao J, Lu H, Wong S, Lu M, Xiao P, Stenzel MH (2017) Influence of nanoparticle shapes on cellular uptake of paclitaxel loaded nanoparticles in 2D and 3D cancer models. Polym Chem 8(21):3317–3326. https://doi.org/10.1039/C7PY00385D
141. Cho W, Kim MS, Lee KH, Park SJ, Shin HJ, Lee YJ et al (2020) Ionizing radiation attracts tumor targeting and apoptosis by radiotropic lysyl oxidase traceable nanoparticles. Nanomedicine 24:102141. https://doi.org/10.1016/j.nano.2019.102141
142. Sadri A, Changizi V, Eivazadeh N (2015) Evaluation of glioblastoma (U87) treatment with ZnO nanoparticle and X-ray in spheroid culture model using MTT assay. Radiat Phys Chem 115:17–21. https://doi.org/10.1016/j.radphyschem.2015.05.035
143. Gaspar VM, Baril P, Costa EC, de Melo-Diogo D, Foucher F, Queiroz JA et al (2015) Bioreducible poly(2-ethyl-2-oxazoline)-PLA-PEI-SS triblock copolymer micelles for co-delivery of DNA minicircles and Doxorubicin. J Control Release 213:175–191. https://doi.org/10.1016/j.jconrel.2015.07.011
144. Gaspar VM, Costa EC, Queiroz JA, Pichon C, Sousa F, Correia IJ (2015) Folate-targeted multifunctional amino acid-chitosan nanoparticles for improved cancer therapy. Pharm Res 32(2):562–577. https://doi.org/10.1007/s11095-014-1486-0
145. Wu J, Feng S, Liu W, Gao F, Chen Y (2017) Targeting integrin-rich tumors with temoporfin-loaded vitamin-E-succinate-grafted chitosan oligosaccharide/d-α-tocopheryl polyethylene glycol 1000 succinate nanoparticles to enhance photodynamic therapy efficiency. Int J Pharm 528(1–2):287–298. https://doi.org/10.1016/j.ijpharm.2017.06.021
146. Jin G, He R, Liu Q, Dong Y, Lin M, Li W et al (2018) Theranostics of triple-negative breast cancer based on conjugated polymer nanoparticles. ACS Appl Mater Interfaces 10(13):10634–10646. https://doi.org/10.1021/acsami.7b14603
147. Kumari P, Jain S, Ghosh B, Zorin V, Biswas S (2017) Polylactide-based block Copolymeric micelles loaded with Chlorin e6 for photodynamic therapy: in vitro evaluation in monolayer and 3D spheroid models. Mol Pharm 14(11):3789–3800. https://doi.org/10.1021/acs.molpharmaceut.7b00548
148. Madsen SJ, Christie C, Hong SJ, Trinidad A, Peng Q, Uzal FA et al (2015) Nanoparticle-loaded macrophage-mediated photothermal therapy: potential for glioma treatment. Lasers Med Sci 30(4):1357–1365. https://doi.org/10.1007/s10103-015-1742-5
149. Cheng X, Li D, Sun M, He L, Zheng Y, Wang X et al (2019) Co-delivery of DOX and PDTC by pH-sensitive nanoparticles to overcome multidrug resistance in breast cancer. Colloids Surf B Biointerfaces 181:185–197. https://doi.org/10.1016/j.colsurfb.2019.05.042
150. Du AW, Lu H, Stenzel MH (2015) Core-cross-linking accelerates antitumor activities of paclitaxel-conjugate micelles to prostate multicellular tumor spheroids: a comparison of 2D and 3D models. Biomacromolecules 16(5):1470–1479. https://doi.org/10.1021/acs.biomac.5b00282
151. Lu H, Utama RH, Kitiyotsawat U, Babiuch K, Jiang Y, Stenzel MH (2015) Enhanced transcellular penetration and drug delivery by crosslinked polymeric micelles into pancreatic multicellular tumor spheroids. Biomater Sci 3(7):1085–1095. https://doi.org/10.1039/C4BM00323C
152. Malarvizhi GL, Retnakumari AP, Nair S, Koyakutty M (2014) Transferrin targeted core-shell nanomedicine for combinatorial delivery of doxorubicin and sorafenib against hepatocellular carcinoma. Nanomedicine 10(8):1649–1659. https://doi.org/10.1016/j.nano.2014.05.011
153. Sarisozen C, Dhokai S, Tsikudo EG, Luther E, Rachman IM, Torchilin VP (2016) Nanomedicine based curcumin and doxorubicin combination treatment of glioblastoma with scFv-targeted micelles: in vitro evaluation on 2D and 3D tumor models. Eur J Pharm Biopharm 108:54–67. https://doi.org/10.1016/j.ejpb.2016.08.013
154. Cheheltani R, Ezzibdeh RM, Chhour P, Pulaparthi K, Kim J, Jurcova M et al (2016) Tunable, biodegradable gold nanoparticles as contrast agents for computed tomography and photoacoustic imaging. Biomaterials 102:87–97. https://doi.org/10.1016/j.biomaterials.2016.06.015

155. Xiao Y-D, Paudel R, Liu J, Ma C, Zhang Z-S, Zhou S-K (2016) MRI contrast agents: classification and application (review). Int J Mol Med 38(5):1319–1326. https://doi.org/10.3892/ijmm.2016.2744
156. Cole LE, Ross RD, Tilley JM, Vargo-Gogola-T, Roeder RK (2015) Gold nanoparticles as contrast agents in x-ray imaging and computed tomography. Nanomedicine (Lond) 10(2):321–341. https://doi.org/10.2217/nnm.14.171
157. Miyamoto Y, Koshidaka Y, Noguchi H, Oishi K, Saito H, Yukawa H et al (2013) Observation of positively charged magnetic nanoparticles inside HepG2 spheroids using electron microscopy. Cell Med 5(2–3):89–96. https://doi.org/10.3727/215517913x666530
158. Peng X, Wang B, Yang Y, Zhang Y, Liu Y, He Y et al (2019) Liver tumor spheroid reconstitution for testing mitochondrial targeted magnetic hyperthermia treatment. ACS Biomater Sci Eng 5(3):1635–1644. https://doi.org/10.1021/acsbiomaterials.8b01630
159. Yohan D, Cruje C, Lu X, Chithrani D (2015) Elucidating the uptake and distribution of nanoparticles in solid tumors via a multilayered cell culture model. Nanomicro Lett 7(2):127–137. https://doi.org/10.1007/s40820-014-0025-1
160. Tutty MA, Movia D, Prina-Mello A (2022) Three-dimensional (3D) liver cell models – a tool for bridging the gap between animal studies and clinical trials when screening liver accumulation and toxicity of nanobiomaterials. Drug Deliv Transl Res. https://doi.org/10.1007/s13346-022-01147-0
161. Park J-K, Utsumi T, Seo Y-E, Deng Y, Satoh A, Saltzman WM et al (2016) Cellular distribution of injected PLGA-nanoparticles in the liver. Nanomedicine 12(5):1365–1374. https://doi.org/10.1016/j.nano.2016.01.013
162. Erler JT, Weaver VM (2009) Three-dimensional context regulation of metastasis. Clin Exp Metastasis 26(1):35–49. https://doi.org/10.1007/s10585-008-9209-8
163. Lefranc F, Brotchi J, Kiss R (2005) Possible future issues in the treatment of glioblastomas: special emphasis on cell migration and the resistance of migrating glioblastoma cells to apoptosis. J Clin Oncol 23(10):2411–2422. https://doi.org/10.1200/jco.2005.03.089
164. Vinci M, Box C, Zimmermann M, Eccles SA (2013) Tumor spheroid-based migration assays for evaluation of therapeutic agents. Methods Mol Biol 986:253–266. https://doi.org/10.1007/978-1-62703-311-4_16
165. Weiswald LB, Bellet D, Dangles-Marie V (2015) Spherical cancer models in tumor biology. Neoplasia 17(1):1–15. https://doi.org/10.1016/j.neo.2014.12.004
166. Cattin S, Ramont L, Rüegg C (2018) Characterization and in vivo validation of a three-dimensional multi-cellular culture model to study heterotypic interactions in colorectal cancer cell growth, invasion and metastasis. Front Bioeng Biotechnol 6:97. https://doi.org/10.3389/fbioe.2018.00097
167. Gao Q, Yang Z, Xu S, Li X, Yang X, Jin P et al (2019) Heterotypic CAF-tumor spheroids promote early peritoneal metastasis of ovarian cancer. J Exp Med 216(3):688–703. https://doi.org/10.1084/jem.20180765
168. Almahmoudi R, Salem A, Murshid S, Dourado MR, Apu EH, Salo T et al (2019) Interleukin-17F has anti-tumor effects in Oral tongue cancer. Cancers 11(5):650
169. Liu H, Lu T, Kremers G-J, Seynhaeve ALB, ten Hagen TLM (2020) A microcarrier-based spheroid 3D invasion assay to monitor dynamic cell movement in extracellular matrix. Biol Proced Online 22(1):3. https://doi.org/10.1186/s12575-019-0114-0
170. Conti S, Kato T, Park D, Sahai E, Trepat X, Labernadie A (2021) CAFs and cancer cells co-migration in 3D spheroid invasion assay. Methods Mol Biol 2179:243–256. https://doi.org/10.1007/978-1-0716-0779-4_19
171. Stejskalová A, Fincke V, Nowak M, Schmidt Y, Borrmann K, von Wahlde M-K et al (2021) Collagen I triggers directional migration, invasion and matrix remodeling of stroma cells in a 3D spheroid model of endometriosis. Sci Rep 11(1):4115. https://doi.org/10.1038/s41598-021-83645-8
172. Nazari SS (2020) Generation of 3D tumor spheroids with encapsulating basement membranes for invasion studies. Curr Protoc Cell Biol 87(1):e105. https://doi.org/10.1002/cpcb.105
173. Bell HS, Wharton SB, Leaver HA, Whittle IR (1999) Effects of N-6 essential fatty acids on glioma invasion and growth: experimental studies with glioma spheroids in collagen gels. J Neurosurg 91(6):989–996. https://doi.org/10.3171/jns.1999.91.6.0989
174. Berens EB, Holy JM, Riegel AT, Wellstein A (2015) A cancer cell spheroid assay to assess invasion in a 3D setting. Vis Exp 105:e53409. https://doi.org/10.3791/53409

175. Verma M (2012) Personalized medicine and cancer. J Perinat Med 2(1):1–14. https://doi.org/10.3390/jpm2010001
176. Pauli C, Hopkins BD, Prandi D, Shaw R, Fedrizzi T, Sboner A et al (2017) Personalized in vitro and in vivo cancer models to guide precision medicine. Cancer Discov 7(5):462–477. https://doi.org/10.1158/2159-8290.Cd-16-1154
177. Offit K (2011) Personalized medicine: new genomics, old lessons. Hum Genet 130(1):3–14. https://doi.org/10.1007/s00439-011-1028-3
178. Gilazieva Z, Ponomarev A, Rutland C, Rizvanov A, Solovyeva V (2020) Promising applications of tumor spheroids and organoids for personalized medicine. Cancers (Basel) 12(10). https://doi.org/10.3390/cancers12102727
179. Ong SM, Zhao Z, Arooz T, Zhao D, Zhang S, Du T et al (2010) Engineering a scaffold-free 3D tumor model for in vitro drug penetration studies. Biomaterials 31(6):1180–1190. https://doi.org/10.1016/j.biomaterials.2009.10.049
180. Kobayashi H, Man S, Graham CH, Kapitain SJ, Teicher BA, Kerbel RS (1993) Acquired multicellular-mediated resistance to alkylating agents in cancer. Proc Natl Acad Sci 90(8):3294–3298. https://doi.org/10.1073/pnas.90.8.3294
181. Lemmo S, Atefi E, Luker GD, Tavana H (2014) Optimization of aqueous biphasic tumor spheroid microtechnology for anticancer drug testing in 3D culture. Cell Mol Bioeng 7(3):344–354. https://doi.org/10.1007/s12195-014-0349-4
182. Doublier S, Belisario DC, Polimeni M, Annaratone L, Riganti C, Allia E et al (2012) HIF-1 activation induces doxorubicin resistance in MCF7 3-D spheroids via P-glycoprotein expression: a potential model of the chemo-resistance of invasive micropapillary carcinoma of the breast. BMC Cancer 12(1):4. https://doi.org/10.1186/1471-2407-12-4
183. Pattni BS, Nagelli SG, Aryasomayajula B, Deshpande PP, Kulkarni A, Hartner WC et al (2016) Targeting of micelles and liposomes loaded with the pro-apoptotic drug, NCL-240, into NCI/ADR-RES cells in a 3D spheroid model. Pharm Res 33(10):2540–2551. https://doi.org/10.1007/s11095-016-1978-1
184. Sachs N, de Ligt J, Kopper O, Gogola E, Bounova G, Weeber F et al (2018) A living biobank of breast cancer organoids captures disease heterogeneity. Cell 172(1–2):373–86.e10. https://doi.org/10.1016/j.cell.2017.11.010
185. Nanki K, Toshimitsu K, Takano A, Fujii M, Shimokawa M, Ohta Y et al (2018) Divergent routes toward Wnt and R-spondin niche independency during human gastric carcinogenesis. Cell 174(4):856–69.e17. https://doi.org/10.1016/j.cell.2018.07.027
186. Vlachogiannis G, Hedayat S, Vatsiou A, Jamin Y, Fernández-Mateos J, Khan K et al (2018) Patient-derived organoids model treatment response of metastatic gastrointestinal cancers. Science 359(6378):920–926. https://doi.org/10.1126/science.aao2774
187. Antunes N, Kundu B, Kundu SC, Reis RL, Correlo V (2022) In vitro cancer models: a closer look at limitations on translation. Bioengineering (Basel) 9(4). https://doi.org/10.3390/bioengineering9040166
188. Ekert JE, Deakyne J, Pribul-Allen P, Terry R, Schofield C, Jeong CG et al (2020) Recommended guidelines for developing, qualifying, and implementing complex in vitro models (CIVMs) for drug discovery. SLAS Discov 25(10):1174–1190. https://doi.org/10.1177/2472555220923332
189. Zanoni M, Cortesi M, Zamagni A, Arienti C, Pignatta S, Tesei A (2020) Modeling neoplastic disease with spheroids and organoids. J Hematol Oncol 13(1):97. https://doi.org/10.1186/s13045-020-00931-0
190. Foster AJ, Chouhan B, Regan SL, Rollison H, Amberntsson S, Andersson LC et al (2019) Integrated in vitro models for hepatic safety and metabolism: evaluation of a human Liver-Chip and liver spheroid. Arch Toxicol 93(4):1021–1037. https://doi.org/10.1007/s00204-019-02427-4
191. Walker PA, Ryder S, Lavado A, Dilworth C, Riley RJ (2020) The evolution of strategies to minimise the risk of human drug-induced liver injury (DILI) in drug discovery and development. Arch Toxicol 94(8):2559–2585. https://doi.org/10.1007/s00204-020-02763-w
192. Williams DP, Lazic SE, Foster AJ, Semenova E, Morgan P (2020) Predicting drug-induced liver injury with Bayesian machine learning. Chem Res Toxicol 33(1):239–248. https://doi.org/10.1021/acs.chemrestox.9b00264
193. Wax PM (1995) Elixirs, diluents, and the passage of the 1938 Federal Food, drug and cosmetic act. Ann Intern Med 122(6):456–461. https://doi.org/10.7326/0003-4819-122-6-199503150-00009

194. Administration USFaD (2017) Investigational new drug (IND) application. U.S. Food and Drug Administration
195. Tutty MA (2021) Three-dimensional (3D) hepatic cell culture models to improve the clinical translation of nanobiomaterials (NBMs). Trinity College Dublin
196. Courau T, Bonnereau J, Chicoteau J, Bottois H, Remark R, Assante Miranda L et al (2019) Cocultures of human colorectal tumor spheroids with immune cells reveal the therapeutic potential of MICA/B and NKG2A targeting for cancer treatment. J Immunother Cancer 7(1):74. https://doi.org/10.1186/s40425-019-0553-9
197. Aref AR, Campisi M, Ivanova E, Portell A, Larios D, Piel BP et al (2018) 3D microfluidic ex vivo culture of organotypic tumor spheroids to model immune checkpoint blockade. Lab Chip 18(20):3129–3143. https://doi.org/10.1039/C8LC00322J
198. Jeong SY, Lee JH, Shin Y, Chung S, Kuh HJ (2016) Co-culture of tumor spheroids and fibroblasts in a collagen matrix-incorporated microfluidic Chip mimics reciprocal activation in solid tumor microenvironment. PLoS One 11(7):e0159013. https://doi.org/10.1371/journal.pone.0159013
199. Shao H, Moller M, Wang D, Ting A, Boulina M, Liu ZJ (2020) A novel stromal fibroblast-modulated 3D tumor spheroid model for studying tumor-stroma interaction and drug discovery. J Vis Exp 156. https://doi.org/10.3791/60660
200. Yakavets I, Francois A, Benoit A, Merlin J-L, Bezdetnaya L, Vogin G (2020) Advanced co-culture 3D breast cancer model for investigation of fibrosis induced by external stimuli: optimization study. Sci Rep 10(1):21273. https://doi.org/10.1038/s41598-020-78087-7
201. Korff T, Augustin HG (1998) Integration of endothelial cells in multicellular spheroids prevents apoptosis and induces differentiation. J Cell Biol 143(5):1341–1352. https://doi.org/10.1083/jcb.143.5.1341
202. Upreti M, Jamshidi-Parsian A, Koonce NA, Webber JS, Sharma SK, Asea AA et al (2011) Tumor-endothelial cell three-dimensional spheroids: new aspects to enhance radiation and drug therapeutics. Transl Oncol 4(6):365–376. https://doi.org/10.1593/tlo.11187
203. Heiss M, Hellström M, Kalén M, May T, Weber H, Hecker M et al (2015) Endothelial cell spheroids as a versatile tool to study angiogenesis in vitro. FASEB J 29(7):3076–3084. https://doi.org/10.1096/fj.14-267633

Chapter 4

Disease Animal Models for Cancer Research

Sara Fuochi and Viola Galligioni

Abstract

Despite nonanimal methods (NAMs) are more and more exploited and new NAMs are developed and validated, animal models are still used in cancer research. Animals are used at multiple levels, from understanding molecular traits and pathways, to mimicking clinical aspects of tumor progression, to drug testing. In vivo approaches are not trivial and involve cross-disciplinary knowledge: animal biology and physiology, genetics, pathology, and animal welfare.

The aim of this chapter is not to list and address all animal models used in cancer research. Instead, the authors would like to guide experimenters in the strategies to adopt in both planning and performing in vivo experimental procedures, including the choice of cancer animal models.

Key words Animal models, Refinement, Animal welfare, Humane endpoints, Clinical score sheets, Cancer models

1 Introduction

The paradigm of the 3Rs, replacement, reduction, and refinement, as defined by Russell and Burch [1] in 1959, is embedded in legislation and in guidelines of in vivo research. Replacement means the substitution for conscious living higher animals of insentient material. Reduction means reduction in the number of animals used to obtain information of a given amount and precision. Refinement means any decrease in the incidence or severity of inhumane procedures applied to those animals, which still must be used. When investigating a scientific question, if NAMs are not available, the right animal model should be used, replacing mammals with less sentient animals whenever possible (replacement), experimental design should allow to get significant data with the least number of animals (reduction), and harms due to experimental procedures should be reduced at minimum, ensuring animal welfare and well-being (refinement).

Dania Movia and Adriele Prina-Mello (eds.), *Cancer Cell Culture: Methods and Protocols*,
Methods in Molecular Biology, vol. 2645, https://doi.org/10.1007/978-1-0716-3056-3_4,
© The Author(s), under exclusive license to Springer Science+Business Media, LLC, part of Springer Nature 2023

The 3Rs, at the beginning, were implemented for ethical reasons to shift to more humane research. But the scientific community has now realized that they lead research to another level, involving further Rs: development of ***reproducible*** science, performed with ***rigor*** and leading to ***robust*** data.

2 Ethical Frame of Animal Use in Cancer Research

The development of replacement for animals is a complex, difficult, and slow process, also considering that the validation of such methods and technologies is ununavoidable and time-demanding step of the whole replacement pipeline [2].

As of today, it is still unrealistic to call for full replacement of animal models in all biomedical research fields. On one hand, progresses are made with excellent results, for example, to investigate the malignant cell transformation in vitro [3]. On the other hand, progresses might be frustrated by unsatisfactory procedures involved in approving new test guidelines and deleting existing ones [4].

Advanced cell culture techniques are remarkable examples of partial replacement, systematically applicable to several types of cancers [5]. Furthermore, the development and validation of biologically relevant in vitro tumor models ultimately resulted in improved translational potential, thus enhanced reliability of such models and a reduction in the total number of the animals required in therapeutic drug screening [6].

Since in vivo models cannot be completely replaced yet, maximization of reduction and refinement measures is crucial and must be acknowledged as a guiding principle by animal experimenters and, more widely, by the entire research community. In silico and in vitro models (replacement) must be explored and as well exploited. These topics are well discussed in other chapters of this book.

Proper application of refinement and reduction principles to animal studies is not only ethical and humane toward the animals used but must be seen as an effective modality to assure better science and, equally important in our times, solidify public support [7].

Ethical treatment of animal models used in cancer research proved the existence of some inconsistency worldwide [8], with some geographical areas accepting levels of burden and experimental endpoint that would not be acceptable, and replicable, in different and more strictly regulated geographical areas, potentially leading to post-publication investigations [8]. At time of writing this chapter, an example is the paper by Hu et al. [9], currently reported as under investigation on the journal's website. Under this perspective, it would be highly recommendable the positive contribution of the peer-review processes, requiring elevated and homogeneous standards of welfare safeguards for publication acceptance, independently from the geographical origin—thus local cultural and regulatory context—of the study. A retraction watch website and database[1] were developed since 2010, brilliantly highlighting and tracking a constantly increasing number of retracted papers. To date, poor or inconsistent animal welfare standards are not explicitly included in the potential reasons for retraction. Endorsement of the relevance of poor *animal* ethics with regard to potential retraction would also be highly recommendable.

3 Refinement at Its Best: Securing Animal Welfare of In Vivo Cancer Models

When the use of a live animal model cannot be avoided, the utmost attention must be given to safeguarding its welfare. Clearly, it is widely acknowledged that a certain degree of pain, suffering, or distress can be considered acceptable, if harm/benefit analysis justifies it, and no avoidable harm is induced to the animal.

Avoiding any non-necessary discomfort experienced by the animals must be seen as a proactive task, to be pursuit while designing the experiment, with an accurate preliminary assessment of the potential negative impact on the animals, as well as during the running of the experiment by accurately monitoring the clinical conditions of the animals. Finally, a retrospective assessment of the actual level of pain, suffering, or distress experienced by the animals in the experimental cohorts is recommended, when not already required by Regulations, to highlight potential inconsistencies and to help to correctly identify the severity degree of similar future experiments—and to take preventive measures accordingly.

To do so, experimenters must be familiar with best practices regarding humane endpoints, severity degree and severity scoring, and euthanasia.

Humane endpoints (HE) are a powerful tool to secure animal welfare of experimental models. Guidelines for setting HE in cancer research have been developed, used, and updated since decades [10–13]. Familiarity with such documents is strongly recommended for all the people working with animal models in cancer research.

In principle, the perfect humane endpoint should be easy to monitor, reproducible, predictive, and reliable in tracking illness progression and correlated impairment of the animal clinical conditions [14]. It should also guarantee the maximum possible reduction of pain, suffering, and distress, while safeguarding experimental design and goal. Finally, it should be accepted

[1] https://retractionwatch.com/ last access on July 6, 2022.

transversally by all the involved stakeholders; thus, consensus should be reached by principal investigators or study directors, veterinarians, animal welfare officers, and the competent authorities—both at the institutional and regulatory level.

The choice of most appropriate humane endpoints should be seen as the best possible combination of ethical, scientific, and regulatory considerations. Under the ethical perspective, the guiding question focuses on the level of constraint that the animals are going to experience and if the expected or actual level of suffering is justifiable under the moral perspective. This decision is normally guided by a well-performed harm-benefit analysis. Further relevant considerations falling under the scientific category pertain to the experimental goal. Clearly, never set HE after the achievement of the experimental goal: delaying the euthanasia of a symptomatic animal, which has reached the experimental final timepoint, must be carefully avoided. This is also valid for circumstances when it becomes clear that the experimental goal cannot be achieved. And finally, always anticipate HE when there is a plausible risk of spontaneous death in the cage: autolysis and post-mortem cannibalism impair appropriate collection of the scientific data, while an early euthanasia could allow for samplings of good-quality and reliable scientific data (Humane Endpoints[2]). Check frequency should always be targeted on actual clinical scoring, with frequency increasing upon increasing of severity of clinical signs and potential approaching of final endpoint or timepoint. This is particularly recommended when timeframes of reduced supervision are approaching, for example, overnight or the weekends.

Several authors [15, 16] remarked that the goal of developing humane endpoints in animal experiments is constantly shifting. Consistently with this statement, the commitment and interest of the research community for the constant refinement of HE in cancer research is proven by an increasing number of publications transversally covering different type of cancers [17–19], non-rodent species [20], and new methodologies including automated recording to detect early or new predictive signs of welfare impairment [21] or imaging to better define endpoints [22].

Severity assessment is strictly related to the level of pain, suffering, and distress experienced by the animal model involved. Relevant factors to be considered include not only the type of tumor model but also the site or target organ, since different levels of pain can be associated with different tissue damage [23]. For example, malignant bone tumors cause one of the most painful conditions in both humans and animals [24], and this aspect should also be considered for severity assessment of models used to study other primary tumors (e.g., breast and prostate) with high affinity

[2] https://www.humane-endpoints.info/en/why-humane-endpoints, Last access on July 6, 2022.

to metastasize to bones. When making hypothesis about the potential pain experienced by animal models of cancer, users must also consider that benignity is not necessarily associated with the absence of pain: a remarkable example are benign, but painful, neuromas [24].

When considering the distress inflicted on such models, also the level of impairment of natural and spontaneous behavior must be considered. As Rollin [7] remarkably highlighted, the satisfaction of their nature might even be more important to animals than pain control. When animals themselves are called to perform a harm/benefit analysis, for example, self-mutilation and removal of a trapped body parts (e.g., legs and/or tail) might be preferable to forced confinement, thus freedom to escape overweighs the drawback of the experienced pain [25–27]. Furthermore, standardized housing conditions and environmental enrichment programs are universally acknowledged as pillars of good laboratory animal management, and they can interestingly foster or delegitimize the reliability and reproducibility of cancer models. As examples, mice living in an enriched housing environment may or may not show reduced tumor growth and increased remission in melanoma and colon cancer studies [28, 29], while housing temperature can affect tumor growth in mice [30].

Scoring is the natural consequence of welfare—or distress—follow-up during in vivo studies, and it is the practical instrument that allows to quantify actual severity and apply humane endpoints. A crucial aspect of a successful and reliable scoring process is its reproducibility. To secure scoring reproducibility, meaning that each scorer will assign the same value to the same finding, subjectivity must be reduced to the minimum levels. This goal can be achieved using project-targeted scoresheets, developed based on anticipated clinical signs, allowing for accurate recording of findings, associated with scores and subsequent actions. A user-friendly and more intuitive version of scoresheets can associate pictures obtained from real clinical cases to corresponding description of clinical findings. Such forms, or visual scoresheets, can facilitate the correct recognition of concurring signs and severity, reducing interpersonal assessment biases [31]. It is good to mention that scoresheets are not carved in stone and can evolve over time as some anticipated signs may never occur and others might need to be added based on experience [32]. To maximize the reliability of the scorers and the reproducibility of each assessment, all the involved personnel, including animal caretakers, researchers, PhD students, or PostDocs, must be appropriately trained to recognize clinical signs and to use the scoresheet accordingly [31].

A remarkable example of nonreliable scoring is the use of weight, and weight loss specifically, as an indicator of poor clinical conditions in cancer animal models. Animal suffering from cancer may lose weight, and at the same time, the tumor itself can grow; thus, the total weight of the animal is not necessarily significantly changed. Body condition scoring, along with signs of emaciation, can conversely provide a reliable assessment of the progressive deterioration of clinical conditions of the animal [33]. Further examples of clinical signs and scoring systems transversal to multiple species are extensively available in literature [10, 31, 34]. Where such a resource is implemented, scoring can also be performed with the use of a colony management software. Such IT supports can help improving archiving clinical data, facilitating retrospective assessments, and post-approval monitoring. Nowadays, several colony management platforms are available on the market, but many institutions have also set up their own internal colony management software programs to address facility- or institute-specific situations.

Finally, a conclusive remark regarding euthanasia, moribund state, or even death as endpoints should always be regarded as unethical and unacceptable—unless strong and well-documented scientific needs require such conditions to be achieved. All measures, including correct application of humane endpoints and accurate, increased frequency control of end-of-stage animals, should be diligently applied to prevent spontaneous death from occurring. Remarkably, such incidents, or more appropriately called animal welfare events, are potentially associated with immediate assignment of the animal to the highest severity degree level [35] relevant for official reporting to the competent authority and statistical analysis.

Provided that the large majority of well-developed, nonregulatory experimental designs do not require such an extreme endpoint and that termination of the animal is proactively performed, attention must be given in granting that, when euthanasia is performed, the literal meaning of this word—that is, good death—is translated into action. This means that the method of choice must be defined together with veterinary staff and animal welfare officers, based on experimental needs, best veterinary practices, species, age, clinical status, and drug or device availability. Also training, skills, and disposition of involved personnel must be accurately considered. For no reason, the method should be simply decided for mere practical—for example, consolidated but old, out-to-date habits—or economic reasons. Safeguarded national ad hoc laws and requirements, were existing, a comprehensive overview of most recommended euthanasia methods, and criteria selection are regularly updated and released in dedicated Guidelines by the American Veterinary Medical Association [36].

4 Choice of the Animal Model

> Among the most important variables in the determination of procedures is that of the species of animal to be used. It is just because it includes a wide knowledge of the special advantages of particular species for particular purposes, that a formal or informal training in zoology has again and again proved its mettle in the progress of medical research. (One need only consider how many years the progress of neurophysiology would have been retarded, if a zoologist called J.Z. Young had not gone to Naples in the thirties to study the comparative cytology of cephalopods.) As Pantin has stressed (1952), nature has been inexhaustibly prodigal in supplying us animals specially fitted for almost any conceivable experimental purpose. Where these rich ores are concerned, it becomes vital to know where to tap.
>
> Russell and Burch (1959)

Detailed characteristics of most common tumor animal models are available in the literature and continuously updated [37–39], but before choosing the animal species, the scientific question or hypothesis should be properly formulated: what we want to measure and what we consider significant should be clear. The worse approach would be "to correct the mistaken choice of a wrong species by forcing it to conform to the requirements of the investigation. This results in just those roundabout methods we should guard against, and is all too liable to end in gross inhumanity" [1].

What do we want from our animal model? Which characteristics do we want to analyze?

- Do we want to understand the consequences at the molecular level of a defined genetic mutation?
- Do we want to study the tumor microenvironment?
- Do we want to analyze clinical lesions, phenotype, and pathological features deriving from cancer development and progression? Which are the limits of the phenotype compared to cancer features in humans?
- Do we want to develop or validate a therapy for target mutations or targeting specific pathways?
- Do we want to understand connections between infectious agents (e.g., *Helicobacter pylori*, HBV) and cancer development?
- Do we want to understand the influence of environment (e.g., diet) on the pathological mechanism?

There is no such animal model that would recapitulate all these features. Each animal model could have one or more attributes that resemble the human condition of interest, but it is important to have clear from the beginning, which human feature we want to resemble in our experiment. Of course, to build a reliable model, a multidisciplinary approach is needed, with different expertises cross-interacting. According to which aspect we want to study, clinicians, pathologists, and molecular biologists define the feature or the behavioral dysfunction they want to be modeled and should elaborate the criteria for model evaluation. Animal biologists, animal care staff, and veterinarians can then guide through different species or strains available, address expected adverse effects, and identify welfare issues and humane endpoints.

Once the scientific hypothesis has been formulated, scientific validity of the model should be evaluated. In fact, one of the causes of poor reproducibility and translational failure is the lack of plausibility and consistency of the model, that is, its validity for a specific use or scientific question. The validation process of a model for a specific disease, such as cancer, includes reliability and replicability (internal validity), construct validity, predictive validity, face validity, and external validity of the model [40].

The internal validity refers to experimental design, how confounding variables are controlled, and how consistent an assessment, testing device, or method is. It includes an assessment of use of appropriate control groups, rigor of measurements, randomization, blinding, and statistical analysis.

Construct validity is the theoretical rationale for using that model. It reflects the degree of fitting the theoretical rationale and of modelling the true nature of the features (e.g., symptoms) to be mimicked by the animal model. Construct validity is the one that addresses the robustness of the theory of using that model, since it provides the tool to interpret data generated by that model.

By predictive validity, it is meant the ability of the model to predict behavior in the situation it is supposed to model. The experimental procedure allows extrapolation of data to other conditions (testing timepoint, age of the animal, other laboratories) and to other species, including humans.

Face validity is the degree of descriptive similarity with what is modeled between animal species and human, like similarity of symptoms. However, the very same behavioral dysfunction or feature may be the expression of underlying physiological states.

External validity provides the tool by which data and findings from an animal model can be generalized and applied to different population, environments, and, eventually, species.

When choosing the appropriate model, it is important to consider how genetic diversity—both of human patients and of the animal model—might impact on the correct understanding of the type of cancer under investigation.

It is well documented, in fact, that animal models with a higher level of genetic variability, compared to typical inbred animals, can improve the predictive value of mouse models in cancer biology and research on the effects of carcinogens and therapeutics in humans [41].

Among such mouse models, a couple of outstanding examples should be mentioned.

The collaborative cross is a panel of inbred mouse lines that captures 90% of the genetic diversity of the mouse and can help identify regions of the genome that are responsible for variation in cancer phenotypes across the population [41]. Collaborative crosses, in fact, displayed a wide range in the incidence and types of spontaneous tumors [42], and exploiting their genetic polymorphism and population diversity allows to more accurately model cancer in animal systems and to improve cancer outcomes [43].

Moving from inbred to outbred, another option to mimic human's genetic polymorphism is the so-called diversity outbred strain. Diversity outbred mice are an evolution of the collaborative cross project and were developed by random outcross mating of collaborative cross recombinant inbred mouse lines, by continued random mating that avoid crosses between siblings. This colony maintenance strategy retains the widest possible genetic diversity in each mouse [44, 45].

Each diversity outbred mouse is genetically unique, and they can effectively approximate the genetic diversity found in human populations, making them excellent models for human cancer [46].

For mice species, there are several resources that can help to identify the best model for human diseases, according to the feature we want to recapitulate:

- The Mouse Genome Informatics[3] hosts the human-mouse: Disease Connection[4] (HMDC) that enables searches for potential mouse models, candidate genes, and similarity between mouse models and humans.
- The Mouse Phenome Database[5] allows to explore phenotypes of mice strains and to compare their biological and physiological characteristics. For example, if you are interested in studying metastases and mammary tumors, you can find a plot collecting data from Lifsted et al. [47] where it's shown how the very same genetic modification on 28 inbred background strain will induce mammary tumor at different age according to the strain.

It is also available the Rat Genome Database,[6] which allows to explore genes, strains, and phenotypes in rats, and hosts a dedicated page for cancer diseases. Interestingly, the rat genome database directly links and interacts with other species-specific NCBI Datasets including, among others, dog, pig, bonobo, African green monkey, naked mole rat, and others.

[3] http://www.informatics.jax.org/, last access on July 6, 2022.

[4] www.informatics.jax.org/humanDisease.shtml, last access on July 6, 2022.

[5] https://phenome.jax.org/, last access on July 6, 2022.

[6] https://rgd.mcw.edu/rgdweb/homepage/, last access on July 6, 2022.

Possible Pitfalls of Animal Models and Reasons for Misleading Results

When choosing the animal model, we also need to understand its limits and the possible pitfalls related to the model or to the experimental procedures performed on it.

- Many cancer models fail to recapitulate patient's tumors [48].
- Genetically heterogeneity of human cancers is not reflected in genetically engineered mouse models of cancer [49].
- Patient-derived tumor xenografts (PDTXs) are transplanted subcutaneously (SC) or orthotopically in immunodeficient mice. Efficiency of engraftment depends on patient tumor and PDTXs may undergo mouse-specific tumor evolution [50, 51].
- PDTXs could host pathogens (e.g., *Mycoplasma* spp., Epstein-Barr virus, murine viruses) that could add confoundable variables and affect translatability of the model [52].
- Genetic evolution of the strain used. Genetic quality assurance, including both genetic monitoring of inbred strains and background characterization of genetically altered (GA) animal models, should be an essential component. There could be spontaneous mutations, genetic drift in inbred strains, and undesirable passenger mutations in the generation of GA strains [53].
- Influence of the genetic background on the genetic modification and phenotype: *Trp53* null mutant mice (allele not designated) developed high rates of mammary tumors when bred on the BALB/cMed background but not C57BL/6 or 129/Sv congenic backgrounds, indicating a more complicated genetic predisposition toward mammary tumorigenesis than only mutations in this one gene [54].
- Genetic modification could trigger development of another tumor. In pancreatic cancer research, engineered mouse models with oncogenic $Kras^{G12D}$ allele are used. The oncogene is activated in the pancreas by *Cre*-mediated recombination, and *Cre* is expressed as *Pdx1* promoter-driven transgene. These mice should develop pancreatic intraductal lesions and eventually metastatic adenocarcinoma. However, *Pdx1* activation leads also to papilloma and cutaneous carcinomas [55, 56].
- Site of human tumor implant. Subcutaneous (SC) tumor implants have the major drawback of not being able to reproduce characteristics and features of primary site human cancers nor sites of metastasis, for example, breast cancers or colon cancers [57]. Therefore, they lack prediction for clinical activity. For these reasons, orthotopic models have been developed.

- In the case of homograft models, the site of tumor implantation (orthotopic vs SC) has consequences. Orthotopic models implanted in physiologically relevant sites have more biological and pharmacological relevance to human disease. They allow to monitor patient-relevant responses and targeting metastatic mechanisms and to study interaction of complex microenvironment. In orthotopic models, tumor growth often slows down in the organ of origin and thus corresponds more closely to the clinical picture.
- Choosing mice of one single strain, single sex, and defined age in genetic terms is the equivalent of an infinite number of a single individual. Therefore, it cannot mirror the heterogeneity of human adult population, including both sexes and/or mixed ages (in case of sex-specific tumors, e.g., prostate, or cervical tumors).

5 The Role of Pathology Misinterpretation in Failure of the Model

The use of optimal pathology analysis and reporting methods enhances the quality of data acquired from many preclinical studies. A comparative pathologist would help to increase quality and reproducibility of data at multiple levels: in planning tissue collection, incorporation of pathology methods in written protocols, conducting pathology procedures according to best practices, and, obviously, pathology data interpretation [58].

By involvement of comparative pathologists, misinterpretation and publication of erroneous pathology data will be avoided, and knowledge of spontaneous diseases of the mouse genetic background will be gained.

Reporting erroneous data (lack of pathology expertise) could lead to failure of the model, because (i) what is happening at the biological level is not understood and (ii) nomenclature of pathological findings/lesions is wrong, and therefore interpretation of the results is misleading (referring to a lesion as malignant when it may be benign, hyperplastic, or normal in some mutant mice strains). Classic example reported by Ward et al. [59] is the inflammatory response caused by bacteria that have been reported as neoplastic lesion in some papers or other types of inflammation. Another example is reported by Chichiowski and Hale [60]: chronic intestinal inflammation was initially thought to be part of phenotype of genetically engineered mice strains. Instead, it was proven to be a consequence of *Helicobacter* spp. infection that was interfering with the study. For this reason, development of tools such as Pathbase[7] is very useful to the research community.

[7] http://pathbase.net/, las access on July 6, 2022.

Pathbase is a database that stores images of the abnormal histology associated with spontaneous and induced mutations of both embryonic and adult mice including those produced by genetic modification.

Mice are not the only species subject to misinterpretation of information. Kent and colleagues [61] reported that *Pseudocapillaria tomentosa* infection confounded a zebrafish carcinogenesis experiment: inflammatory changes and aggressive neoplasms of the intestinal were seen in zebrafish research facilities. Kent identified these lesions were caused by a nematode that can easily spread in zebrafish facilities.

These examples in mice and zebrafish, therefore, lead to the importance of knowing the health status of animals we use as models, since factors such as infections and infestations can interfere with the results and, therefore, mine data robustness.

6 Think Outside the Box: Wild Animal Models

As described above, common animal models have their own limitations, but knowledge of biology, physiology, and pathology of other animal species could help in choosing a model that has a construct validity.

We report below some animal species that are used in cancer research, to answer specific scientific hypothesis. Of course, the use of these models could be limited and not easy to manage, but that does not justify the use of less reliable models.

6.1 North American Woodchuck (*Marmota monax*)

Woodchucks can be naturally or experimentally infected with the woodchuck hepatitis virus (WHV), with consequent progression to cancer. As in humans, liver cancer develops as result of persistent viral infection, mimicking human hepatitis B virus-induced hepatocellular carcinoma (HCC) development.

Many antiviral treatments of chronic HBV infections in humans have already been FDA approved by using the woodchuck model to evaluate effectiveness of the treatment [62, 63].

Engineered mouse models can be used in liver cancer research, can have intact immunological background, and (not in the case of xenografts) can develop spontaneous liver tumors. However, the rodent model does not recapitulate the chronic hepatitis B infection, which is the main cause of HCC in humans.

Despite it does not mirror some human clinical symptoms such as cirrhosis and ascites, the woodchuck model is important to understand mechanisms of human hepatocarcinogenesis. In detail, the woodchuck model highlights how the viral infection develops in cancer: it includes virus integration; chronic necroinflammation, which damages hepatocytes and their cycle of regeneration; and all signal alterations that occur during progression of viral hepatitis-induced HCC.

Because of pathological relevance of the model, it can be used to investigate therapeutic approaches in HCC, such as the efficacy of nanoparticles uptake in the liver for drug delivery [64].

6.2 Naked Mole-Rat (Heterocephalus glaber)

Naked mole-rat is famous for its resistance to cancer and its longevity. It lives underground, in hypoxia conditions, that resemble the tumor environment. Spontaneous neoplasia is very rare in naked mole-rat, and even induced chemical mutagenesis has not been successful yet (for a review, see [65]).

It appears to possess few but efficient modifications in its translation apparatus that might contribute to achieving its famous high resistance to cancer. The main genetic features involved in cancer resistance are 28S rRNA fragmentation, RPL26 and eIF4G overexpression, global downregulation of mTOR pathway, specific amino acid residues in RAPTOR (P908) and RICTOR (V1695), and the absence of 4E-BP3 [66]. Genomic and transcriptomic data have been investigated, finding that hypoxia decreases translation through mTORC1 signaling pathway, which represents a central node to control cancer onset and progression. *H. glaber* mTORC1 signaling is globally downregulated and, along with the translational features above described, appears to be among the breaks that stop cancer development.

6.3 Chicken (Gallus domesticus)

Another example of unusual model to study cancer pathophysiology and assays related to medullary thyroid cancer drug testing [67] is the use of avian chorioallantoic membrane (CAM). The CAM provides a microenvironment that supports tumor cells with blood vessels and connective tissue, which is an advantage compared to in vitro methods. Hence, this method allows investigation of invasion and angiogenesis. Moreover, at the biological level, this method could be used to identify expression of signaling molecules and possible therapeutic targets, especially antiangiogenic molecules, because of the lack of the immune system. However, even though the chick embryo has an immature immune system at the time of tumor grafts placement (6–10 days post-incubation) and conditions for rejection have not been established, nonspecific inflammatory reaction can occur after 15 days of incubation. CAM is also used in neuroblastoma research for tumor angiogenesis and metastasis development [68]. However, also this method has its limitations: the test can be performed in a short period, before immunocompetency of the chick develops; tumor cells in direct contact with the CAM survive and grow, whereas tumor cells in upper layers of xenograft often don't have protective tissues, and their fate is necrosis [69].

Chicken has also been used as model of spontaneous ovarian cancer (OVC) for epidemiological and molecular studies: Cancer antigen 125 (CA-125) and E-cadherin expressions are similar to human OVC [70]. In case of scientific question related to epidemiology, such as environmental factors affecting reproductive life cycle, OVC, and its molecular pathogenesis, the chicken model would be more robust and translatable than some mice models.

6.4 Tasmanian Devil (*Sarcophilus harrisii*)

Another interesting model of cancer is the Tasmanian devil, the largest marsupial carnivorous in Australia. This species is a remarkable model for transmissible tumors. It is in fact currently listed as endangered species also in consideration of the high incidence in the wild population of a transmissible cancer known as devil facial tumor disease (DFTD). This kind of cancer was first considered associated with low genetic diversity of the population of devils enabling the allograft transmission of DFTD, while later evidence revealed that genetically diverse animals succumb to the disease as well [71]. What is particularly interesting in this model, beyond the population transmission dynamics, is the spontaneous regression. The specific mechanisms underlying tumor regression without treatment in humane patients are not well understood, but Tasmanian devils are used to understand comparative and functional genomics to identify tumor genetic variation associated with tumor regression, with potential positive impacts on therapeutic treatment options through mechanism for tumor inhibition [72], as well as vaccines [71].

7 Pets as Cancer Animal Models

Along with the evolution of veterinary medical practice in the field of oncology, the process of treating companion and wild animals afflicted with cancer opened new possibilities to study oncological diseases in spontaneous models.

Protocol design and review, approvals, and reporting might vary at national levels; nevertheless, in general, studies should be conducted in the spirit of good clinical practice [73, 74].

As of today, pets, particularly cats and dogs, but occasionally also wild species, are not only considered relevant models for cancer research but are consistently used in preclinical and clinical trials focusing on new anti-neoplastic drug research and development [75, 76].

Extensive literature on canine and feline models for cancer biology and treatment is available; a partial selection highlighting potentials for comparative oncology and some interesting breed and genome-based aspects is included in the chapter bibliography [75, 77–81].

The use of this kind of models, integrated between the preclinical laboratory animal model and human clinical trials, shows many advantages including a more accurate etiology potential, as exposure to environmental carcinogens, including air, food, and water is also common between people and their pets [82].

A striking example is the use of dog as model for prostate cancer [83]. The canine prostate gland shares morphological and functional similarities with the human prostate. Dogs are the only animals to present with a significant incidence of spontaneous prostate cancer, with clinical features, including late age at onset and metastatic patterns resembling those in humans. An important difference between human and canine prostate cancer is the role of androgen. Canine prostate cancer usually does not respond to androgen deprivation therapy or surgical castration. Since dogs have a shorter life span than humans, clinical trials using dogs can be conducted in a relatively shorter period. Thus, the canine prostate cancer model could be informative to study the pathogenesis of advanced prostate cancer, especially metastatic castration-resistant prostate cancer (mCRPC), and to assess biomarkers and therapeutics.

Other positive aspects of such models include the nurture of a "one health" or "one oncology" approach, implementing collaboration and sharing of knowledge between preclinical research scientists, veterinary, and human oncologists [82]. This comparative oncology virtuous circle already contributed to the creation of cooperative strategies for drug screening and developments such as the Comparative Oncology and Genomics Consortium [84, 85].

A very unique and peculiar characteristic of this specific field of in vivo research is its positive outlook not only on pet owners, but more widely on the lay public [7], somehow presenting in a more understandable and immediate way that research with animals saves lives and also to animals.

8 Conclusions

Consistently with the nature of this chapter, we would like to conclude with a sort of unorthodox vision of the "more-than-3" Rs guiding animal testing.

Respect Back to basics, when no alternatives are available and live animals must be used in research, respecting each animal as a sentient being, independently from the species' evolution, is the very first rule to be observed. Respect includes a wide range of activities, safeguards, and attitudes: not only securing to the maximum possible extent the principle of the five freedoms but going beyond with the most appropriate welfare management [86]. Environmental and husbandry management conditions, well-designed studies and appropriate humane endpoints, pain management, environmental enrichment programs, trained and competent staff: they all concur to a respectful use of live animal models.

Rigor of the scientific method: Well-designed experiments, including expected adverse events, appropriate scoring, and interpretation of biological data will minimize suffering and maximize **robustness** of experimental outcomes. To address a scientific question, understanding which models and methods are available is the basis to plan how to test the hypothesis and analyze the generated data.

Reporting of the study: Poor reporting of study details, measures to reduce bias (e.g., randomization and blinding), and statistical analyses is unfortunately still common. Poor reporting can affect research, rendering external validity, internal validity, and statistical robustness and power largely unclear [87]. The limitations found within studies are further exacerbated by reporting biases such as publication bias and selective outcome reporting. Tools such as systematic reviews of animal studies ([88], for more information see SYRCLE[8] and CAMARADES[9]) and preclinical registration ([89], see Preclinicaltrials.eu,[10] PROSPERO[11]) help in generating high-quality evidence, avoid duplication studies, and increase transparency.

Recycling This concept might be applied to several aspects of in vivo cancer research. As previously mentioned, pets with naturally occurring cancers provide researchers with insights about human cancer; thus, their clinical cases and veterinary research can be recycled as relevant trials for humane therapies. New technologies allow longitudinal measurements recycling the same animal for multiple time points with no cumulative effects negatively impacting on severity degree. Imaging technologies are widely described in literature; selected references pertaining to cancer models [90, 91] range from positron emission tomography (PET), magnetic resonance imaging (MRI), and optical imaging, comprising bioluminescence imaging (BLI) and fluorescence imaging (FLI). Less described, yet increasingly used, are home-cage monitoring devices, applied to cancer models [92]. An interesting perspective with home-cage monitoring systems is that the huge amount of data obtained by home-cage recording might always be re-analyzed (again, recycled) at a later stage to answer new questions arising (e.g., during peer reviewed processes), without the need to use further animals, thus concurring to a general **reduction**.

[8] https://www.syrcle.network/, last access on July 6, 2022.

[9] https://www.ed.ac.uk/clinical-brain-sciences/research/camarades, last access on July 6, 2022.

[10] https://preclinicaltrials.eu/, last access on July 6, 2022.

[11] https://www.crd.york.ac.uk/prospero/, last access on July 6, 2022.

References

1. Russell WMS, Burch RL (1959) The principles of humane experimental technique. Methuen, London. https://caat.jhsph.edu/principles/the-principles-of-humane-experimental-technique
2. Festing S, Wilkinson R (2007) The ethics of animal research. Talking point on the use of animals in scientific research. EMBO Rep 8(6):526–530. https://doi.org/10.1038/sj.embor.7400993
3. Mascolo MG, Perdichizzi S, Vaccari M, Rotondo F, Zanzi C, Grilli S, Paparella M, Jacobs MN, Colacci A (2018) The transformics assay: first steps for the development of an integrated approach to investigate the malignant cell transformation in vitro. Carcinogenesis 39(7):955–967. https://doi.org/10.1093/carcin/bgy037
4. Combes RD (2018) The replacement of animal tests. Altern Lab Anim 46(2):103–104. https://doi.org/10.1177/026119291804600202
5. Lovitt CJ, Shelper TB, Avery VM (2014) Advanced cell culture techniques for cancer drug discovery. Biology (Basel) 3(2):345–367. https://doi.org/10.3390/biology3020345
6. Pampaloni F, Reynaud EG, Stelzer EH (2007) The third dimension bridges the gap between cell culture and live tissue. Nat Rev Mol Cell Biol 8:839–845. https://doi.org/10.1038/nrm2236
7. Rollin BE (2019) The ethics of animal use in cancer research. In: Bernicker E (ed) Cancer and society. Springer, Cham. https://doi.org/10.1007/978-3-030-05855-5_11
8. Bik, Elisabeth. Science Integrity Digest (2020) Animal ethics misconduct: mice with very large tumors. Science Integrity Digest. https://scienceintegritydigest.com/2020/05/07/animal-ethics-misconduct-mice-with-very-large-tumors/. Accessed 3 Mar 2022
9. Hu H, Qiu Y, Guo M, Huang Y, Fang L, Peng Z, Ji W, Xu Y, Shen S, Yan Y, Huang X, Zheng J, Su C et al (2015) Targeted Hsp70 expression combined with CIK-activated immune reconstruction synergistically exerts antitumor efficacy in patient-derived hepatocellular carcinoma xenograft mouse models. Oncotarget 6:1079–1089. Under investigation on https://www.oncotarget.com/article/2835/text/. Accessed 3 March 2022
10. Workman P, Aboagye E, Balkwill F et al (2010) Guidelines for the welfare and use of animals in cancer research. Br J Cancer 102:1555–1577. https://doi.org/10.1038/sj.bjc.6605642
11. Wallace J (2000) Humane endpoints and cancer research. ILAR J 41(2):87–93. https://doi.org/10.1093/ilar.41.2.87
12. Canadian Council on Animal Care (1998) Choosing an appropriate endpoint in experiments using animals for research, teaching and testing. https://www.ccac.ca/Documents/Standards/Guidelines/Appropriate_endpoint.pdf. Accessed on 18 Fe. 2022
13. Expert Working Group of the EU Commission (2012) Working document on a severity assessment framework, en.pdf (europa.eu). Accessed 2 Mar 2022
14. 3Rs-Centre Utrecht Life Sciences (2016) Humane endpoints – why humane endpoints? https://www.humane-endpoints.info/en/why-humane-endpoints#. Accessed 18 Feb 2022
15. Hendriksen CFM, Morton DB (eds) (1999) Humane endpoints in animal experiments for biomedical research; proceedings of the international conference; 22–25 November 1998. The Royal Society Medical Press, Ziest
16. National Research Council (US) Committee on Recognition and Alleviation of Pain in Laboratory Animals (2009) Recognition and alleviation of pain in laboratory animals. National Academies Press (US)
17. Silva-Reis R, Faustino-Rocha AI, Gonçalves M et al (2021) Refinement of animal model of colorectal carcinogenesis through the definition of novel humane endpoints. Animals (Basel) 11(4):985. Published 2021 Apr 1. https://doi.org/10.3390/ani11040985
18. Faustino-Rocha AI, Ginja M, Ferreira R, Oliveira PA (2019) Studying humane endpoints in a rat model of mammary carcinogenesis. Iran J Basic Med Sci 22(6):643–649. https://doi.org/10.22038/ijbms.2019.33331.7957
19. Oliveira M, Nascimento-Gonçalves E, Silva J, Oliveira PA, Ferreira R, Antunes L, Arantes-Rodrigues R, Faustino-Rocha AI (2017) Implementation of humane endpoints in a urinary bladder carcinogenesis study in rats. In vivo (Athens, Greece) 31(6):1073–1080. https://doi.org/10.21873/invivo.11172
20. Solin SL, Shive HR, Woolard KD, Essner JJ, McGrail M (2015) Rapid tumor induction in zebrafish by TALEN-mediated somatic inactivation of the retinoblastoma1 tumor suppressor rb1. Sci Rep 5:13745. https://doi.org/10.1038/srep13745

21. Winn CB, Hwang SK, Morin J et al (2021) Automated monitoring of respiratory rate as a novel humane endpoint: a refinement in mouse metastatic lung cancer models. PLoS One 16(9):e0257694. Published 2021 Sep 20. https://doi.org/10.1371/journal.pone.0257694
22. O'Farrell A, Shnyder S, Marston G, Coletta P, Gill J (2013) Non-invasive molecular imaging for preclinical cancer therapeutic development. Br J Pharmacol 169(4):P719–P735. https://doi.org/10.1111/bph.121552013
23. Pacharinsak C, Beitz A (2008) Animal models of cancer pain. Comp Med 58(3):220–233
24. Wacnik PW, Pacharinsak C, Beitz AJ (2011) Animal models of cancer pain. In: Ma C, Zhang JM (eds) Animal models of pain. Neuromethods, vol 49. Humana Press, Totowa. https://doi.org/10.1007/978-1-60761-880-5_8
25. Rollin BE (2015) Beyond pain – controlling suffering in laboratory animals. Bioscience 65(12):1113–1114. https://doi.org/10.1093/biosci/biv148
26. Fleming PA, Muller D, Bateman PW (2007) Leave it all behind: a taxonomic perspective of autotomy in invertebrates. Biol Rev Camb Philos Soc 82(3):481–510. https://doi.org/10.1111/j.1469-185X.2007.00020.x
27. Baker SE, Ayers M, Beausoleil NJ, Belmain SR, Berdoy M, Buckle AP, Cagienard C, Cowan D, Fearn-Daglish J, Goddard P, Golledge HDR, Mullineaux E, Sharp T, Simmons A, Schmolz E (2022) An assessment of animal welfare impacts in wild Norway rat (Rattus norvegicus) management. Animal welfare (South Mimms, England) 31:51–68. https://doi.org/10.7120/09627286.31.1.005
28. Cao L, Liu X, Lin EJ, Wang C, Choi EY, Riban V, Lin B, During MJ (2010) Environmental and genetic activation of a brain-adipocyte BDNF/leptin axis causes cancer remission and inhibition. Cell 142(1):52–64. https://doi.org/10.1016/j.cell.2010.05.029
29. Westwood JA, Darcy PK, Kershaw MH (2013) Environmental enrichment does not impact on tumor growth in mice [version 1; peer review: 2 approved]. F1000Research 2:140. https://doi.org/10.12688/f1000research.2-140.v1
30. Kokolus KM, Capitano ML, Lee CT, Eng JW, Waight JD, Hylander BL, Sexton S, Hong CC, Gordon CJ, Abrams SI, Repasky EA (2013) Baseline tumor growth and immune control in laboratory mice are significantly influenced by subthermoneutral housing temperature. Proc Natl Acad Sci U S A 110(50):20176–20181. https://doi.org/10.1073/pnas.1304291110
31. Pacharinsak C, Sharp P, Zintzsch A, Fuochi S (2022) Recognition of pain, distress, and suffering. In: Practical handbook on the 3Rs in the context of the directive 2010/63/EU. Academic Press, pp 181–205
32. Fentener van Vlissingen J, Borrens M, Girod A, Lelovas P, Morrison F, Torres YS (2015) The reporting of clinical signs in laboratory animals: FELASA Working Group Report. Lab Anim 49(4):267–283. https://doi.org/10.1177/0023677215584249
33. Ullman-Culleré MH, Foltz CJ (1999) Body condition scoring: a rapid and accurate method for assessing health status in mice. Comp Med 49(3):319–323
34. Orellana-Muriana JM (2012) Animal models in cancer research: assessment of severity and the application of humane endpoints. In: Martínez Murillo R, Martínez A (eds) Animal models of brain Tumors. Neuromethods, vol 77. Humana Press, Totowa. https://doi.org/10.1007/7657_2012_59
35. Federal Department of Home Affairs FDHA, Federal Food Safety and Veterinary Office FSVO Animal Welfare No 1.04_e (2018) https://www.blv.admin.ch/dam/blv/en/dokumente/tiere/publikationen-und-forschung/tierversuche/klassifikation-schweregrad-tv.pdf.download.pdf/116104_EN.pdf. Accessed 2 Mar 2022
36. AVMA Guidelines for the euthanasia of animals (2020). https://www.avma.org/sites/default/files/2020-02/Guidelines-on-Euthanasia-2020.pdf Accessed on 18 Feb 2022
37. Teicher BA (ed) (2010) Tumor models in cancer research. Springer
38. Li Z, Zheng W, Wang H et al (2021) Application of animal models in cancer research: recent Progress and future prospects. Cancer Manag Res 13:2455–2475. Published 2021 Mar 15. https://doi.org/10.2147/CMAR.S302565
39. Harman RM, Das SP, Bartlett AP et al (2021) Beyond tradition and convention: benefits of non-traditional model organisms in cancer research. Cancer Metastasis Rev 40:47–69. https://doi.org/10.1007/s10555-020-09930-6
40. van der Staay FJ, Arndt SS, Nordquist RE (2009) Evaluation of animal models of neurobehavioral disorders. Behav Brain Funct 5(1):1–23
41. Reilly KM (2016) Using the collaborative cross to study the role of genetic diversity in cancer-related phenotypes. Cold Spring Harb Protoc 2016(3):pdb.prot079178. https://doi.org/10.1101/pdb.prot079178

42. Wang P, Wang Y, Langley SA et al (2019) Diverse tumour susceptibility in Collaborative Cross mice: identification of a new mouse model for human gastric tumourigenesis Gut 68:1942–1952. https://gut.bmj.com/content/68/11/1942
43. Hunter K (2012) Mouse models of cancer: does the strain matter? Nat Rev Cancer 12: 144–149. https://doi.org/10.1038/nrc3206
44. The Jackson Laboratory (2022) Diversity Outbred (J/DO) 009376 – Strain Details (jax.org). Accessed 2 Mar 2022
45. Churchill GA, Gatti DM, Munger SC, Svenson KL (2012) The diversity outbred mouse population. Mamm Genome 23(9–10):713–718. https://doi.org/10.1007/s00335-012-9414-2
46. Yang C, Wang Y, Xu W, Liu Z, Zhou S, Zhang M, Cui D (2019) Genome-wide association study using diversity outcross mice identified candidate genes of pancreatic cancer. Genomics 111(6):1882–1888.,ISSN 0888-7543. https://doi.org/10.1016/j.ygeno.2018.12.011
47. Lifsted T, Le Voyer T, Williams M, Muller W, Klein-Szanto A, Buetow KH, Hunter KW (1998) Identification of inbred mouse strains harboring genetic modifiers of mammary tumor age of onset and metastatic progression. Int J Cancer 77(4):640–644
48. Kamb A (2005) What's wrong with our cancer models? Nat Rev Drug Discov 4:161–165
49. Cheon DJ, Orsulic S (2011) Mouse models of cancer. Annu Rev Pathol 6:95–119
50. Ben-David U, Ha G, Tseng YY, Greenwald NF, Oh C, Shih J, McFarland JM, Wong B, Boehm JS, Beroukhim R, Golub TR (2017) Patient-derived xenografts undergo mouse-specific tumor evolution. Nat Genet 49(11): 1567–1575
51. Byrne AT, Alférez DG, Amant F, Annibali D, Arribas J, Biankin AV, Bruna A, Budinská E, Caldas C, Chang DK, Clarke RB (2017) Interrogating open issues in cancer precision medicine with patient-derived xenografts. Nat Rev Cancer 17(4):254–268
52. Yuan Z, Fan X, Zhu JJ et al (2021) Presence of complete murine viral genome sequences in patient-derived xenografts. Nat Commun 12: 2031. https://doi.org/10.1038/s41467-021-22200-5
53. Benavides F, Rülicke T, Prins JB, Bussell J, Scavizzi F, Cinelli P, Herault Y, Wedekind D (2020 Apr) Genetic quality assurance and genetic monitoring of laboratory mice and rats: FELASA Working Group Report. Lab Anim 54(2):135–148
54. Kuperwasser C, Hurlbut GD, Kittrell FS et al (2000) Development of spontaneous mammary tumors in BALB/c p53 heterozygous mice. A model for Li-Fraumeni syndrome. Am J Pathol 157(6):2151–2159
55. Gades NM et al (2008) Spontaneous vulvar papillomas in a colony of mice used for pancreatic cancer research. Comp Med 58(3): 271–275
56. Mazur PK et al (2010) Identification of epidermal Pdx1 expression discloses different roles of Notch1 and Notch2 in murine KrasG12D-induced skin carcinogenesis in vivo. PLoS One 5(10):e13578
57. Bibby M (2004) Orthotopic models of cancer for preclinical drug evaluation: advantages and disadvantages. Eur J Cancer 40:852–857
58. Everitt JI et al (2018) Pathology study design, conduct, and reporting to achieve rigor and reproducibility in translational research using animal models. ILAR J 59(1):4–12
59. Ward JM, Shonfield PN, Sundberg J (2017) Reproducibility of histopathological findings in experimental pathology of the mouse: a sorry tail. LabAnimal 46(4):146–151
60. Chichlowski M, Hale L (2009) Effects of Helicobacter infection on research: the case for eradication of Helicobacter from rodent research colonies. Comp Med 59(1):10–17
61. Kent ML, Bishop-Stewart JK, Matthews JL, Spitsbergen JM (2002) Pseudocapillaria tomentosa, a nematode pathogen, and associated neoplasms of zebrafish (Danio rerio) kept in research colonies. Comp Med 52(4): 354–358
62. Korba BE, Cote P, Hornbuckle W, Schinazi R, Gangemi JD, Tennant BC, Gerin JL (2000) Enhanced antiviral benefit of combination therapy with lamivudine and alpha interferon against Whv replication in chronic carrier woodchucks. Antivir Ther 5:95–104
63. Menne S, Cote PJ, Korba BE, Butler SD, George AL, Tochkov IA, Delaney WET, Xiong S, Gerin JL, Tennant BC (2005) Antiviral effect of oral administration of tenofovir disoproxil fumarate in woodchucks with chronic woodchuck hepatitis virus infection. Antimicrob Agents Chemother 49:2720–2728
64. Liu LY (2020) Nanoparticle uptake in a spontaneous and immunocompetent woodchuck liver cancer model. ACS Nano 14:4698–4715
65. Hadi F, Smith ESJ, Khaled WT (2021) Named Mole-rats: resistant to developing cancer or good at avoiding it? In: Buffenstein REA (ed) The extraordinary biology of the Naked Mole-Rat. Springer, Cham, pp 1033–1070

66. Maldonado G, Hernández G (2020) Translational control in the naked mole-rat as a model highly resistant to cancer. BBA – Rev Cancer. https://doi.org/10.1016/j.bbcan.2020.188455
67. Ghaffari-Tabrizi-Wizsy N et al (2019) The avian chorioallantoic membrane as an alternative tool to study medullary thyroid cancer. Endocr Connect 8(5):462–467
68. Ribatti D, Tamma R (2018) The chick embryo chorioallantoic membrane as an in vivo experimental model to study human neuroblastoma. J Cell Physiol:1–6
69. Endo Y (2019) The history of the development of chick embryo tumor xenograft models. The Enzymes 46:11–22
70. Hawkridge AM (2014) The chicken model of spontaneous ovarian cancer. Proteomics Clin Appl 8:689–699
71. Pinfold TL, Brown GK, Bettiol SS, Woods GM (2014) Mouse model of devil facial tumour disease establishes that an effective immune response can be generated against the cancer cells. Front Immunol 5:251. https://doi.org/10.3389/fimmu.2014.00251
72. Margres MJ, Ruiz-Aravena M, Hamede R, Chawla K, Patton AH, Lawrance MF, Fraik AK, Stahlke AR, Davis BW, Ostrander EA, Jones ME (2020 Aug 1) Spontaneous tumor regression in Tasmanian devils associated with RASL11A activation. Genetics 215(4): 1143–1152
73. Khanna C, London C, Vail D, Mazcko C, Hirschfeld S (2009) Guiding the optimal translation of new cancer treatments from canine to human cancer patients. Clin Cancer Res 15(18):5671–5677. https://doi.org/10.1158/1078-0432.CCR-09-0719
74. European Medicines Agency (EMA) (2001) Committee VS. Good clinical practice VICH GL9 final guidance. https://www.ema.europa.eu/en/documents/scientific-guideline/vich-gl9-good-clinical-practices-step-7_en.pdf. Accessed 19 Feb 2022
75. Di Cerbo A, Palmieri B, De Vico G, Iannitti T (2014) Onco-epidemiology of domestic animals and targeted therapeutic attempts: perspectives on human oncology. J Cancer Res Clin Oncol 140(11):1807–1814. https://doi.org/10.1007/s00432-014-1664-9
76. Vail DM, Thamm DH (2004) Spontaneously occurring Tumors in companion animals as models for drug development. In: Teicher BA, Andrews PA (eds) Anticancer drug development guide. Cancer drug discovery and development. Humana Press, Totowa. https://doi.org/10.1007/978-1-59259-739-0_12
77. De Vico G, Maiolino P (2008) Canine and feline models for cancer. In: Conn PM (ed) Sourcebook of models for biomedical research. Humana Press. https://doi.org/10.1007/978-1-59745-285-4_70
78. MacEwen EG (1990) Spontaneous tumors in dogs and cats: models for the study of cancer biology and treatment. Cancer Metastasis Rev 9(2):125–136. https://doi.org/10.1007/BF00046339
79. McNiel EA (2001) Vaccine-associated sarcomas in cats: a unique cancer model. Clin Orthop Relat Res 382:21–27. https://doi.org/10.1097/00003086-200101000-00005
80. Davis BW, Ostrander EA (2014) Domestic dogs and cancer research: a breed-based genomics approach. ILAR J 55(1):59–68. https://doi.org/10.1093/ilar/ilu017
81. Schiffman JD, Matthew B (2015) Comparative oncology: what dogs and other species can teach us about humans with cancer. Phil Trans R Soc B37020140231201402 31. https://doi.org/10.1098/rstb.2014.0231
82. Giuliano A (2021) Companion animal model in translational oncology; feline oral squamous cell carcinoma and canine oral melanoma. Biology 11(1):54. https://doi.org/10.3390/biology11010054
83. Maeda S et al (2022) Anti-CCR4 treatment depletes regulatory T cells and leads to clinical activity in a canine model of advanced prostate cancer. J Immunother Cancer 10(e003731)
84. Gordon I, Paoloni M, Mazcko C, Khanna C (2009) The Comparative Oncology Trials Consortium: using spontaneously occurring cancers in dogs to inform the cancer drug development pathway. PLoS Med 6(10): e1000161. https://doi.org/10.1371/journal.pmed.1000161
85. Pinho SS, Carvalho S, Cabral J, Reis CA, Gärtner F (2012) Canine tumors: a spontaneous animal model of human carcinogenesis. Transl Res 159(3):165–172. https://doi.org/10.1016/j.trsl.2011.11.005
86. Mellor DJ (2016) Updating animal welfare thinking: moving beyond the "five freedoms" towards "a life worth living". Animals 6(3):21. https://doi.org/10.3390/ani6030021
87. van der Naald M, Chamuleau SAJ, Menon JML, et al (2021) A 3-year evaluation of preclinicaltrials.eu reveals room for improvement in preregistration of animal studies. PLoS Biol 19(9):e3001397. Published 2021 Sep 9. https://doi.org/10.1371/journal.pbio.3001397

88. Pound P, Ritskes-Hoitinga M (2020) Can prospective systematic reviews of animal studies improve clinical translation? J Transl Med 18(1):15. Published 2020 Jan 9. https://doi.org/10.1186/s12967-019-02205-x
89. van der Naald M, Wenker S, Doevendans PA, Wever KE, Chamuleau SA (2020) Publication rate in preclinical research: a plea for preregistration. BMJ Open Sci 4(1):e100051
90. Adiseshaiah PP, Patel NL, Ileva LV, Kalen JD, Haines DC, McNeil SE (2014) Longitudinal imaging of cancer cell metastases in two preclinical models: a correlation of noninvasive imaging to histopathology. Int J Mol Imaging 2014:102702. https://doi.org/10.1155/2014/102702
91. Puaux A-L, Ong LC, Jin Y, Teh I, Hong M, Chow PKH, Golay X, Abastado J-P (2011) A comparison of imaging techniques to monitor tumor growth and cancer progression in living animals. Int J Mol Imaging 2011., Article ID 321538, 12 pages. https://doi.org/10.1155/2011/321538
92. Nwagwu CD, Defensor E, Jiang MY et al (2020) Endpoint in ovarian cancer xenograft model predicted by nighttime motion metrics. Lab Anim 49:227–232. https://doi.org/10.1038/s41684-020-0594-1

Part II

Specific Methods for Cancer Cell Culture: Step-by-Step Methodologies

Chapter 5

Generation of Radioresistant Prostate Cancer Cells

Laure Marignol

Abstract

The development of in vitro isogenic models of radioresistance through exposure to fractionated radiation is an increasingly used approach to investigate the mechanisms of radioresistance in cancer cells. Owing to the complex nature of the biological effect of ionizing radiation, the generation and validation of these models requires the careful consideration of radiation exposure protocols and cellular endpoints. This chapter presents a protocol we used to derive and characterize an isogenic model of radioresistant prostate cancer cells. This protocol may be applicable to other cancer cell lines.

Key words Ionizing radiation, Clonogenic survival, Cell death, DNA damage

1 Introduction

With over 50% of all cancer patients receiving radiotherapy, this treatment has earned its place as an efficient anticancer treatment [1]. Ionizing radiation provides DNA double-strand breaks (DSBs) production that is random, high yield and "dirty." Its unique physical characteristics allow the induction of DSBs within a few minutes of exposure [2]. Approximatively 30% of these DSBs are produced in conjunction with other types of DNA damage (i.e., base damage, single-strand breaks) from clustered energy deposition [3]. These "dirty" DSBs are repaired less efficiently than isolated damage and uniquely provide ionizing radiation with its lethal properties [4]. In recent years, radiotherapy treatment outcomes have improved because the rapid evolution of radiation therapy techniques has driven progressive radiation dose prescriptions escalations [5], fractionated radiation dose timing optimizations, and, more recently, increased accessibility to alternate radiation types with a surge in carbon ions and proton therapy centers [6, 7].

Dania Movia and Adriele Prina-Mello (eds.), *Cancer Cell Culture: Methods and Protocols*,
Methods in Molecular Biology, vol. 2645, https://doi.org/10.1007/978-1-0716-3056-3_5,

© The Author(s), under exclusive license to Springer Science+Business Media, LLC, part of Springer Nature 2023

DSB production is effective against the expansion and invasion of cancer cells including prostate cancer [8]. This is demonstrated by the high probability of tumor control achieved by radiation therapy in men with prostate cancer. This probability, called tumor control probability (TCP), is highly dependent on the dose of radiation, the proportionate production of DSBs [9], and the subsequent inactivation of proliferating cells within the tumor cell population [10]. Increases in dose and hence DSBs production are dependent on mitigating the risk of normal tissue complications, by controlling the dose delivered to these organs with careful treatment planning, the use of scarce radiation protective agents [11], the insertion of protective hydrogel spacers [12], or the upcoming use of restorative strategies [13]. Despite these improvements and the further advances that combination with chemotherapy, hormone therapy, targeted therapies, and more recently immunotherapy have brought [14–16], some tumors will not respond to radiotherapy treatment. Characterizing this lack of response or radioresistance is complicated by a molecular radiation response that affects all aspects of cancer cell biology and by the need to eradicate cancer stem cells [17, 18].

Exposure of cancer cells to fractionated radiation schedules can select a cancer subpopulation with modified cell fate in response to subsequent radiation exposure and affect tumor control probability [19]. This selection process can be reproduced in vitro to investigate the molecular response of cancer cells and guide the development of novel biomarkers of radiotherapy failure (as reviewed in [20]). These isogenic models of radioresistance are thus cells of the same origin that differ in radiosensitivity following continuous fractionated radiation protocols. In this context, we have developed a 22Rv1 prostate cancer model that has been proven to be a clinically relevant model to elucidate the potential molecular determinants of the radiation response [21]. The radioresistant cell subline we developed maintained, in fact, proliferating activity despite radiation exposure, as proven by the relationship between clonogenic survival and radiation dose [22]. A change in this relationship is often linked with both reduced induction of DSBs and increased DNA repair capacity using a modified alkaline comet assay and γ-H2AX detection.

The repair of DSBs is influenced by many parameters such as the type of DSBs, spatial distribution, cell cycle phase, and the cell line itself [23, 24]. DSBs are predominantly repaired by two main pathways: homologous recombination and nonhomologous end-joining (NHEJ) [25–27]. In response to the generation of large amounts of DSBs, as is the case following radiation exposure, the majority of the damage is repaired by the NHEJ within 3 h [28]. Decades of research into the processing of DSBs following radiation exposure identify that the choice of DSBs repair pathway may be dependent on the degree of chromatin destabilization generated [23, 29]. The prediction of outcomes is further complicated by the random induction of DSBs by radiation and the stochastic nature of subsequent cell lethality [30, 31]. Chromosome rearrangements incompatible with mitosis can lead to mitotic cell death and contribute to the antitumor effects of ionizing radiation alongside apoptosis, autophagy, or senescence [32]. This was evident during the creation of our radioresistant isogenic cell line. The emergence of radioresistant cells was associated with collective modifications of mitotic cell death, apoptosis, and senescence susceptibilities [33].

The introduction of novel diagnostic tools that can predict the response of each individual patient to a specific radiation therapy protocol remains the most likely innovation to transform the practice of radiation oncology [34, 35]. Isogenic models of radioresistance can support the search for candidate biomarkers predictive of radiotherapy outcome. In prostate cancer, these models enabled screening for miRNAs associated with radioresistance [36]. Upon combination with protein profiling analysis, these studies identified and later validated the presence of the YB-1 protein in biopsy specimens of patients diagnosed with disease recurrence and radiotherapy failure [21]. Nevertheless, there is a lack of consensus in the literature for the best way to generate and characterize isogenic models of radioresistance. Heterogeneity in doses per fraction, number of fractions, overall treatment time, and inter-fraction recovery times is evident. Inconsistencies in the end points used to determine the presence and quantify the magnitude of radioresistance are also reported [20]. Furthermore, the discrimination between genetic alterations due to the age of the cell lines, rather than radiation induced genetic instability as a result of long overall treatment times is not often examined in the literature. This chapter aims to provide a road map to the generation and validation of novel isogenic radioresistance cancer models. We define the generation of a radioresistant (RR) subline from a parent, wild-type (WT) cell line. We advocate the generation of an age-matched control (AMC): a wild-type cell line kept in culture over the same overall treatment time as the irradiated line. Comparing these three sub-cell lines ensures that other causes for radiation resistance in the isogenic cell line such as abnormalities in karyotype and changes in global gene expression patterns due to age can be ruled out [37, 38].

2 Materials

2.1 Cell Culture

1. Human prostate cancer cell line 22Rv1.
2. Growth medium: RPMI cell culture medium, l-glutamine, 10% fetal bovine serum (FBS), 1% penicillin/streptavidin.
3. 5% Trypsin-EDTA.
4. Cell culture incubator.

2.2 Irradiation

1. Cabinet irradiator (*see* **Note 1**).

2.3 Clonogenic Assays

1. 6-well plates.
2. Phosphate buffer saline (PBS).
3. Trypan blue.
4. Cell counting slide and cover slips.
5. Brightfield microscope.
6. Colony fixing solution: 70% methanol, 0.05% crystal violet, PBS.
7. Colony counter (*see* **Note 2**).

2.4 Flow Cytometry

1. PBS.
2. Distilled water.
3. Annexin V-FITC.
4. Fixation solution: 70% ethanol, distilled water.
5. Permeabilization solution: 0.1% v/v Triton X-100, PBS.
6. 1× binding buffer: 0.1 M HEPES, 1.4 M NaCl, 25 mM $CaCl_2$, pH 7.4.
7. Cell cycle staining solution: 0.02 mg/mL PI, 0.2 mg/ml RNase A, 0.5 mL Triton X-100.
8. Apoptosis staining solution: Annexin V-FITC, PBS.
9. Bafilomycin A1: 100 μM sterile-filtered Bafilomycin A1, PBS.
10. C12FDG: 2 mM 5-Dodecanoylaminofluorescein di-β-D-galactopyranoside, methanol.
11. pH meter.
12. Flow cytometer.

3 Method

3.1 Cell Culture and Irradiation

1. Grow cells to approximately 70–80% confluency in vented 75 cm^2 culture flasks at 37 °C and 5% CO_2.
2. Deliver 2Gy radiation x-rays dose (250 keV, 15 mA) to one flask using a cabinet irradiator. Keep culturing the other flask in parallel as age-matched controls (AMC-22Rv1).
3. Incubate the irradiated cells in the vented 75 cm^2 culture flasks at 37 °C and 5% CO_2 until they reach 70–80% confluency.
4. Deliver 2Gy radiation x-rays dose (250 keV, 15 mA) using the cabinet irradiator.
5. Repeat this process 30 times or up to total radiation dose of 60Gy (*see* **Note 3**).
6. Keep culturing the other flask in parallel as age-matched controls (AMC-22Rv1).

7. Subculture the lines once to twice weekly to maintain exponential growth.
8. To monitor the successful generation of the isogenic radiation-resistant sub-cell line, perform the clonogenic assay (*see* Subheading 3.2) every 3 weeks.
9. Once an increase in clonogenic survival is detected, confirm that radiation resistance with the apoptosis (*see* Subheading 3.3), senescence (*see* Subheading 3.4) assays, and cycle analysis (*see* Subheading 3.5) at 4, 8, 12, 24, and 30 h post irradiation.
10. Once the radiation-resistant sub-cell line is generated, keep the cells in culture for a month and repeat the clonogenic assay (*see* Subheading 3.2) to confirm that the radioresistant phenotype has remained stable (*see* **Note 8**).

3.2 Clonogenic Assay

1. Harvest cells in exponential growth phase (approximately 70% confluency) by trypsinization.
2. Count to obtain a single cell suspension of 1×10^6cells/mL (*see* **Note 4**).
3. Seed cells into 6-well plates at a density of $1 \times 10^3 - 1 \times 10^4$ cells/well and allow cells to adhere to plates overnight in the incubator at 37 °C in 95% humidified air containing 5% CO_2 (*see* **Note 5**).
4. Irradiate cells (2, 4, 6, 8, or 10 Gy single doses) and incubate for 7–10 days in 95% humidified air containing 5% CO_2.
5. At the selected time-point (*see* **Note 6**), remove the cell culture media, rinse each well carefully with PBS.
6. Add 1 mL/well of the colony fixing solution (*see* **Note** 7).
7. Incubate for 1 min at room temperature.
8. Remove the colony fixing solution.
9. Turn the plate upside down and leave to dry overnight.
10. Count colonies using a colony counter.
11. Calculate the platting efficiency (PE) in the unirradiated plates using the equation below:

$$\text{PE} = \frac{\text{Number of colonies counted}}{\text{Number of cells seeded}}$$

12. Calculate then the surviving fraction at each radiation dose, using the equation below:

$$\text{Surviving fraction} = \frac{\text{Number of colonies}}{(\text{Number of cells seeded} \times \text{PE})}$$

13. Plot the radiation dose in the *x*-axis and the log of the surviving fraction on the *y*-axis, to generate a radiation survival curve (*see* **Note 9**).

3.3 Apoptosis Assay by Flow Cytometry

1. Seed cells in 4 × 6-well plates at a density of 1–2 × 10^5 cells/well.
2. Prepare four addition 6-well plates and label Mock irradiated, 2Gy, 4Gy, and 8Gy.
3. Allow to adhere overnight.
4. Irradiate each plate with either 2Gy, 4Gy, or 8Gy, as appropriate.
5. Return the cells to the incubator along with mock irradiated controls for 24 h.
6. At the end of the 24-h period, harvest the cell media containing non-adherent, dead cells.
7. Trypsinize the adherent cells and combine them with the supernatant containing non-adherent cells.
8. Centrifuge the cells at 180× *g* for 3 min and decant the supernatant.
9. Resuspend cell pellets in ice cold 1× binding buffer, making sure that there are no clumps.
10. Add 3 μL of apoptosis staining solution.
11. Incubate at 4 °C for 1 h.
12. Centrifuge the cells at 180× *g* for 3 min and decant the supernatant.
13. Add ice-cold PBS and resuspend the cell pellet.
14. Centrifuge the cells at 180× *g* for 3 min and decant the supernatant.
15. Resuspend cells in 1× binding buffer.
16. Analyze the cells using a flow cytometer and quantify the percentage (%) of apoptotic cells.

3.4 Senescence Assay by Flow Cytometry

1. Seed cells in 4 × 6-well plates at a density of 1–2 × 10^5 cells/well.
2. Prepare four additional 6-well plates and label Mock irradiated, 2Gy, 4Gy, and 8Gy.
3. Allow to adhere overnight.
4. Irradiate each plate with either 2Gy, 4Gy, or 8Gy, as appropriate.
5. Return the cells to the incubator along with mock irradiated controls for 24 h.
6. At the end of the 24-h period, wash the cells with PBS.
7. Trypsinize and centrifuge at 180× *g* for 3 min. Remove the supernatant carefully.
8. Resuspend the cells in growth culture media, making sure that there are no clumps by pipetting very slowly.

9. Add Bafilomycin A1 and adjust pH to a value of 6.
10. Incubate for 1 h at 37 °C, 5% CO_2.
11. Add C12FDG.
12. Incubate for 1 h at 37 °C, 5% CO_2.
13. When ready for analysis, centrifuge the cells 180× *g* for 3 min, discard the supernatant, and resuspend the cell pellet in 1 ml PBS.
14. Analyze immediately using a flow cytometer and extrapolate the mean fluorescence intensity.

3.5 Cell Cycle Analysis by Flow Cytometry

1. Harvest cells at 4, 8, 12, 24, and 30 h post irradiation, by trypsinization.
2. Resuspend the pellet with fixation solution, and add dropwise 4 mL of ice-cold permeabilization solution, making sure that no clumps form.
3. Allow to fix for a minimum of 2 h at 4 °C.
4. Centrifuge the fixed cells at 180 × g for 3 min and decant the supernatant.
5. Wash each sample in PBS and centrifuge at 180 × g for 3 min.
6. Resuspend unstained fixed controls in PBS and each irradiated sample in 0.5 mL of cell cycle staining solution.
7. Incubate samples at 37 °C for 30 min and read immediately by flow cytometry.
8. Determine the percentage (%) of cells in each phase of the cell cycle.

4 Notes

1. In our lab, we use a RS225 cabinet irradiator (X-Strahl, Surrey, UK) set at 250 keV, 15 mA, to deliver a dose rate of 3.25Gy/min photon x-rays. Cabinet irradiators from other manufacturers can be used.
2. In our experiments, a ColCount® colony counter (Oxford Optronix Ltd., Oxford, UK) was used. Other manufacturers supply colony counters with same performances.
3. It is recommended to freeze stock of cells on a regular basis, to avoid issues if the main culture becomes compromised.
4. Make sure to fully disaggregate cell clusters.
5. Once the cells are added to the cell culture media, it is recommended to swirl the plate gently to enable even spread of the cells within the well.

6. The unirradiated plate should be used as a control. Once clones are visible to the naked eye on that plate, all plates prepared for the experiment should be stained at the same time.
7. The colony fixing solution is toxic and will stain all surfaces permanently. Wear protective clothing, protect the lab bench, and do not pour down the sink.
8. Once the radiation-resistant sub-line is generated, you will need to generate frozen stock. When you defrost a stock, it is recommended that you regularly perform a clonogenic assay at 0Gy, 2Gy, 4Gy, and 6Gy to confirm that the surviving fractions that you obtain remain similar to those at the time of freezing.
9. If you detect a statistically significant decrease in surviving fraction overtime, your cells are losing their radiation-resistant phenotype, and they will require further irradiation. Return to **step 4** in Subheading 3.1. Give three more 2Gy fractions and repeat **steps 6–10**.

References

1. Lievens Y, Borras JM, Grau C (2020) Provision and use of radiotherapy in Europe. Mol Oncol 14(7):1461–1469. https://doi.org/10.1002/1878-0261.12690
2. Friedland W, Dingfelder M, Kundrat P, Jacob P (2011) Track structures, DNA targets and radiation effects in the biophysical Monte Carlo simulation code PARTRAC. Mutat Res 711(1–2):28–40. https://doi.org/10.1016/j.mrfmmm.2011.01.003
3. Ward JF (1990) The yield of DNA double-strand breaks produced intracellularly by ionizing radiation: a review. Int J Radiat Biol 57(6): 1141–1150. https://doi.org/10.1080/09553009014551251
4. Shikazono N, Noguchi M, Fujii K, Urushibara A, Yokoya A (2009) The yield, processing, and biological consequences of clustered DNA damage induced by ionizing radiation. J Radiat Res 50(1):27–36. https://doi.org/10.1269/jrr.08086
5. Pinkawa M, Piroth MD, Holy R, Klotz J, Djukic V, Corral NE et al (2012) Dose-escalation using intensity-modulated radiotherapy for prostate cancer – evaluation of quality of life with and without (18)F-choline PET-CT detected simultaneous integrated boost. Radiat Oncol 7:14. https://doi.org/10.1186/1748-717X-7-14
6. Walsh S, Roelofs E, Kuess P, Lambin P, Jones B, Georg D et al (2016) A validated tumor control probability model based on a meta-analysis of low, intermediate, and high-risk prostate cancer patients treated by photon, proton, or carbon-ion radiotherapy. Med Phys 43(2):734–747. https://doi.org/10.1118/1.4939260
7. Sidaway P (2016) Prostate cancer: proton therapy delays progression. Nat Rev Urol 13(4): 181. https://doi.org/10.1038/nrurol.2016.50
8. Tamponi M, Gabriele D, Maggio A, Stasi M, Meloni GB, Conti M et al (2019) Prostate cancer dose-response, fractionation sensitivity and repopulation parameters evaluation from 25 international radiotherapy outcome data sets. Br J Radiol 92(1098):20180823. https://doi.org/10.1259/bjr.20180823
9. Dugle DL, Gillespie CJ, Chapman JD (1976) DNA strand breaks, repair, and survival in x-irradiated mammalian cells. Proc Natl Acad Sci U S A 73(3):809–812
10. Krause M, Yaromina A, Eicheler W, Koch U, Baumann M (2011) Cancer stem cells: targets and potential biomarkers for radiotherapy. Clin Cancer Res 17(23):7224–7229. https://doi.org/10.1158/1078-0432.CCR-10-2639
11. Devine A, Marignol L (2016) Potential of Amifostine for chemoradiotherapy and radiotherapy-associated toxicity reduction in advanced NSCLC: a meta-analysis. Anticancer Res 36(1):5–12
12. Leiker AJ, Desai NB, Folkert MR (2018) Rectal radiation dose-reduction techniques in prostate cancer: a focus on the rectal spacer.

Future Oncol 14(26):2773–2788. https://doi.org/10.2217/fon-2018-0286
13. Coppes RP, van der Goot A, Lombaert IM (2009) Stem cell therapy to reduce radiation-induced normal tissue damage. Semin Radiat Oncol 19(2):112–121. https://doi.org/10.1016/j.semradonc.2008.11.005
14. Wengner AM, Scholz A, Haendler B (2020) Targeting DNA damage response in prostate and breast cancer. Int J Mol Sci 21(21). https://doi.org/10.3390/ijms21218273
15. Quinn DI, Sandler HM, Horvath LG, Goldkorn A, Eastham JA (2017) The evolution of chemotherapy for the treatment of prostate cancer. Ann Oncol 28(11):2658–2669. https://doi.org/10.1093/annonc/mdx348
16. Madan RA, Gulley JL (2017) Prostate cancer immunotherapy: the path forward. Curr Opin Support Palliat Care 11(3):225–230. https://doi.org/10.1097/SPC.0000000000000278
17. Boss MK, Bristow R, Dewhirst MW (2014) Linking the history of radiation biology to the hallmarks of cancer. Radiat Res 181(6): 561–577. https://doi.org/10.1667/RR13675.1
18. Schulz A, Meyer F, Dubrovska A, Borgmann K (2019) Cancer stem cells and radioresistance: DNA repair and beyond. Cancers (Basel) 11(6). https://doi.org/10.3390/cancers11060862
19. Zaider M, Hanin L (2011) Tumor control probability in radiation treatment. Med Phys 38(2):574–583. https://doi.org/10.1118/1.3521406
20. McDermott N, Meunier A, Lynch TH, Hollywood D, Marignol L (2014) Isogenic radiation resistant cell lines: development and validation strategies. Int J Radiat Biol 90(2): 115–126. https://doi.org/10.3109/09553002.2014.873557
21. Inder S, Bates M, Ni Labhrai N, McDermott N, Schneider J, Erdmann G et al (2019) Multiplex profiling identifies clinically relevant signalling proteins in an isogenic prostate cancer model of radioresistance. Sci Rep 9(1):17325. https://doi.org/10.1038/s41598-019-53799-7
22. Franken NA, Rodermond HM, Stap J, Haveman J, van Bree C (2006) Clonogenic assay of cells in vitro. Nat Protoc 1(5): 2315–2319. https://doi.org/10.1038/nprot.2006.339
23. Schipler A, Iliakis G (2013) DNA double-strand-break complexity levels and their possible contributions to the probability for error-prone processing and repair pathway choice. Nucleic Acids Res 41(16):7589–7605. https://doi.org/10.1093/nar/gkt556
24. Arnould C, Legube G (2020) The secret life of chromosome loops upon DNA double-strand break. J Mol Biol 432(3):724–736. https://doi.org/10.1016/j.jmb.2019.07.036
25. Jasin M, Haber JE (2016) The democratization of gene editing: insights from site-specific cleavage and double-strand break repair. DNA Repair (Amst) 44:6–16. https://doi.org/10.1016/j.dnarep.2016.05.001
26. Shibata A, Jeggo PA (2020) Roles for 53BP1 in the repair of radiation-induced DNA double strand breaks. DNA Repair (Amst) 93: 102915. https://doi.org/10.1016/j.dnarep.2020.102915
27. Jeggo PA, Geuting V, Lobrich M (2011) The role of homologous recombination in radiation-induced double-strand break repair. Radiother Oncol 101(1):7–12. https://doi.org/10.1016/j.radonc.2011.06.019
28. Leatherbarrow EL, Harper JV, Cucinotta FA, O'Neill P (2006) Induction and quantification of gamma-H2AX foci following low and high LET-irradiation. Int J Radiat Biol 82(2): 111–118. https://doi.org/10.1080/09553000600599783
29. Iliakis G, Mladenov E, Mladenova V (2019) Necessities in the processing of DNA double strand breaks and their effects on genomic instability and cancer. Cancers (Basel) 11(11). https://doi.org/10.3390/cancers11111671
30. Sia J, Szmyd R, Hau E, Gee HE (2020) Molecular mechanisms of radiation-induced cancer cell death: a primer. Front Cell Dev Biol 8:41. https://doi.org/10.3389/fcell.2020.00041
31. Hill MA (2020) Radiation track structure: how the spatial distribution of energy deposition drives biological response. Clin Oncol (R Coll Radiol) 32(2):75–83. https://doi.org/10.1016/j.clon.2019.08.006
32. Eriksson D, Stigbrand T (2010) Radiation-induced cell death mechanisms. Tumour Biol 31(4):363–372. https://doi.org/10.1007/s13277-010-0042-8
33. McDermott N, Meunier A, Mooney B, Nortey G, Hernandez C, Hurley S et al (2016) Fractionated radiation exposure amplifies the radioresistant nature of prostate cancer cells. Sci Rep 6:34796. https://doi.org/10.1038/srep34796
34. Yaromina A, Krause M, Baumann M (2012) Individualization of cancer treatment from radiotherapy perspective. Mol Oncol 6(2): 211–221. https://doi.org/10.1016/j.molonc.2012.01.007

35. Baumann M, Krause M, Overgaard J, Debus J, Bentzen SM, Daartz J et al (2016) Radiation oncology in the era of precision medicine. Nat Rev Cancer 16(4):234–249. https://doi.org/10.1038/nrc.2016.18
36. McDermott N, Meunier A, Wong S, Buchete V, Marignol L (2017) Profiling of a panel of radioresistant prostate cancer cells identifies deregulation of key miRNAs. Clin Transl Radiat Oncol 2:63–68. https://doi.org/10.1016/j.ctro.2017.01.005
37. Ruutu M, Johansson B, Grenman R, Syrjanen K, Syrjanen S (2004) Effect of confluence state and passaging on global cancer gene expression pattern in oral carcinoma cell lines. Anticancer Res 24(5A):2627–2631
38. Wenger SL, Senft JR, Sargent LM, Bamezai R, Bairwa N, Grant SG (2004) Comparison of established cell lines at different passages by karyotype and comparative genomic hybridization. Biosci Rep 24(6):631–639. https://doi.org/10.1007/s10540-005-2797-5

Chapter 6

Generation and Characterization of an Isogenic Cell Line Model of Radioresistant Esophageal Adenocarcinoma

Aoife Cannon, Stephen G. Maher, and Niamh Lynam-Lennon

Abstract

Radiation therapy is a cornerstone of cancer treatment worldwide. Unfortunately, in many cases, it does not control tumor growth, and many tumors display treatment resistance. The molecular pathways leading to treatment resistance in cancer have been subject to research for many years. Isogenic cell lines with divergent radiosensitivities are an extremely useful tool to study the molecular mechanisms that underpin radioresistance in cancer research, as they reduce the genetic variation that is present in patient samples and cell lines of different origin, thus allowing the elucidation of molecular determinants of radioresponse. Here, we describe the process of generating an in vitro isogenic model of radioresistant esophageal adenocarcinoma by chronic irradiation of esophageal adenocarcinoma cells with clinically relevant doses of X-ray radiation. We also characterize cell cycle, apoptosis, reactive oxygen species (ROS) production, DNA damage and repair in this model to investigate the underlying molecular mechanisms of radioresistance in esophageal adenocarcinoma.

Key words Cancer, Radiation, Isogenic cell line, Radioresistance, Clonogenic assay, DNA repair, Cell cycle, Apoptosis, Reactive oxygen species

1 Introduction

Radiation therapy is one of the most effective treatments for cancer, but resistance to radiation is a significant obstacle in the successful outcome of this treatment. Failure of a tumor to respond to treatment can lead to tumor growth and disease progression [1]. However, the exact molecular mechanisms underlying radioresistance remain poorly elucidated. In part, this is due to the complexity of studying radioresistance in vivo. For this reason, an in vitro isogenic model of radioresistance is an exemplar model for studying the genetic and cellular processes that contribute to radioresistance, as it reduces the genetic variation that is present in patient samples and cell lines of different origin, thus allowing the elucidation of molecular determinants of radioresponse.

Dania Movia and Adriele Prina-Mello (eds.), *Cancer Cell Culture: Methods and Protocols*, Methods in Molecular Biology, vol. 2645, https://doi.org/10.1007/978-1-0716-3056-3_6,
© The Author(s), under exclusive license to Springer Science+Business Media, LLC, part of Springer Nature 2023

A number of isogenic cell lines have been described in the literature for various human cancers including esophageal [2, 3], lung [4, 5], breast [6], pancreatic [7], and prostate [8]. In this chapter, we describe how to create an in-house radioresistant OE33 esophageal adenocarcinoma (EAC) cell line. In establishing the methodology, OE33 cells were exposed to clinically relevant fractionated doses of 2 Gy X-ray radiation until a radioresistant subline (OE33 R) was generated. Parental OE33 cells (OE33 P) were handled identically but were mock-irradiated, thus providing an age- and passage-matched control [2]. Radiosensitivity was determined using the gold standard clonogenic assay [9]. This assay tests the ability of each cell in a population to retain its reproductive integrity following exposure to cytotoxic insults. The isogenic model of radioresistant EAC was characterized in terms of parameters frequently implicated in radioresistance such as altered cell cycling [10], apoptosis [11, 12], production of reactive oxygen species (ROS) [13], and DNA damage and repair (as measured by the accumulation of γH2AX foci) [14, 15] to identify underlying molecular mechanisms underpinning radioresistance in EAC.

Cell cycle analysis was performed by propidium iodide (PI) staining and flow cytometry [16]. Apoptosis was measured using annexin V-FITC and PI staining [17] and assessed by flow cytometry. Intracellular ROS levels were also measured by flow cytometry using 5-(-6)-chloromethyl-2′,7′-dichlorodihydrofluorescein diacetate acetyl ester (CM-H_2DCFDA), a cell permeant, nonfluorescent molecule.

Our novel in vitro model provides a unique tool with which to investigate the molecular basis of radioresistance in the EAC cancer subtype. Radiation-induced DNA damage and kinetics of DNA repair can in fact be investigated by assessment of phosphorylated histone 2AX (γH2AX) by immunofluorescence-based microscopy. Radiation exposure induces a plethora of DNA damage, with DNA double-strand breaks (DSB) reportedly being the most lethal to cells [18]. Histone 2AX is rapidly phosphorylated following radiation-induced strand breaks (in particular DSB), forming foci at these sites of DNA damage [19], and therefore allows assessment of radiation-induced DNA damage. In addition, the loss of γH2AX foci at sites of DNA DSB is consistent with their repair [20]. Assessment of γH2AX foci at different time points post radiation enables investigation of the kinetics of DNA repair following radiation-induced DNA damage. In this model, γH2AX foci was assessed using a GE InCell analyzer 1000, a microscope-based screening platform capable of high-throughput large-scale analysis of fluorescently labeled cells, using automated image acquisition, data management, and multi-parametric analysis.

2 Materials

Prepare and store all reagents at 4 °C (unless indicated otherwise). Warm media, trypsin, and phosphate buffer saline (PBS) at 37 °C in a water bath for 1 h before use. Closely follow health and safety regulations when disposing of waste materials.

2.1 Cell Culture

1. Grade II laminar air flow cabinet.
2. OE33 cell line.
3. Complete growth medium: Roswell Park Memorial Institute (RPMI) 1640 medium, 10% (v/v) heat-inactivated fetal bovine serum (FBS).
4. PBS: 13.8 mM NaCl, 2.7 mM KCl, pH 7.4.
5. Trypsin ethylene-diamine tetra-acetic acid (EDTA): 0.05% (w/v) trypsin, 0.02% (w/v) EDTA.
6. Tissue culture flasks (25 cm^2 and 75 cm^2, vented).
7. Sterile serological pipettes and pipette aid.
8. Sterile 3.5 mL Pasteur pipettes.
9. Sterile 15 mL tubes.
10. Centrifuge with buckets to hold 15 mL tubes and capable of speed up to 180 × *g*.
11. Light microscope with 10× magnification.
12. Cell incubator, 5% CO_2/95% humidified air.

2.2 Cell Counting and Viability Assessment

1. 0.4% (w/v) trypan blue solution
2. 96-well flat-bottom plate
3. Neubauer chamber hemocytometer (chamber depth of 0.1 mm).
4. Light microscope with 40× magnification.

2.3 X-Ray Irradiation

1. Medical X-ray generator (*see* **Note 1**).

2.4 Clonogenic Assay

2.4.1 Colony Staining

1. Staining/fixing solution: 0.05% w/v crystal violet, 25% v/v methanol.
2. Sterile 6-well flat-bottom plates.
3. Water bath.

2.4.2 Colony Counting

1. Light microscope or colony counter and accompanying counting software (*see* **Note 2**).

2.5 PI Cell Cycle Analysis

1. PBS.
2. Fixation/permeabilization solution: 70% v/v ice-cold ethanol, 30% v/v PBS.
3. Staining solution: 0.1% v/v Triton X-100, 0.02 mg/mL propidium iodide, 0.2 mg/mL RNase A, PBS.
4. 25 cm^2 tissue culture flasks.
5. 5 mL flow cytometry tubes.
6. 5 mL Pasteur pipettes.
7. Centrifuge with buckets to hold 5 mL flow cytometry tubes and capable of speed up to 180 × g.
8. Vortex mixer.
9. Flow cytometer.

2.6 Annexin-V-FITC/PI Apoptosis Assay

1. PBS.
2. Complete growth medium (Subheading 2.1).
3. 1× binding buffer: 0.01 M HEPES, NaOH, pH 7.4, 0.14 M NaCl, 2.5 mM $CaCl_2$.
4. PI solution: 0.25 μg propidium iodide, 1× binding buffer.
5. Annexin V-FITC.
6. Sterile 3.5 mL Pasteur pipettes.
7. Sterile 15 mL tubes.
8. 5 mL flow cytometry tubes.
9. Sterile 6-well flat-bottom plates.
10. Centrifuge with buckets to hold 5 and 15 mL tubes and capable of speed up to 180 × *g*.
11. Vortex mixer.
12. Flow cytometer.

2.7 Detection of ROS

1. PBS.
2. CM-H_2DCFDA: 10 μM 5-(−6)-chloromethyl-2′,7 ′-dichlorodihydrofluorescein diacetate acetyl ester, PBS.
3. PI stain solution: 1 mg/mL propidium iodide, PBS.
4. Positive control: H_2O_2 (100 μM).
5. 5 mL flow cytometry tubes
6. Centrifuge with buckets to hold 5 mL flow cytometry tubes and capable of speed up to 180 × *g*.
7. Flow cytometer.

2.8 Analysis of DNA Damage by γH2AX Detection

1. PBS.
2. Fixative solution: 4% v/v paraformaldehyde, PBS.
3. Blocking buffer: 5% goat serum, 3% Triton X-100, PBS.
4. Antibody dilution buffer: 1% bovine serum albumin (BSA), 0.3% Triton X-100, PBS.
5. Primary antibody: anti-human γH2AX (ser139).
6. Secondary antibody: Alexafluor 488-labeled secondary antibody.
7. Nuclear stain: 3 μg/mL Hoechst 33342, PBS.
8. High content analyzer/fluorescent microscope.
9. Half-area 96-well microplate (0.15 cm^2 well area).
10. Parafilm™.

3 Methods

3.1 Subculturing of Cells

Subculturing should be performed when cells reach 80–90% confluency.

1. Decant the growth medium to waste and wash cells with 5 mL warm, sterile PBS to remove residual complete growth medium.
2. Add a volume of 0.5 mL (25 cm^2 flask) or 1 mL (75 cm^2 flask) trypsin-EDTA to the flask.
3. Incubate flasks at 37 °C in 5% CO_2/95% humidified air for approximately 5 min, to allow cells to detach from the flask surface.
4. Via microscope, visually confirm detachment of 95–100% of cells and then add an equal volume of complete growth medium to flasks to inactivate the trypsin (*see* **Note 3**).
5. Decant the contents to a sterile 15 mL tube.
6. Centrifuge at 180 × *g* for 3 min, and decant the supernatant.
7. Resuspend the cell pellet in 1 mL complete growth medium to obtain a cell suspension.
8. Use this suspension to seed fresh flasks in complete growth medium.
9. Maintain cells at 37 °C in 5% CO_2/95% humidified air.

3.2 Generation of an Isogenic Model of Radioresistant EAC

1. Cells are irradiated in 75 cm^2 tissue culture flasks. Perform irradiation when cells are at 50% confluency.
2. Place the flask containing cells into the irradiator (*see* **Note 4**).
3. Irradiate cells with the desired dose of X-ray radiation at the center of the beam, at a focus to skin distance of 40 cm with a

beam at a dose rate of 3.25 Gy/min (in this case produced by 200 kilovolts (kV) and 15 milliamperes (mA)).

4. Ensure a uniform delivery of ionizing radiation to cells plated in flasks with the generation of a broad radiation beam (12.5 cm × 12.5 cm).
5. Following irradiation, maintain cells at 37 °C in 5% CO_2/95% humidified air, allowing them to recover.
6. Subculture the cells when 90% confluency has been reached.
7. Repeat irradiation when cells are at 50% confluency (*see* **Note 5**).
8. For the parental "P" cells, handle the cells identically but mock-irradiate by placing them in the irradiator for the same length of time as irradiated cells without administering a dose of X-ray radiation. These cells will act as an age- and passage-matched radiosensitive control for comparison (*see* **Note 6**).

3.3 Assessing Radioresistance: Clonogenic Assay

1. Harvest cells in the exponential growth phase by trypsinization and obtain a single cell suspension.
2. Count the cells and adjust the cell suspension to the required concentration density using complete growth medium. This will vary from cell line to cell line and will require optimization. Optimize cell seeding densities to ensure at least 100 viable colonies (consisting of at least 50 cells) per radiation dose are present at the end of the clonogenic incubation period. Optimized cell seeding densities for this model are outlined in Table 1.
3. Seed the cells at optimized seeding densities into 6-well plates to a final volume of 2 mL complete growth medium.
4. Allow the cells to adhere to the plates overnight at 37 °C in 5% CO_2/95% humidified air.
5. The next day, irradiate plates with 0, 2, 4, or 6 Gy X-ray radiation. Mock irradiate the control plates.
6. Incubate all plates at 37 °C in 5% CO_2/95% humidified air for 7–14 days to allow surviving cells to proliferate and reach maximum density without allowing the merging of colonies.
7. Remove complete growth medium from wells and add 0.5 mL crystal violet staining/fixing solution to each well.
8. Incubate the plates for 30 min at room temperature.
9. Decant the staining/fixing solution to waste. Residual unbound crystal violet can be removed by gently submerging plates in a water bath containing dH_2O (*see* **Note 7**).
10. Allow the plates to air dry at room temperature overnight.

Table 1
Optimum clonogenic cell seeding densities for OE33 cells

Treatment	Cell seeding density (cells/well) (6-well plate)
Control	1.5×10^3
2 Gy	3×10^3
4 Gy	6×10^3
6 Gy	1.2×10^4

11. Image the colonies using wide-angled optics. Electroluminescent trans-illumination is used to create maximal contrast with the crystal violet stain. The application of static counting parameters eliminates bias and ensures objective results.
12. Calculate the plating efficiency (PE), which is the fraction of colonies established from untreated cells, using the formula below:

$$PE = \frac{\text{Number of colonies}}{\text{Number of cells seeded}}$$

13. Calculate the surviving fraction (SF), which is the number of colonies produced after treatment, expressed in terms of PE, using the formula below:

$$SF = \frac{\text{Number of colonies}}{(\text{Number of cells seeded} \times \text{PE})}$$

14. Survival curves can be constructed by plotting SF on a log scale on the Y-axis and radiation dose on the X-axis. *See* Fig. 1.

3.4 Assessing Molecular Features of an Isogenic Model of Radioresistant EAC

3.4.1 PI Cell Cycle Analysis

1. Harvest cells in the exponential growth phase by trypsinization.
2. Count and seed cells in a 25 cm^2 flask at an appropriate concentration density. This will vary from cell line to cell line. In this model, cells were seeded at a density of 3.7×10^5 cells in 25 cm^2 flasks and allowed to adhere overnight at 37 °C in 5% CO_2/95% humidified air.
3. Irradiate the cells with X-ray radiation, while mock-irradiating the control flasks.
4. Collect cells at several time points post irradiation by trypsinization and transfer to 5 mL flow cytometry tubes. In this model, cells were collected 6, 10, 24, and 30 h post irradiation.
5. Centrifuge the cells at $180 \times g$ for 3 min and discard the supernatant.

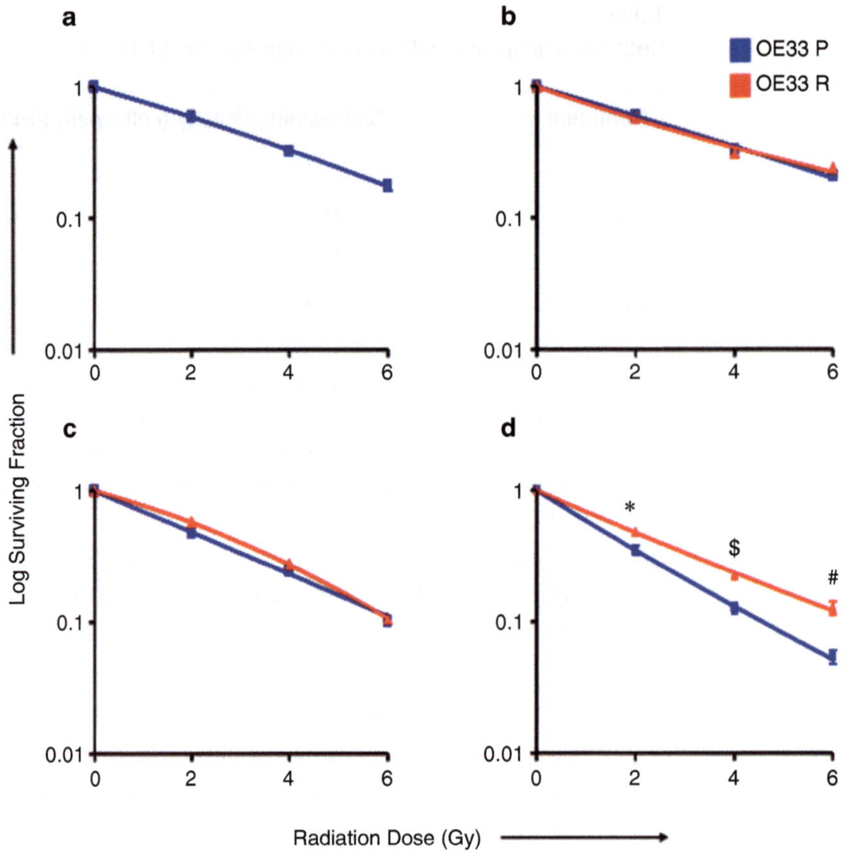

Fig. 1 Generation of a radioresistant cell line. The survival of OE33 cells to the cytotoxic effects of X-ray radiation was measured by clonogenic assay. (**a**) The inherent radiosensitivity of OE33 cells at 2, 4, and 6 Gy bolus dose of radiation. (**b**) The survival of OE33 P and OE33 R7, which had received 7 cycles of 2 Gy radiation (cumulative dose 14 Gy). Survival of OE33 P and OE33 R7 is similar at 2, 4, and 6 Gy bolus dose. (**c**) The survival of OE33 P and OE33 R15, which had received 15 cycles of 2 Gy radiation (cumulative dose 30 Gy). Survival of OE33 P and OE33 R15 is similar at 2, 4, and 6 Gy bolus dose. (**d**) Survival of OE33 P and OE33 R25, which had received 25 cycles of 2 Gy radiation (cumulative dose 50 Gy). OE33 R25 cells are significantly more radioresistant than OE33 P cells at 2, 4, and 6 Gy bolus dose. Data are presented as mean ± SEM of at least three independent experiments. Analysis was performed using a paired two-tailed Student's t-test. *$p = 0.0037$ vs OE33 P, $ $p = 0.0009$ vs OE33 P, # $p = 0.0015$ vs OE33 P. (Figure adapted from Lynam-Lennon et al. 2010 [2])

6. Fix and permeabilize the cell pellet by dropwise addition of 4.5 mL fixation/permeabilization solution while vortexing to avoid the formation of aggregates (*see* **Note 8**).
7. Store the cells in the fixation/permeabilization solution for at least 2 h at 4 °C.
8. Centrifuge the fixed cells at 180 × *g* for 3 min and discard the supernatant.
9. Wash cells with 1 mL PBS and centrifuge at 180 × *g* for 3 min. Discard the supernatant.

10. Resuspend the cell pellets in 0.5 mL staining solution, except for appropriate controls (*see* **Note 9**).
11. Incubate the samples at 37 °C for 30 min and then at room temperature for 1.5 h in staining solution. Unstained control samples used for instrumental setup are incubated in 1 mL PBS.
12. Measure DNA content using a flow cytometer (*see* **Note 10**).

3.4.2 Annexin-V-FITC/PI Apoptosis Assay

1. Harvest the cells in the exponential growth phase by trypsinization.
2. Count and seed the cells in 6-well plates at an appropriate concentration density. This will vary from cell line to cell line. In this model, cells were seeded at a density of 1.5×10^5 cells/well in 6-well plates, and allowed to adhere overnight at 37 °C in 5% CO_2/95% humidified air.
3. Wash the cells with 1 mL warmed PBS.
4. Add 1.5 mL warmed complete growth medium.
5. Irradiate cells with X-ray radiation, while mock-irradiating control plates.
6. At 24 h and 48 h post irradiation, collect supernatants (containing non-adherent/dead cells) in 5 mL flow cytometry tubes.
7. Harvest adherent cells by trypsinization and combine with their respective supernatants in a 15 mL tube.
8. Pellet the cells by centrifugation at 180 × *g* for 3 min.
9. Discard the supernatants and resuspend the cell pellets with 0.5 mL PBS. Centrifuge at 180 × *g* for 3 min.
10. Discard the supernatants and resuspend the cell pellet by vortexing in 1 mL 1× binding buffer.
11. Pellet the cells by centrifugation at 180 × *g* for 3 min and discard the supernatant.
12. Add 3 μL annexin V-FITC to each sample, with the exception of appropriate controls.
13. Vortex the samples and incubate for 15–20 min in the dark at room temperature.
14. Resuspended the cells in 500 μL PI solution, with the exception of appropriate controls. Controls are resuspended in 0.5 mL 1× binding buffer.
15. Vortex the tubes and measure fluorescence using a flow cytometer (*see* **Note 11**).

3.4.3 Detection of ROS

1. Harvest cells in the exponential growth phase by trypsinization.
2. Pellet the cells by centrifugation at 180 × *g* for 3 min.
3. Resuspend the cells in 100 μL PBS at an appropriate concentration density. This will vary from cell line to cell line. In this model, cells were resuspended at a concentration of 1.5×10^5 cells/tube in 5 mL tubes.
4. Load the cells with 0.5 mL CM-H_2DCFDA, except for appropriate unstained controls, and incubate for 30 min at 37 °C in 5% CO_2/95% humidified air, in the dark. Unstained control samples are incubated in PBS.
5. Pellet the cells by centrifugation at 180 × *g* for 3 min and resuspend in 0.5 mL PBS.
6. Add a volume of 5 μL PI stain solution to all samples, except appropriate unstained controls.
7. Irradiate appropriate samples with X-ray radiation, and place the samples on ice in the dark (*see* **Note 12**).
8. Assess fluorescence immediately using a flow cytometer (*see* **Note 13**).

3.4.4 Analysis of DNA Damage and Repair by γH2AX Detection

1. Harvest cells in the exponential growth phase by trypsinization.
2. Seed the cells in triplicate in a half-area 96-well microplate (0.15 cm^2 well area) at an appropriate concentration density. This will vary from cell line to cell line. In this model, cells were seeded at a density of 7×10^3 cells/well in a half-area 96-well microplate and allowed to adhere overnight at 37 °C in 5% CO_2/95% humidified air. Use a separate plate for each treatment.
3. Irradiate cells with X-ray radiation, while mock-irradiating control plates.
4. At desired time points post irradiation, remove complete growth medium from the wells and wash the cells with 100 μL PBS. Desired time points can be determined by optimization as they will vary from cell line to cell line. In this model, 20 min, 6 h and 24 h were used.
5. Remove PBS from wells and add 100 μL fixative solution to each well.
6. Seal the plates with Parafilm™, and incubate at 37 °C for 30 min.
7. Remove the fixative solution and wash the cells three times with 100 μL PBS for 5 min each time.
8. Add 50 μL blocking buffer to each well and incubate for 1 h at room temperature.
9. Remove the blocking buffer and incubate the cells with 25 μL primary antibody in antibody dilution buffer. Incubate the cells at 4 °C overnight.

10. Remove the primary antibody and wash the cells three times with 100 μL PBS for 5 min each time.
11. Incubate the cells with 25 μL secondary antibody for 1 h at room temperature in the dark.
12. Remove the secondary antibody and wash the cells three times with 100 μL PBS for 5 min each time.
13. Incubate the cells with a Hoechst-based nuclear stain for 30 min at 37 °C. Wash the cells three times with 100 μL PBS for 5 min each time.
14. Add fresh PBS and image cells immediately (*see* **Note 14**).

4 Notes

1. In our lab, irradiation was performed using a Gulmay Medical X-ray generator, model (RS225) (Gulmay Medical, Surrey, UK), at a dose rate of 3.25 Gray per minute. Dosimetry was performed by a qualified medical physicist, using ionization chambers and thermoluminescent dosimeters. Other medical X-ray generators can be used, but a similar dose rate should be employed.
2. In our lab, a PC software-operated colony counter (Col-Count™, Oxford Optronix Ltd., Oxford, UK) was used. Other commercially available systems can be utilized.
3. Visually confirm detachment under a light microscope. Cells will become rounded in appearance and will be suspended in the liquid phase and not attached to the surface of the flask. Intermittently tap the side of the flask gently to encourage cell detachment.
4. Prior to irradiating cells, ensure any necessary warm-ups are conducted on the X-ray irradiator.
5. In our methodology, cells were repeatedly exposed to a clinically relevant dose of 2 Gy X-ray radiation until they had received a cumulative dose of 50 Gy (over ~13 months).
6. When carrying out downstream assays, ensure at least a 10-day period between the last irradiation and conducting the assay.
7. For cells that are weakly adherent, cells can be washed gently by addition and removal of dH_2O using a 3.5 mL Pasteur pipette to prevent loss of colonies during the washing process.
8. Permeabilization aids the access of propidium iodide to DNA.
9. The addition of RNase-A ensures any RNA present is digested, preventing any interference with the DNA signal. Make sure there are instrumental controls performed in parallel with the samples, to define the boundaries of each population during

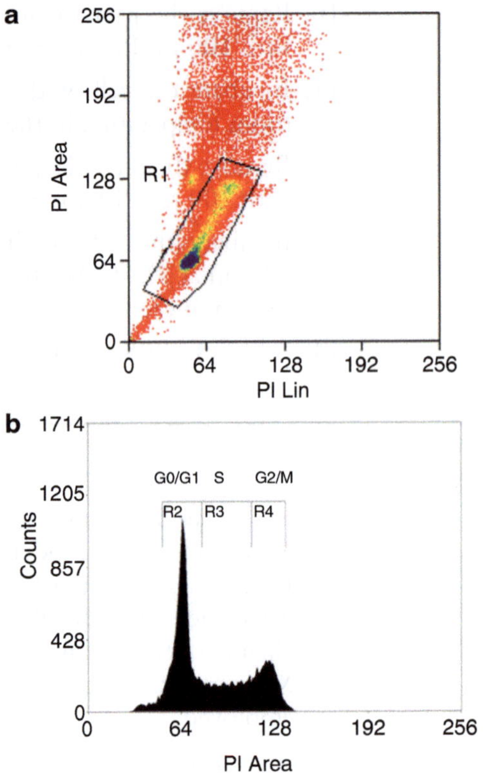

Fig. 2 PI staining for cell cycle analysis. (**a**) Doublet exclusion. (**b**) PI staining, cell cycle phase markers

analysis on the flow cytometer. These include an unstained control and single-stained cells with PI only.

10. Ensure the collection of a minimum of 10,000 events on the flow cytometer, and exclude doublets from analysis using doublet discrimination. *See* Fig. 2 for representative cell cycle distribution plot.

11. Ensure instrumental controls are set up in parallel with the samples to define the boundaries of each population during analysis on the flow cytometer. These controls include unstained cells, single-stained cells with annexin V-FITC only and PI only. Collect a minimum of 10,000 events on the flow cytometer.

12. Ensure instrumental controls are set up in parallel with the samples, to define the boundaries of each population during analysis on the flow cytometer. These include unstained controls, single-stained cells with CM-H_2DCFDA only, single-stained cells with a PI only, and a positive control. Positive controls are obtained by treating samples with 100 μM H_2O_2 for 30 min.

13. Collect a minimum of 10,000 events. Assess mean fluorescence intensity (MFI) as a measure of ROS.
14. Acquire 15 fields of view per well using a 20× objective. Nuclear staining with Hoechst is detected using an excitation filter of 360 nm and emission filter of 460 nm. Alexafluor-488 is detected using an excitation filter of 480 nm and an emission filter of 535 nm. Calculate nuclear MFI as a measure of γH2AX.

Acknowledgments

A.C. and N.L.L. are supported by the Health Research Board, grant numbers ILP-POR-2017-007 and EIA-2017-020.

References

1. Domina EA, Philchenkov A, Dubrovska A (2018) Individual response to ionizing radiation and personalized radiotherapy. Crit Rev Oncog 23(1-2):69–92. https://doi.org/10.1615/CritRevOncog.2018026308
2. Lynam-Lennon N, Reynolds JV, Pidgeon GP, Lysaght J, Marignol L, Maher SG (2010) Alterations in DNA repair efficiency are involved in the radioresistance of esophageal adenocarcinoma. Radiat Res 174(6):703–711. https://doi.org/10.1667/rr2295.1
3. Fukuda K, Sakakura C, Miyagawa K, Kuriu Y, Kin S, Nakase Y et al (2004) Differential gene expression profiles of radioresistant oesophageal cancer cell lines established by continuous fractionated irradiation. Br J Cancer 91(8):1543–1550. https://doi.org/10.1038/sj.bjc.6602187
4. Wang L, Li S, Zhu X (2020) Construction of radiation surviving/resistant lung cancer cell lines with Equidifferent gradient dose irradiation. Dose-Response 18(4):1559325820982421. https://doi.org/10.1177/1559325820982421
5. Li Y, Li H, Peng W, He XY, Huang M, Qiu D et al (2015) DNA-dependent protein kinase catalytic subunit inhibitor reverses acquired radioresistance in lung adenocarcinoma by suppressing DNA repair. Mol Med Rep 12(1):1328–1334. https://doi.org/10.3892/mmr.2015.3505
6. Gray M, Turnbull AK, Ward C, Meehan J, Martínez-Pérez C, Bonello M et al (2019) Development and characterisation of acquired radioresistant breast cancer cell lines. Radiat Oncol (Lond, Engl) 14(1):64. https://doi.org/10.1186/s13014-019-1268-2
7. Lee JU, Hosotani R, Wada M, Doi R, Kosiba T, Fujimoto K et al (1999) Role of Bcl-2 family proteins (Bax, Bcl-2 and Bcl-X) on cellular susceptibility to radiation in pancreatic cancer cells. Eur J Cancer (Oxf, Engl: 1990) 35(9):1374–1380. https://doi.org/10.1016/s0959-8049(99)00134-3
8. McDermott N, Meunier A, Mooney B, Nortey G, Hernandez C, Hurley S et al (2016) Fractionated radiation exposure amplifies the radioresistant nature of prostate cancer cells. Sci Rep 6:34796. https://doi.org/10.1038/srep34796
9. Franken NA, Rodermond HM, Stap J, Haveman J, van Bree C (2006) Clonogenic assay of cells in vitro. Nat Protoc 1(5):2315–2319. https://doi.org/10.1038/nprot.2006.339
10. Pawlik TM, Keyomarsi K (2004) Role of cell cycle in mediating sensitivity to radiotherapy. Int J Radiat Oncol Biol Phys 59(4):928–942. https://doi.org/10.1016/j.ijrobp.2004.03.005
11. Brown JM, Attardi LD (2005) The role of apoptosis in cancer development and treatment response. Nat Rev Cancer 5(3):231–237. https://doi.org/10.1038/nrc1560
12. Meyn RE, Stephens LC, Milas L (1996) Programmed cell death and radioresistance. Cancer Metastasis Rev 15(1):119–131. https://doi.org/10.1007/bf00049491
13. Diehn M, Cho RW, Lobo NA, Kalisky T, Dorie MJ, Kulp AN et al (2009) Association of

reactive oxygen species levels and radioresistance in cancer stem cells. Nature 458(7239): 780–783. https://doi.org/10.1038/nature07733

14. Taneja N, Davis M, Choy JS, Beckett MA, Singh R, Kron SJ et al (2004) Histone H2AX phosphorylation as a predictor of radiosensitivity and target for radiotherapy. J Biol Chem 279(3):2273–2280. https://doi.org/10.1074/jbc.M310030200
15. Kuwahara Y, Li L, Baba T, Nakagawa H, Shimura T, Yamamoto Y et al (2009) Clinically relevant radioresistant cells efficiently repair DNA double-strand breaks induced by X-rays. Cancer Sci 100(4):747–752. https://doi.org/10.1111/j.1349-7006.2009.01082.x
16. Darzynkiewicz Z, Juan G, Li X, Gorczyca W, Murakami T, Traganos F (1997) Cytometry in cell necrobiology: analysis of apoptosis and accidental cell death (necrosis). Cytometry 27(1):1–20
17. Vermes I, Haanen C, Steffens-Nakken H, Reutelingsperger C (1995) A novel assay for apoptosis. Flow cytometric detection of phosphatidylserine expression on early apoptotic cells using fluorescein labelled Annexin V. J Immunol Methods 184(1):39–51. https://doi.org/10.1016/0022-1759(95)00072-i
18. Radford IR (1985) The level of induced DNA double-strand breakage correlates with cell killing after X-irradiation. Int J Radiat Biol Relat Stud Phys Chem Med 48(1):45–54. https://doi.org/10.1080/09553008514551051
19. Fernandez-Capetillo O, Chen HT, Celeste A, Ward I, Romanienko PJ, Morales JC et al (2002) DNA damage-induced G2-M checkpoint activation by histone H2AX and 53BP1. Nat Cell Biol 4(12):993–997. https://doi.org/10.1038/ncb884
20. Banáth JP, Klokov D, MacPhail SH, Banuelos CA, Olive PL (2010) Residual gammaH2AX foci as an indication of lethal DNA lesions. BMC Cancer 10:4. https://doi.org/10.1186/1471-2407-10-4

Chapter 7

Developing an In Vitro Isogenic Model of Chemotherapy-Resistant Lung Cancer

Martin P. Barr

Abstract

Despite improvements in therapies available for small subsets of patients affected by non-small cell lung cancer (NSCLC), the chemotherapy drug cisplatin is still one of the most commonly used treatments for advanced NSCLC patients in the absence of oncogenic driver mutations or immune checkpoints. Unfortunately, as in the case of many solid tumors, acquired drug resistance is a common phenomenon in NSCLC and presents a significant clinical challenge for oncologists. In order to study and elucidate the cellular and molecular mechanisms implicated in the development of drug resistance in cancer, the use of isogenic models provides a valuable in vitro tool for investigating novel biomarkers and the identification of potential targetable pathways involved in drug-resistant cancers.

Key words Cisplatin, Chemotherapy, Drug resistance, Isogenic, In vitro models, NSCLC

1 Introduction

In advanced NSCLC, treatment with cisplatin doublet chemotherapy has been the standard of care for many patients for a considerable number of years in the absence of molecular targetable mutations or immune checkpoints [1]. Following the discovery of cisplatin (cis-diamminedichloroplatinum (II)) by Perone in the 1840s [2], further studies by Rosenberg and colleagues in the 1970s [3] reported the ability of platinum compounds to inhibit cell division in *E. coli*, which subsequently led to the discovery of their anti-tumor activity [4]. Cisplatin is a small, square planar complex, composed of a platinum ion core and four ligands consisting of two amine and two chloride groups (Fig. 1). Cisplatin enters the cells through a combination of passive and active transport systems via membrane-bound transporter proteins. Within the cell, cisplatin undergoes a series of spontaneous aqueous-based reactions, in which the chloride ions are displaced by a water molecule that unlocks the cytotoxic potential of the drug

Dania Movia and Adriele Prina-Mello (eds.), *Cancer Cell Culture: Methods and Protocols*,
Methods in Molecular Biology, vol. 2645, https://doi.org/10.1007/978-1-0716-3056-3_7,
© The Author(s), under exclusive license to Springer Science+Business Media, LLC, part of Springer Nature 2023

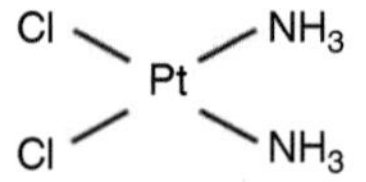

Fig. 1 Cisplatin is an inorganic water-soluble platinum compound composed of two ammine and two chloro ligands which are oriented in a *cis* planar configuration around a platinum ion. Following hydrolysis, cisplatin reacts with cellular DNA to produce both intra- and inter-strand crosslinks. These crosslinks impair replication and transcription of DNA resulting in cell apoptosis

[5]. Since its FDA approval in 1978, cisplatin represents a major landmark in the history of successful anti-cancer drugs and has changed the therapeutic management of several human solid tumors, including lung cancer. While treatment response to cisplatin can vary across different cancer types, the development of acquired drug resistance is a significant clinical challenge in the treatment of patients [6]. Drug resistance is responsible for more than 90% of deaths in patients with cancer receiving conventional chemotherapeutic drugs [7]. Studies have reported that in the context of cisplatin-based therapies, resistance is inferred when the clinical dose of cisplatin required for anti-tumor activity is increased 50–100-fold as part of intensive drug-based treatments [8]. While a plethora of resistance mechanisms to platinum agents such as cisplatin have been reported in cancer cells [9], further studies are warranted to identify novel biomarkers and elucidate pathways underlying this drug-resistant phenotype and to facilitate the prediction of clinical response to treatment with this chemotherapy drug in patients with cancer. The development and extensive characterization of clinically relevant, isogenic in vitro cisplatin-resistant cell lines offer a unique model for the study of resistance mechanisms and the preclinical evaluation of novel targeted therapies in the context of acquired drug resistance in cancer. In this chapter, we describe a method to generate an isogenic in vitro model of cisplatin-resistant NSCLC.

2 Materials

All general laboratory chemicals and reagents should be of analytical grade and stored at room temperature or according to manufacturer's instructions. Prepare buffers and solutions in deionized water and autoclave. Weigh all solid reagents using a calibrated electronic balance. Transfer liquid volumes up to 1 mL using calibrated pipettes. Dispense volumes greater than 1 mL using 10 mL serological disposable pipettes. Use a graduated cylinder for measuring volumes greater than 10 mL.

2.1 Cell Line Generation

1. Cells (NSCLC cell lines; H460, A549, SKMES-1) (*see* **Note 1**).
2. Class IIB laminar air flow unit.
3. Cell culture incubator for maintaining cell lines in a humidified atmosphere of 5% CO_2 in air at 37 °C.
4. Water bath, set at 37 °C for pre-heating cell culture media and reagents prior to use.
5. Centrifuge with insert buckets for spinning 15 mL and 50 mL tubes.
6. Supplemented cell culture medium: RPMI-1640 (for H460 cells) or Hams-F12 (for A549 cells) or EMEM (for SKMES-1 cells) cell culture medium, 10% fetal bovine serum (FBS), 100 U/mL penicillin, 100 μg/mL streptomycin.
7. Phosphate buffered saline (PBS) without calcium chloride & magnesium chloride, sterile-filtered for cell culture use.
8. Sterile-filtered 1× Trypsin-EDTA: 0.5 g porcine trypsin, 0.2 g EDTA.
9. Fixation solution: 70% v/v ethanol, distilled water.
10. Sterile plastics (cell culture flasks, sterile 15 mL and 50 mL tubes, sterile individually wrapped 10 mL pipettes). Use vented flasks for adherent cell lines.
11. Bright-Line Hemacytometer for cell counting.
12. 0.4% Trypan Blue solution (sterile-filtered for cell culture use).
13. Bright-field light microscope.
14. Cisplatin: commercially available 3.3 mM *cis*-diammineplatinum(II) dichloride.
15. 0.15 M NaCl.

2.2 Mycoplasma and Agarose Gel Electrophoresis

1. Sterile Eppendorf tubes.
2. Bench-top microfuge.
3. Class II laminar flow hood for PCR.
4. DreamTaq Green Master mix (*see* **Note 2**).
5. Forward and reverse mycoplasma primers (*see* **Note 3**).
6. Nuclease-free Molecular Biology Grade sterile water.
7. Thermal Cycler.
8. Agarose gel: 2% agarose gel, 1× TAE (Tris-Acetate EDTA) buffer (*see* **Note 4**).
9. DNA molecular weight ladder (1 kb).
10. SYBR Safe DNA gel stain: 1:10,000 SYBR Safe DNA gel stain, 1× TAE buffer.
11. UV imaging system.

2.3 MTT Assay

1. MTT staining solution: 5 mg/mL 3-[4,5-dimethylthiazol-2-yl]-2,5-diphenyltetrazolium bromide), PBS. Store the MTT reagent in the dark at 4 °C.
2. Sterile 96-well culture plates.
3. Molecular Biology Grade DMSO (dimethyl sulfoxide).
4. Microplate reader.

2.4 Apoptosis Assay

1. Sterile 6-well plates.
2. Polystyrene (5 mL) flow cytometry tubes.
3. 1× binding buffer: 0.01 M HEPES, 140 mM NaCl, 2.5 mM $CaCl_2$, PBS. Store at 4 °C
4. Commercially available Annexin V-FITC fluorescent conjugate.
5. 1 mg/mL Propidium iodide (1:4000), 1× binding buffer.
6. Flow cytometer (BD FACSCanto™ II).

2.5 Clonogenic Assay

1. Sterile 6-well culture plates.
2. Crystal violet stain/fixation solution: 0.05% (w/v) crystal violet, 25% (v/v) methanol, distilled water. Store at room temperature.
3. Activated charcoal powder.
4. Colony counter (ColCount™ Oxford Optronix Ltd, UK).

3 Methods

3.1 Mycoplasma Screen

1. Mycoplasma screening should be performed prior to the generation of drug-resistant cell lines.
2. Culture cells in a vented flask until they reach at least 80% confluency.
3. Remove 1 mL of culture supernatant and place in a clean sterile Eppendorf tube.
4. Centrifuge supernatants at 140× *g* for 1 min to pellet any cellular debris.
5. Thaw all RT-PCR reagents (DreamTaq Green PCR Master mix, forward and reverse mycoplasma primers) on ice prior to use. UV-treat the PCR hood for 15 min prior to use.
6. Prepare an RT-PCR master mix consisting of (per sample), 25 μL DreamTaq Green PCR Master Mix, 1 μL forward Mycoplasma primer, 1 μL reverse Mycoplasma primer and 22 μL Molecular Biology Grade nuclease-free water. Mix well by vortexing and place on ice.

7. Add 1 μL of each test cell culture supernatant into clean sterile Eppendorf tubes.
8. For negative and positive controls, add 1 μL of Molecular Biology Grade nuclease-free water and 1 μL of a positive infected culture supernatant, respectively.
9. To each test and control sample, add 49 μL of RT-PCR Master Mix, mix well by vortexing and place on ice. Briefly spin tubes in a picofuge and load on Thermal Cycler.
10. Carry out RT-PCR using PCR cycling conditions consisting of 95 °C for 5 min, followed by 40 cycles of DNA denaturation at 94 °C for 30 s, primer annealing at 55 °C for 30 s, primer extension at 72 °C for 1 min, followed by a final extension at 72 °C for 10 min.
11. Run PCR products on a 2% agarose gel containing SYBR Safe DNA gel stain (1:10,000) diluted in 1× TAE buffer.
12. Carefully load 10 μL of sample or control into each well of the gel, in parallel with a 1 kb DNA molecular weight marker. Run the gel by electrophoresis at 80 v for 30 min.
13. Visualize PCR products under UV light using the appropriate imaging system. A PCR product of 270 bp is indicative of mycoplasma infection and similar in size to that observed in the positive control. Negative controls should show no band in this region.

3.2 Drug Solution Preparation

1. Prepare 0.15 M NaCl in PBS by weighing 2.922 g of NaCl. Transfer the solution to a clean 50 mL Falcon tube and add 50 mL of PBS pre-heated at 37 °C and mix thoroughly. Use this as the buffer for dissolving cisplatin to a final stock concentration of 1 mg/mL (3.30 mM).
2. Filter-sterilize the drug solution by gently passing it through a sterile syringe & barrel attached to a 0.22 μM syringe filter into a clean sterile 50 mL tube. Pipette 1 mL aliquots of the drug into sterile Eppendorf tubes and store at −80 °C for long-term storage. Further dilutions of the stock drug can be made as required at the time of experiments.

3.3 Cisplatin Dose-Response Studies

1. Pre-sterilize a laminar flow hood using UV light for 15 min prior to the experiment.
2. Seed *wild*-type (parental) cells (2.5×10^3) in a 96-well plate, in triplicate, using supplemented cell culture medium (200 μL per well) and incubate overnight in a humidified CO_2 incubator at 37 °C to allow cells to adhere. Include wells, in triplicate, containing supplemented cell culture medium only for assay background and vehicle (0.15 M NaCl) controls.

3. In order to generate cisplatin-resistant sublines from *wild*-type cells, prepare increasing concentrations of cisplatin in 15 mL tubes ranging from 0 to 100 μM (0, 10 μM, 20 μM, 40 μM, 60 μM, 80 μM, 100 μM) in supplemented cell culture medium. Mix well by vortexing.
4. Remove cell culture media from control (background) wells and replace with 200 μL of fresh complete medium. For vehicle control, prepare 0.15 M NaCl in fresh supplemented cell culture medium and add 200 μL to each vehicle control well.
5. Gently remove the supplemented cell culture medium from each well and add 200 μL of each cisplatin concentration (0–100 μM) into each well, in triplicate.
6. Incubate plate for 72 h in a humidified CO_2 incubator at 37 °C.
7. Analyze cell viability using the MTT assay (*see* Subheading 3.4) [10].

3.4 MTT Cell Viability Assay

1. Add 10 μL of MTT staining solution directly to the supplemented cell culture medium in each well (except background controls) of the 96-well plate.
2. Incubate for 4 h in a humidified CO_2 incubator at 37 °C.
3. Confirm the formation of purple formazan crystals in each well under light microscopy. If there is no evidence of purple formazan crystals after this time, incubate for longer, or overnight.
4. Add 100 μL of DMSO to each well and agitate gently on a shaker to dissolve formazan crystals.
5. Measure absorbance at 595 nm with a reference wavelength of 650 nm using a microplate reader.
6. Calculate the average absorbance for each sample and subtract the average background absorbance from that of each treatment or control.
7. Set the untreated (control) absorbance at 100% and calculate the percentage cell viability for each cisplatin concentration using the following equation:

$$\text{Cell viability (\%)} = \frac{\text{Average absorbance of test sample}}{\text{Average absorbance of control}} \times 100$$

8. Extrapolate the cisplatin half-maximal inhibitory concentration (IC_{50}) for the *wild*-type cell line, by constructing sigmodal dose-response curves using GraphPad software or other statistical software package.

3.5 Generation of Cisplatin-Resistant Cell Sublines

1. Seed (from 1×10^6 to 2×10^6 cells/mL) *wild*-type (parental) cells in two T75 cm^2 culture flasks containing 12 mL of supplemented cell culture medium each.
2. Allow cells to adhere overnight in a humidified CO_2 incubator at 37 °C. Grow in culture until cells are at least 70% confluent.
3. In flask #1, remove all the supplemented cell culture medium and add 12 mL of fresh supplemented cell culture medium. Label this flask as "parental" cells. Prepare 12 mL of a cisplatin solution at its IC_{50} in supplemented cell culture medium, mix well by vortexing and add it to flask #2. Label this flask as "cisplatin-resistant" cells.
4. Culture both flasks for 72 h in a humidified CO_2 incubator at 37 °C.
5. Remove the supernatant from both flasks and replace it with 12 mL of fresh supplemented cell culture medium and allow cells to recover for a further 72 h.
6. Check cell morphology and confluency every 2–3 days using a light microscope.
7. After 72 h, passage both flasks of cells (parental and cisplatin-resistant) and repeat **steps 3–6** until a resistant phenotype in flask #2 is confirmed.
8. To monitor for the development of resistance, perform the MTT cell viability assay (*see* Subheading 3.4) every four weeks on both the parental and cisplatin-resistant cell lines and compare the cisplatin IC_{50} for the two sublines.
9. Repeat for several months until such time as a drug resistance phenotype becomes apparent (Fig. 2). A log-fold change in IC_{50} concentrations between parental and the cisplatin-resistant derived subline, or significant differences in cell viability in response to different cisplatin concentrations between the matched cell lines, is indicative of a resistant phenotype.
10. Further functional characterization may be carried out using apoptosis (*see* Subheading 3.6) and clonogenic (*see* Subheading 3.7) assays.

3.6 Apoptosis (Annexin V-Propidium Iodide) Assay by Flow Cytometry

1. Seed cells (1×10^5 cells/mL) in 6-well plates in 2 mL of supplemented cell culture medium and allow to adhere overnight in a humidified CO_2 incubator at 37 °C. Prepare sufficient wells for treatments and untreated controls. Experimental controls should also be included for unstained cells, Annexin-V only, and PI only.
2. Prepare increasing concentrations of cisplatin ranging from 0 to 100 μM (0, 10 μM, 20 μM, 40 μM, 60 μM, 80 μM, 100 μM) in supplemented cell culture medium. Mix them well by vortexing.

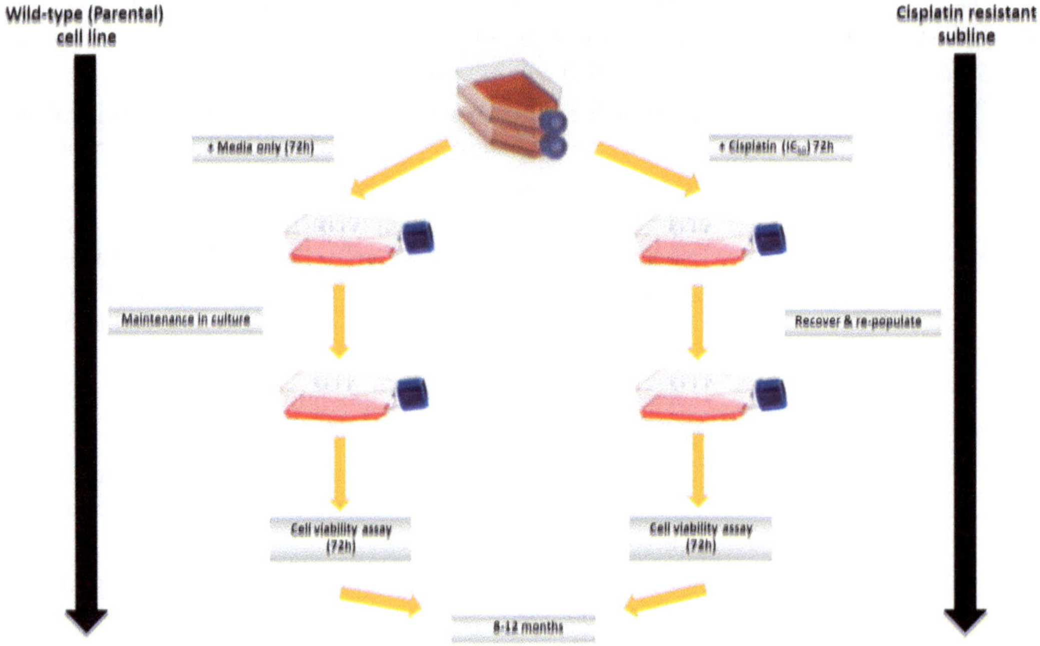

Fig. 2 Drug-resistant sublines are generated from original *wild*-type (parental) cells in response to treatment with IC_{50} concentrations of cisplatin over a period of 8–12 months. Development of the resistant phenotype is monitored every four weeks using cisplatin dose-response studies in functional assays that include cell viability, apoptosis, and clonogenic analysis

3. Add 2 mL of each cisplatin concentration to the 6-well plate and 2 mL of supplemented cell culture medium to control wells.
4. Incubate in a humidified CO_2 incubator at 37 °C for 72 h.
5. After 72 h, collect any dead/floating cells in the cell culture supernatants from each well using a Pasteur pipette and transfer into clean, labelled flow cytometry tubes.
6. Trypsinize adherent cells and transfer the cell suspension into flow cytometry tubes containing cell culture supernatants.
7. Spin the tubes by centrifugation at 360× *g* for 3 min.
8. Discard supernatants and re-suspend the cell pellets in 1 mL of 1× binding buffer and vortex.
9. Centrifuge at 360× *g* for 3 min and discard supernatant.
10. Add 3 μL of Annexin-V staining solution to each test sample, but not to the appropriate controls (PI only and unstained), and vortex.
11. Place tubes on ice, cover with foil, and incubate at 4 °C for 20 min.

12. Suspend samples in 1 ml of 1× binding buffer, vortex and centrifuge at 360× *g* for 3 min. Discard supernatants and suspend cell pellets in 500 μL of 1× binding buffer.
13. Add 500 μL of PI staining solution and vortex.
14. Acquire test samples (10,000 events) on the flow cytometer following compensation and gating using the appropriate controls. A significant increase in cisplatin-mediated lung cancer cell apoptosis should be observed in parental cells relative to their matched cisplatin resistant counterparts.

3.7 Clonogenic Assay

The clonogenic assay, otherwise known as the colony formation assay, is a gold standard in vitro cell survival assay for assessing the reproductive integrity of cells following exposure to ionizing radiation or in response to a cytotoxic insult [11]. In the context of cisplatin resistance, the clonogenic assay can be used to measure the survival ability of both parental and drug-resistant cells in response to cisplatin.

1. Harvest parental and cisplatin-resistant sublines in the exponential phase of cell growth by trypsinization.
2. Centrifuge at 360× *g* for 3 min to pellet cells.
3. Resuspend the cell pellets in supplemented cell culture medium and seed cells in 6-well plates prior to treatment with cisplatin using optimized seeding densities of 500 (untreated), 1000 (0.1 μM), 2000 (1 μM), 10,000 (5 μM) and 50,000 (10 μM). Allow cells to adhere overnight in a humidified CO_2 incubator at 37 °C.
4. Following overnight incubation, remove the supernatant from each well and treat cells for 72 h with 2 mL of increasing concentrations of cisplatin (from 0.1 μM to 10 μM) prepared in supplemented cell culture medium. Mock treat all control cells with 2 mL of supplemented cell culture medium.
5. After 72 h, remove supernatants from each well and add supplemented cell culture medium. Incubate in a humidified CO_2 incubator at 37 °C for at least 10 days to allow growth of single discrete clones. Clones are considered to represent viable cells if they contain in excess of 50 cells.
6. After 10 days, or thereafter, when control dishes have formed sufficiently large colonies, remove the supernatant from each well and wash with 2 mL of PBS.
7. Add 1 mL of fixation/staining solution to each well and leave at room temperature for 30 min.
8. Wash plates by submerging gently in a water bath containing tap water to remove any residual fixation/staining solution from each well.

9. Leave plates to dry at room temperature overnight.
10. Image and count the number of stained colonies using a colony counter. For imaging, use where possible, wide-angled optics to provide visual coverage of the entire 6-well plate and electroluminescent trans-illumination to create maximal contrast with the crystal violet stain.
11. Determine the plating efficiency (PE) of control cells. This is the fraction of colonies from control cells (those that are not exposed to the treatment). This is calculated using the formula below:

$$\mathrm{PE} = \frac{\text{Number of colonies formed}}{\text{Number of cells seeded}}$$

12. Calculate the surviving fraction (SF) of cells following treatment using the formula below, where this takes into account the PE of untreated or control cells:

$$\mathrm{SF} = \frac{\text{Number of colonies formed after treatment}}{(\text{Number of cells seeded} \times \mathrm{PE})}$$

13. Construct survival curves by plotting the SF on the Y-axis and treatment concentrations on the X-axis. Compare differences in surviving colonies between parental and cisplatin-resistant sublines at each cisplatin concentration by statistical analysis.

4 Notes

1. This protocol can be used for generating cisplatin-resistant sublines from any cancer cell type. Optimization of seeding concentrations may be required.
2. The use of DreamTaq Green PCR Master Mix (2×) is recommended. This is a ready-to-use solution containing DreamTaq DNA polymerase, DreamTaq Green buffer, $MgCl_2$ and dNTPs. The mix contains two tracking dyes and a density reagent that allows for direct loading of PCR products on a gel. The dyes used in the master mix do not interfere with PCR performance and retain all features of DreamTaq DNA polymerase. It can be used for robust amplification up to 6 kb from genomic DNA and up to 20 kb from viral DNA.
3. Prepare a working stock (10 μM) of forward and reverse primers for Mycoplasma RT-PCR:

   Forward: 5′-TGCACCATCTGTCACTCTGTTAACCTC-3′

   Reverse: 5′-GGGAGCAAACAGGATTAGATACCCT-3′.
4. For a 50× stock solution of 1× TAE buffer, weigh 242 g of Tris-base (MW = 121.14 g/mol) and dissolve in 700 mL of

deionized water. Add 57.1 mL of 100% glacial acetic acid and 100 mL of 0.5 M EDTA (pH 8.0). Adjust the pH of the solution to 8.5 following calibration of the pH meter using pH 7.0 and pH 10.0 calibration buffers. Make up to a final volume of 1 L and store at room temperature.

References

1. Ardizzoni A, Boni L, Tiseo M, Fossella FV, Schiller JH, Paesmans M, Radosavljevic D, Paccagnella A, Zatloukal P, Mazzanti P, Bisset D, Rosell R, CISCA (CISplatin versus CArboplatin) Meta-analysis Group (2007) Cisplatin- versus carboplatin-based chemotherapy in first-line treatment of advanced non-small-cell lung cancer: an individual patient data meta-analysis. J Natl Cancer Inst 99:847–857
2. Peyrone M (1844) Ueber die Einwirkung des Ammoniaks auf Platinchlorür. Ann Chemie Pharm 51:1–29
3. Rosenberg B, Vancamp L, Krigas T (1965) Inhibition of cell division in Escherichia Coli by electrolysis products from a platinum electrode. Nature 205:698–699
4. Rosenberg B, VanCamp L, Trosko JE, Mansour VH (1969) Platinum compounds: a new class of potent antitumour agents. Nature 222: 385–386
5. Goodsell DS (2006) The molecular perspective: cisplatin. Stem Cells 24:514–515
6. Ko T, Li S (2019) Genome-wide screening identifies novel genes and biological processes implicated in cisplatin resistance. FASEB J 33: 7143–7154
7. Bukowski K, Kciuk M, Kontek R (2022) Mechanisms of multidrug resistance in cancer chemotherapy. Int J Mol Sci 21:3233
8. Hagopian GS, Mills GB, Khokhar AR, Bast RC Jr, Siddik ZH (1999) Expression of p53 in cisplatin-resistant ovarian cancer cell lines: modulation with the novel platinum analogue (1R, 2R-diaminocyclohexane)(trans-diacetato)(dichloro)-platinum(IV). Clin Cancer Res 5: 655–663
9. Zhou J, Kang Y, Chen L, Wang H, Liu J, Zeng S, Yu L (2020) The drug-resistance mechanisms of five platinum-based antitumor agents. Front Pharmacol 11:343
10. Mosmann T (1983) Rapid colorimetric assay for cellular growth and survival: application to proliferation and cytotoxicity assays. J Immunol Methods 65:55–63
11. Franken NA, Rodermond HM, Stap J, Haveman J, van Bree C (2006) Clonogenic assay of cells in vitro. Nat Protoc 1:2315–2319

Chapter 8

Culturing Human Lung Adenocarcinoma Cells in a Serum-Free Environment

Aline Chary

Abstract

The human lung adenocarcinoma cell line A549 is commonly used in cancer research as a model of malignant alveolar type II epithelial cells. A549 cells are frequently cultured in Ham's F12K (Kaighn's) or Dulbecco's Modified Eagle's Medium (DMEM), supplemented with glutamine and 10% fetal bovine serum (FBS). However, the use of FBS presents significant scientific concerns, such as the presence of undefined components and batch-to-batch variation leading to possible reproducibility issues in experiments and readouts. This chapter describes how to transition A549 cells to FBS-free medium and gives some insights on the further characterizations and functionality assays that would be necessary to perform for the validation of the cultured cells.

Key words Fetal bovine serum, Cell culture, Serum-free media, Chemically defined media, Xeno-free media, Cell lines, A549, Air–liquid interface

1 Introduction

For the last decades, cell culture has been used to study many aspects of cell biology, for example, with the development of in vitro models for the assessment of toxicological and pharmacological effects of chemicals and pharmaceuticals. The human lung adenocarcinoma cell line A549 is commonly used in cancer research as a model of the malignant alveolar type II epithelial cells [1]. A549 cells are able to produce lung surfactant and can thus be cultured at the Air–Liquid Interface (ALI), allowing to mimic in vivo conditions for respiratory epithelial cells [2]. A549 cells are frequently cultured in Ham's F12K (Kaighn's) or Dulbecco's Modified Eagle's Medium (DMEM), supplemented with glutamine and 10% Fetal Bovine Serum (FBS).

FBS has been, for a long time, considered as the gold standard for cells culture, but this is now raising concerns. In applied research, the validation of an in vitro method in a regulatory

Dania Movia and Adriele Prina-Mello (eds.), *Cancer Cell Culture: Methods and Protocols*,
Methods in Molecular Biology, vol. 2645, https://doi.org/10.1007/978-1-0716-3056-3_8,
© The Author(s), under exclusive license to Springer Science+Business Media, LLC, part of Springer Nature 2023

context is crucial. The robustness, reproducibility, and predictive value of the in vitro model must be proven and the inter-laboratory reproducibility represents a key stage in the process, where the responses of a cellular model need to be consistent between different laboratories [3]. Recently, an inter-laboratory study was performed by two laboratories on A549 cells cultured in submerged and at ALI conditions following the same detailed protocol. Experimental variability in the behavior of the cells was reported, related to the use of different batches of FBS, limiting the inter-laboratory reproducibility of the model [4]. Also, it has become evident that cells behave differently as they are actually doing in the human body when they are cultured in an FBS-enriched environment originally optimized for the development of calves. A549 cells transitioned and cultured in two different FBS-free media have showed, in fact, more sensitivity to toxicants and a differential genes expression as compared to the same cells cultured in an FBS-enriched environment. Therefore, the use of FBS-free medium may change the output of the A549 cell model, and responses closest to the in vivo human situations may be obtained by using FBS-free media [5].

Several FBS-free media exist on the market, with different degrees of chemical definition. A universal FBS-free medium suitable for all cell types may not be feasible due to the requirements requisite for each cell type. Likewise, not all FBS substitutes may be suitable for a particular cell type. In our experiments, on four selected FBS substitutes, only two were suitable to grow the A549 cell line [5].

Different approaches are possible to adapt a cell line to FBS-free medium (*see* Table 1) [6]. Basically, cells need to be gradually adapted to lower FBS concentrations until attaining FBS-free conditions, following a progressive weaning process of FBS in the medium. A direct replacement of FBS-containing media with FBS-free media is generally unsuccessful.

Table 1
Available approaches for the adaptation of cells to an FBS-free environment (SFM: Serum Free Media)

1 – Reduction of serum content	2 – Sequential adaptation
Cultivation of cells in normal medium (DMEM + 10%FBS)	Cultivation of cells in 100% normal medium + 0% SFM
Subcultivation in SFM + 5% FBS	Subcultivation in 80% normal medium + 20% SFM
Subcultivation in SFM + 2% FBS	Subcultivation in 60% normal medium + 40% SFM
Subcultivation in SFM + 1% FBS	Subcultivation in 40% normal medium + 60% SFM
Subcultivation in SFM + 0.5% FBS	Subcultivation in 20% normal medium + 80% SFM
Subcultivation in SFM + 0.1% FBS	Subcultivation in 0% normal medium + 100% SFM
Subcultivation in SFM	

Adapted and modified from Ref. [6]

In this chapter, we describe how the reduction of serum content and the sequential adaptation to an FBS-free media can be successfully performed for the A549 cell line.[5]

2 Materials

2.1 Equipment

1. Cell culture incubator with controlled humidified environment and 5% CO_2 (5%)
2. Laminar flow hood
3. Water bath
4. Light microscope
5. Centrifuge
6. Refrigerator (2–8 °C) and freezer (−20 °C)
7. Liquid nitrogen tank
8. Automated cell counter
9. Sterile plasticware
10. Cell freezing container

2.2 Culture and Cryopreservation Materials

1. Human cell line A549 (*see* **Notes 1** and **2**)
2. Serum Free Medium (SFM): CnT-Prime Airway Epithelial Proliferation Medium (CnT-PR-A) or XVIVO™ 10 medium (Lonza) (*see* **Note 3**)
3. Dimethyl Sulfoxide (DMSO)
4. 70% Ethanol: 70% Ethanol, 30% water
5. Dulbecco's Modified Eagle's Medium (DMEM)
6. Fetal Bovine Serum (FBS) Superior
7. Complete culture medium containing FBS (CCM-F): 50 mL of FBS Superior, 450 mL of DMEM (*see* **Note 4**)
8. Phosphate buffered saline (PBS)
9. ProFreeze™ (Lonza) (*see* **Note 5**)
10. Recombinant cell-dissociation enzyme TrypLE (*see* **Note 6**)
11. Mycoplasma Detection Kit
12. Freezing medium: 85% of ProFreeze™, 15% of DMSO

3 Methods

All the operations should be carried out under strict sterile condition; it is highly recommended that protective gloves and clothing always be used.

3.1 A549 Cell Culture in DMEM

1. Upon receipt of the cell line from the supplier, place a T75 flask containing 15 mL of CCM-F into the incubator for at least 15 min, to allow the medium to reach its normal pH (7.0–7.6).
2. Thaw the vial by gentle agitation in a 37 °C water bath for 2 min. To reduce the possibility of contamination, keep the cap out of the water. Remove the vial from the water bath as soon as the content is thawed and decontaminate by spraying with 70% ethanol.
3. Transfer the vial contents to a centrifuge tube containing 9 mL of CCM-F and centrifuge at 300× *g* for 5 min.
4. Resuspend cell pellet with CCM-F from the flask and transfer the cell suspension back to the flask.
5. Check the confluency of the cells every day under a light microscope. When cells reach approximately 80–90% of confluency, the adaptation of the cells can begin.

3.2 Method 1 – Reducing the Serum Content in the Cell Medium Step-by-Step

1. Once A549 cells reach 80–90% of confluency in CCM-F, wash cells with 5 mL of PBS and add 1–2 mL of TrypLE to the flask.
2. Incubate for 10 min at 37 °C in a cell incubator.
3. Transfer the cell suspension in 8 mL of SFM in a 15 mL falcon tube, and centrifuge at 300× *g* for 5 min.
4. Resuspend the pellet in 1 mL of SFM, and count cells. For counting, 10 μL of cell suspension is diluted in 390 μL of PBS and mixed.
5. Seed cells in a new flask containing 15 mL of SFM supplemented with 750 μL of FBS (5% of FBS). For CNT-PR-A medium, 1×10^6 cells are seeded per flask. For X-VIVO medium, 0.5×10^6 cells are seeded per flask (*see* **Note 7**).
6. Place the flask in the cell incubator.
7. Assess the morphology and confluency of the cells daily. Change medium every other day.
8. When cells reach 80–90% of confluency, repeat **steps 1–4**.
9. Seed the cells in a new flask containing 15 mL of SFM supplemented with 300 μL of FBS (2% of FBS).
10. Place the flask in the cell incubator.
11. Assess the morphology and confluency of the cells daily. Change medium every other day.
12. When cells reach 80–90% of confluency, repeat **steps 1–4**.
13. Seed cells in a new flask containing 15 mL of SFM supplemented with 150 μL of FBS (1% of FBS).
14. Place the flask in the cell incubator.

15. Assess the morphology and confluency of the cells daily. Change medium every other day.
16. When cells reach 80–90% of confluency, repeat **steps 1–4**.
17. Seed cells in a new flask containing 15 mL of SFM with 75 μL of FBS (0.5% of FBS).
18. Place the flask in the cell incubator.
19. Assess the morphology and confluency of the cells daily. Change medium every other day.
20. When cells reach 80–90% of confluency, repeat **steps 1–4**.
21. Seed cells in a new flask containing 15 mL of SFM with 15 μL of FBS (0.1% of FBS).
22. Place the flask in the cell incubator.
23. Assess the morphology and confluency of the cells daily. Change medium every other day.
24. When cells reach 80–90% of confluency, repeat **steps 1–4**.
25. Seed cells in a new flask containing 15 mL of SFM (*see* **Note 8**).
26. Place the flask in the cell incubator. A549 cells are now adapted to the SFM medium.
27. Passage cells for at least 3 further passages by repeating **steps 24–26** (*see* **Note 9**). At the third passage, cells can be either used for further experiments (*see* **Note 10**) or cryopreserved (*see* Subheading 3.4).

3.3 Method 2 – Sequential Adaption

1. Once A549 cells reach 80–90% of confluency in CCM-F, wash cells with 5 ml of PBS and add 1–2 mL of TrypLE to the flask.
2. Incubate for 10 min at 37 °C in a cell incubator.
3. Transfer the cell suspension in 8 mL of CCM-F in a 15 mL falcon tube, and centrifuge at 300× *g* for 5 min.
4. Resuspend the pellet in 1 mL of CCM-F, and count cells. For counting, 10 μL of cell suspension is diluted in 390 μL of PBS and mixed.
5. Seed cells in a new flask containing 12 mL (80%) of CCM-F and 3 mL (20%) of SFM. For CNT-PR-A medium, 1×10^6 cells are seeded per flask. For X-VIVO medium, 0.5×10^6 cells are seeded per flask (*see* **Note 7**).
6. Place the flask in the cell incubator.
7. Assess the morphology and confluency of the cells daily. Change medium every other day.
8. When cells reach 80–90% of confluency, repeat **steps 1–4**.

9. Seed cells in a new flask containing 9 mL (60%) of CCM-F and 6 mL (40%) of SFM.
10. Place the flask in the cell incubator.
11. Assess the morphology and confluency of the cells daily. Change medium every other day.
12. When cells reach 80–90% of confluency, repeat **steps 1–4**.
13. Seed the cells in a new flask containing 6 mL (40%) of CCM-F and 9 mL (60%) of SFM.
14. Place the flask in the cell incubator.
15. Assess the morphology and confluency of the cells daily. Change medium every other day.
16. When cells reach 80–90% of confluency, repeat **steps 1–4**.
17. Seed the cells in a new flask containing 3 mL (20%) of CCM-F and 12 mL (80%) of SFM (80%).
18. Place the flask in the cell incubator.
19. Assess the morphology and confluency of the cells daily. Change medium every other day.
20. When cells reach 80–90% of confluency, repeat **steps 1–4**.
21. Seed the cells in a new flask containing 15 mL of SFM (100%).
22. Place the flask in the cell incubator. A549 cells are now adapted to the SFM medium.
23. Passage cells for at least 3 further passages by repeating **steps 20–22** (*see* **Note 9**). At the third passage, cells can be either used for further experiments (*see* **Note 10**) or cryopreserved (*see* Subheading 3.4).

3.4 Cell Cryopreservation in Serum-Free Conditions

1. Once A549 cells reach 80–90% of confluency in CCM-F, wash cells with 5 mL of PBS and add 1–2 mL of TrypLE to the flask.
2. Incubate for 10 min at 37 °C in a cell incubator.
3. Transfer the cell suspension in 8 mL of CCM-F in a 15 mL falcon tube, and centrifuge at 300× *g* for 5 min.
4. Resuspend the pellet in 1 mL of SFM, and count cells.
5. Dilute cells to a concentration of 1×10^6 cells per 500 μL in SFM.
6. Add 500 μL of cell suspension to each cryovial and place them on ice.
7. Slowly add 500 μL of freezing medium to each vial.
8. Cool the vial to −80 °C at a rate of −1 °C/min in a cell freezing container overnight, and then transfer them to a liquid nitrogen tank.

4 Notes

1. Upon receipt of the cell line from the supplier, it is recommended to multiply cells and freeze a new batch of cells before beginning the procedure of adaptation. Use A549 cells between 1 and 20 passages; then a new cryovial of cells should be thawed.
2. To ensure reproducibility, reliability, credibility, and acceptance of any results produced by this protocol, it is recommended to check the quality of the cells regularly. The identity of the A549 cell line must be confirmed before and after the adaptation to Serum Free Medium (SFM) via human short tandem repeat (STR) profiling cell authentication. A549 cells (adapted to SFM or not) must be tested every 6 months for mycoplasma contamination using a mycoplasma detection kit.
3. Two types of SFM were validated in our protocol: The CnT-Prime Airway Epithelial Proliferation Medium which is a chemically defined medium, and the X-VIVO^{TM}10, which is a xeno-free medium. We, therefore, recommend using these.
4. Store at +4 °C. To reduce the possibility of contamination, it is recommended to transfer 10 mL of CCM-F to a 25 cm^2 flask and incubate it for 48–72 h at 37 °C in a cell incubator. Determine sterility by visual inspection.
5. It is recommended to use ProFreeze^{TM}, as this is a cryopreservation animal-free medium which is chemically defined.
6. The use of a recombinant cell-dissociation enzyme such as TrypLE instead of the classical porcine trypsin is encouraged to foster as much the replacement of animal components in cell culture. Preliminary assays showed no differences between the use of porcine trypsin and TrypLE for cellular detachment and dissociation. This alternative can be directly substituted for trypsin without modifying the protocol.
7. The indicated cell densities are based on our studies. If using a different FBS-free medium, it is advised to start with the cell density which is generally used in CCM-F medium. Assess daily the morphology and cellular confluency under a light microscope. If cells are growing slower in the transition/FBS-free medium, more cells can be added to the flask.
8. The use of coating such as human fibronectin may improve the growth of A549 cells cultures in CNT-PR-A [5].
9. Unlike trypsin, TrypLE does not need to be inactivated by adding excess protein, such as serum or soybean protein, and allows for less destructive cell dissociation. Thus, A549 cells

grown in CCM-F or X-VIVO™ can be seeded in SFM without the need to perform an inactivation step. When using CNT-PRA, always centrifuge the cells in order to remove the TrypLE.

10. The growth curve and the doubling time of the cells transitioned to SFM must be determined in order to know the cell density to seed in plates or flasks for further experiments. An unwanted selection of change in the population may occur during the transition. Thus, it is necessary to monitor cellular morphology during the weaning process and to check the performance and functionality of the cells after the adaptation to SFM [6]. The verification of key features that need to be expressed by A549 cells such as the production of surfactant is important as well, especially when the cells should be cultured at the Air–Liquid Interface (ALI) in further experiments. The question regarding which reference for comparison should be used to evaluate the functionality of the cells is still outstanding [7], as one should keep in mind that cells grown in FBS environment express phenotypical changes that may not be human-relevant from a physiological point of view.

Acknowledgments

This work was in part supported by the LIST internal project VEGALI (Esch-sur-Alzette, Luxembourg) and PETA Science Consortium International e.V. (Stuttgart, Germany).

References

1. Lieber M, Todaro G, Smith B, Szakal A, Nelson-Rees W (1976) A continuous tumor-cell line from a human lung carcinoma with properties of type II alveolar epithelial cells. Int J Cancer 17:62–70
2. Lacroix G et al (2018) Air–liquid interface in vitro models for respiratory toxicology research: consensus workshop and recommendations. Appl In Vitro Toxicol 4:91–106
3. Hartung T et al (2004) A modular approach to the ECVAM principles on test validity. Alternat Lab Anim 32:467–472
4. Barosova H et al (2021) Inter-laboratory variability of A549 epithelial cells grown under submerged and air-liquid interface conditions. Toxicol in Vitro 75:105178
5. Chary A et al (2022) Maximizing the relevance and reproducibility of A549 cell culture using FBS-free media. Toxicol in Vitro 105423. https://doi.org/10.1016/j.tiv.2022.105423
6. van der Valk J et al (2010) Optimization of chemically defined cell culture media – replacing fetal bovine serum in mammalian in vitro methods. Toxicol in Vitro 24:1053–1063
7. van der Valk J (2022) Fetal bovine serum – a cell culture dilemma. Science. https://doi.org/10.1126/science.abm1317

Chapter 9

A Method for Culturing 3D Tumoroids of Lung Adenocarcinoma Cells at the Air–Liquid Interface

Dania Movia and Adriele Prina-Mello

Abstract

Three-dimensional (3D) tumor spheroids and tumoroids are among the most exploited cell culture methods in the lung cancer field, finding applications in the investigation of tumor growth and proliferation, invasion, and drug screening. However, 3D tumor spheroids and tumoroids cannot fully mimic the architecture of the human lung adenocarcinoma tissue and, in particular, the direct contact of the lung adenocarcinoma cells with the air, as they lack polarity. Our method allows to overcome this limitation by enabling to grow tumoroids of lung adenocarcinoma cells and healthy lung fibroblasts at the Air–Liquid Interface (ALI). This ensures straightforward access to both the apical and basal surface of the cancer cell culture, with several advantages in drug screening applications.

Key words Lung adenocarcinoma, New approach methodology, Alternatives to animal models, 3D culture, Air–Liquid Interface

1 Introduction

Non-small-cell lung cancer (NSCLC) and its most common subtype, lung adenocarcinoma, are a leading cause of cancer-related mortality in Europe [1] and worldwide [2]. The recent review conducted by the European Commission Joint Research Centre on advanced non-animal models for research into respiratory tract diseases [3] concluded that the lung-cancer field would benefit from a closer association with non-cancer lung models to progress toward animal replacement and more human-predictive research.

In the recent years, our group has successfully started addressing this methodological gap by establishing a protocol to form tumoroids of either lung adenocarcinoma cells [4] or lung adenocarcinoma cells co-cultured with healthy human lung fibroblasts [5] grown at the Air–Liquid Interface (ALI). Unlike tumor spheroids, in our tumoroids the tissue polarity is preserved via the ALI culturing conditions, thus replicating the direct contact of lung

Dania Movia and Adriele Prina-Mello (eds.), *Cancer Cell Culture: Methods and Protocols*,
Methods in Molecular Biology, vol. 2645, https://doi.org/10.1007/978-1-0716-3056-3_9,
© The Author(s), under exclusive license to Springer Science+Business Media, LLC, part of Springer Nature 2023

adenocarcinoma cells with the air. The latter is a key advantage, as culturing lung adenocarcinoma cells in direct contact with the air is essential for successfully mimicking the disease pathogenesis in vitro [6]. The culture polarity enables to replicate the anatomy of the human air–blood barrier affected by lung adenocarcinoma, while allowing for straightforward access to both the apical and basal tissue surfaces. This enables the testing of both inhaled and systemically administered drugs.

Despite the limitations of being a partially separated culture, our in vitro ALI co-culture tumoroid replicates the chemical cancer cell-fibroblast crosstalk, and it is able of mimick some of the human-specific multi-drug resistance and feedback activation loops seen in lung adenocarcinoma patients [5]. Furthermore, such ALI tumoroid has been recognized as a promising new approach methodology (NAM) in the cancer field, by being included in the knowledge base compiled by the European Centre for the Validation of Alternative Methods (ECVAM) of the European Commission [3]. Such knowledge base collects all the promising non-animal models that have been developed for respiratory disease research to date.

2 Materials

2.1 Cell Culture Formation

2.1.1 Cell Components

1. Human lung adenocarcinoma epithelial cells (A549 cell line) (*see* **Note 1**).
2. Human lung fibroblasts (MRC-5 cell line) (*see* **Note 1**).

2.1.2 Cell Substrate

1. Transparent PET membrane (0.4 μm pore size) inserts (BD Falcon cell culture inserts; growth area inserts = 0.3 cm^2; catalog number: 353095) (*see* **Note 2**).
2. 24-well companion plates (BD Falcon) (*see* **Note 2**).

2.1.3 Reagents

1. Complete DMEM: DMEM medium, 1000 mg/l glucose, 5 μg/ml gentamicin, 10% Fetal Bovine Serum (FBS).
2. Complete EMEM: EMEM medium, 10% FBS, 1% penicillin/streptavidin.
3. Agent for the dissociation of adherent mammalian cells from plastic (e.g., trypsin, TryplE, etc).

2.1.4 Equipment

1. Brightfield microscope.

2.2 Immuno-fluorescence Staining for Confocal Imaging Analysis

1. Washing solution: 1X phosphate buffered saline (PBS).
2. Fixation solution: 3.7% w/w paraformaldehyde, 1X PBS.
3. Permeabilization solution: 0.25% Triton-X, 1X PBS.

4. Blocking solution: 5% bovine serum albumin (BSA), 0.1% Triton-X, 1X PBS.
5. Fluorescent stain: Hoechst (1:200), Rhodamine phalloidin (1: 40), 1X PBS.

2.3 Quantitative Evaluation of Cell Viability by Flow Cytometry

1. Agent for the dissociation of adherent mammalian cells from plastic (e.g., trypsin, TryplE, etc).
2. Live/Dead Fixable Dead Cell Stain Kit (*see* **Note 3**).
3. Washing solution: 1X PBS.
4. Fixation solution: 3.7% w/w paraformaldehyde, 1X PBS.
5. Cell suspension buffer: 1% BSA, 1X PBS.

3 Methods

3.1 Cell Culture Formation

3.1.1 Membrane Conditioning

1. Add 0.7 ml of complete EMEM into the wells of the companion plates.
2. Place inserts into the wells with medium.
3. Pipette 0.35 ml of complete EMEM onto the inserts.
4. Incubate inserts overnight in humidified atmosphere at 37 °C and 5% CO_2.

3.1.2 Seeding of Human Lung Fibroblasts

1. Remove medium from both compartments.
2. Place inserts into a new companion 24-well plate and turn the plate with the inserts upside-down.
3. Detach the fibroblasts from the culturing flask and resuspend in complete EMEM at a concentration of 1.5×10^6 cells/ml.
4. Add the fibroblasts' suspension (0.1 ml/insert, equal to 1.5×10^5 cells/insert) on the inverted inserts (final seeding density = 1.5×10^5 cells/cm^2).
5. Use the bottom of the 24-well plate as lid.
6. Incubate overnight in humidified atmosphere at 37 °C and 5% CO_2.
7. Turn the plate with the transwell inserts back to its original orientation.
8. Add 0.7 ml complete EMEM to the basolateral side.
9. Confirm fibroblasts' attachment to the basal side of the inserts via visual inspection by means of a brightfield microscope.

3.1.3 Seeding of Lung Cancer Cells

1. Detach the lung cancer cells from the culturing flask and resuspend in complete DMEM at a concentration of 7.5×10^5 cells/ml.

2. Add the lung cancer cells' suspension (0.2 ml/insert, equal to 1.5×10^5 cells/insert) on the apical side of each insert (final seeding density = 1.5×10^5 cells/cm^2).
3. Incubate for 24 h in humidified atmosphere at 37 °C and 5% CO_2.

3.1.4 Airlifting

1. Check cells for full confluency by visual inspection with a brightfield microscope.
2. If cells are confluent, remove all the complete DMEM from the apical side of the cultures (*see* **Note 4**).
3. Change medium in basolateral side with complete EMEM.
4. Incubate in humidified atmosphere at 37 °C and 5% CO_2 for 72 h.
5. Confirm the 3D structure of the culture by immunofluorescence staining and imaging.
6. Confirm the cell culture viability by flow cytometry.

3.1.5 Cell Culture Maintenance Overtime

1. For long-term experiments, substitute medium in the basolateral compartment with complete EMEM every 3 days.

3.2 Immunofluorescence Staining for Confocal Imaging

3.2.1 Fixation

1. Discard all medium and wash cultures with pre-warmed washing solution.
2. Add 200 μl/insert of fixation solution on the apical compartment of the cultures.
3. Add 700 μl/insert of fixation solution on the basolateral compartment.
4. Incubate for 10 min at room temperature.
5. Wash two times with washing solution.

3.2.2 Fluorescent Staining

1. Add 100 μl/insert and 700 μl/insert of permeabilization solution in the apical and basolateral compartment of the culture, respectively.
2. Incubate for 10 min at room temperature.
3. Discard the permeabilization solution and wash once with the washing solution.
4. Add blocking solution to both sides of the cultures and incubate for 1 h at room temperature.
5. Discard the blocking solution and add the staining solution on the apical side of both sides of the cultures.
6. Incubate for 3 h at room temperature, or overnight at 4 °C, in the dark. To avoid evaporation, keep the plate covered during the incubation.

7. Discard the staining solution and wash three times with the washing solution.
8. Detach the PET membranes from the plastic inserts with a clean scalpel blade, air dry the specimens for one min, and then mount them on glass slides with mounting medium. Seal with nail varnish.
9. Image immediately by confocal microscopy and/or store in the dark at 2–6 °C.
10. Process the confocal projections with the ImageJ analysis software, to confirm the 3D architecture of the culture (*see* **Note 5**).

3.2.3 Quantitative Evaluation of Cell Viability by Flow Cytometry

1. Wash inserts with washing solution.
2. Place inserts in a new 24-well companion plate containing 700 μl/well of washing solution.
3. Add 100 μL of dissociation agent on the apical side of each insert, and incubate for 10 min at 37 °C.
4. Disaggregate the cell cultures by pipetting vigorously and scratching the membrane with the end of a sterile tip.
5. Transfer the cell suspension obtained from each insert in a different Eppendorf tube containing 600 μl of complete DMEM.
6. Centrifuge for 1 min at 845× *g*.
7. Discard the supernatant carefully, and resuspend the cell pellet in 1 ml of pre-warmed washing solution.
8. Add the fluorescent reactive dye at concentration as per manufacturer's instructions, and mix well by pipetting up and down.
9. Incubate at room temperature for 30 min, protected from light.
10. Centrifuge for 1 min at 845× *g*.
11. Discard supernatant and resuspend the cell pellet in 500 μl of fixation solution.
12. Incubate at room temperature for 15 min.
13. Centrifuge for 1 min at 845× *g*.
14. Discard the supernatant carefully and resuspend the cell pellet in 200 μl of cell suspension buffer.
15. Analyze the fixed cell suspension by flow cytometry as per the instructions provided by the manufacturer of the staining kit.

4 Notes

1. In our studies, A549 and MRC-5 cell lines were acquired from the American Type Culture Collection. These can also be obtained commercially from several suppliers (e.g., Sigma-Aldrich).
2. We strongly advise against the use of inserts and companion plates produced by different manufacturers, as this can negatively affect the reproducibility of our protocol. If using inserts and companion plates produced by a manufacturer different than the one indicated, protocol optimization will be needed.
3. We recommend using a fixable stain for flow cytometry (e.g., Invitrogen). This ensures that viability data are not affected by the lengthy sample preparation procedure.
4. If lung cancer cells are not confluent, incubate in humidified atmosphere at 37 °C and 5% CO_2 for further 24 h.
5. After 72 h in culture, lung cancer cells should have formed 2–3 multilayers, as reported in [5].

Acknowledgments

We acknowledge the financial support from Science Foundation Ireland and the Irish Research Council (SFI-IRC Pathway Award to D.M. - grant ID: 21/PATH-S/9320).

References

1. JRC: ECIS – European Cancer Information System. https://ecis.jrc.ec.europa.eu. Accessed February 2021
2. Sung H, Ferlay J, Siegel RL, Laversanne M, Soerjomataram I, Jemal A et al (2021) Global cancer statistics 2020: GLOBOCAN estimates of incidence and mortality worldwide for 36 cancers in 185 countries. CA Cancer J Clin 71(3): 209–249. https://doi.org/10.3322/caac.21660
3. Dura A, Gribaldo L, Deceuninck P, Campia I (2020) EURL ECVAM Review of non-animal models in biomedical research – respiratory tract diseases. http://data.europa.eu/89h/176d71e6-5082-4b29-8472-b719f6bda323: European Commission Joint Research Centre (JRC)
4. Movia D, Bazou D, Volkov Y, Prina-Mello A (2018) Multilayered cultures of NSCLC cells grown at the air-liquid Interface allow the efficacy testing of inhaled anti-cancer drugs. Sci Rep 8(1):12920. https://doi.org/10.1038/s41598-018-31332-6
5. Movia D, Bazou D, Prina-Mello A (2019) ALI multilayered co-cultures mimic biochemical mechanisms of the cancer cell-fibroblast cross-talk involved in NSCLC MultiDrug resistance. BMC Cancer 19(1):854. https://doi.org/10.1186/s12885-019-6038-x
6. Geles KG, Zhong W, O'Brien SK, Baxter M, Loreth C, Pallares D et al (2016) Upregulation of RNA processing factors in poorly differentiated lung cancer cells. Transl Oncol 9(2): 89–98. https://doi.org/10.1016/j.tranon.2016.01.006

Chapter 10

Isolation and Cryopreservation of Mononuclear Cells from Peripheral Blood and Bone Marrow of Blood Cancer Patients

Sarah Brophy, Rebecca Amet, Hayley Foy-Stones, Nicola Gardiner, and Anthony M. McElligott

Abstract

Peripheral blood and bone marrow aspirates are routinely obtained from blood cancer patients for diagnostic investigations and provide an accessible source of patient-specific cancer cells, as well as non-malignant cells, for research proposes. The simple and reproducible method presented here allows isolation of viable mononuclear cells, including malignant cells, from fresh peripheral blood or bone marrow aspirates using density gradient centrifugation. The cells obtained using the protocol described can be further purified for a variety of cellular, immunological, molecular, and functional assays. In addition, these cells can be cryopreserved and bio-banked for future research studies.

Key words MNC, PBMC, Density gradient centrifugation, Cryopreservation, Primary cell culture, Leukemia, Lymphoma, Myeloma, Biobanking

1 Introduction

Blood cancers are a diverse group of malignancies affecting the blood, bone marrow, and secondary lymphoid organs that include various forms of leukemia, lymphoma, and myeloma. Although accounting for only about 10% of all cancers [1], blood cancer research has been at the forefront of the development of new treatment modalities in oncology [2, 3]. This is due, in part, to the accessibility of blood and bone marrow, providing a convenient source of material for research purposes. While blood cancer cell lines are valuable experimental models in blood cancer research [4], the availability of more physiologically relevant patient material has significantly contributed to our understanding of the pathophysiology of blood cancers. Peripheral blood samples and bone marrow aspirates are routinely obtained from blood cancer patients for diagnostic investigations and, following appropriate ethical approval and patient consent, simultaneous samples can be

Dania Movia and Adriele Prina-Mello (eds.), *Cancer Cell Culture: Methods and Protocols*, Methods in Molecular Biology, vol. 2645, https://doi.org/10.1007/978-1-0716-3056-3_10,
© The Author(s), under exclusive license to Springer Science+Business Media, LLC, part of Springer Nature 2023

obtained for research purposes without the need for extra procedures to be performed on the patient.

Over 50 years ago, Böyum [5] described a method of isolation of mononuclear cells from whole blood using density gradient centrifugation that has become the standard technique for isolation of mononuclear cells from peripheral blood and bone marrow. The blood or bone marrow is layered on top of a density gradient media, such as Lymphoprep™, and the various subtypes of cells migrate differentially through the density gradient media during centrifugation. Low-density white blood cells, i.e., mononuclear cells and platelets, concentrate at the interface between the upper plasma layer and the density gradient media. High-density cells, such as erythrocytes and granulocytes, are pelleted at the bottom of the centrifuge tube. Various modifications of this technique have been developed with several density gradient media formulations commercially available, along with specialized centrifugation tubes [6, 7].

Here, we describe a simple and standardized method of isolation and cryopreservation of mononuclear cells from fresh peripheral blood or bone marrow aspirates from blood cancer patients. The mononuclear cells contain blood cancer cells, as well as a heterogeneous population of non-malignant cell types, including lymphocyte subsets (T cells, B cells and NK cells). Both the blood cancer and non-malignant mononuclear cells can be further purified by a variety of methods, including fluorescence-activated cell sorting (FACS) and magnetic cell sorting [8], and can be used for a variety of studies, including cellular, immunological, molecular, and functional analysis [9–14].

2 Materials

2.1 Isolation of Patient Derived Mononuclear Cells

1. Freshly collected peripheral blood or bone marrow aspirates (*see* **Notes 1–3**).
2. Commercially available density gradient media, such as Lymphoprep™ or Ficoll-Paque™.
3. RPMI-1640 medium. Store it at 4 °C and bring it to room temperature before use.
4. Supplemented medium: 90% RPMI-1640 medium, 10% fetal bovine serum (FBS) (*see* **Note 4**). Store it at 4 °C and heat it to 37 °C before use.
5. Sterile phosphate-buffered saline (PBS), pH 7.2.
6. Neubauer hematocytometer.
7. Cell viability stain (*see* **Note 5**).
8. A centrifuge with a horizontal rotor that can accommodate 15 mL and 50 mL conical bottom centrifuge tubes.

2.2 Freezing and Storage of Mononuclear Cells

1. Dimethyl sulfoxide (DMSO).
2. FBS.
3. Freezing medium: 10% DMSO, 40% FBS, RPMI-1640 medium (*see* **Note 6**).
4. Centrifuge.
5. Cell freezing container.
6. 1.8 mL cryogenic vials
7. Liquid nitrogen storage vessel.

2.3 Thawing of Mononuclear Cells

1. 70% ethanol: 70% ethanol, 30% water
2. Supplemented RPMI-1640 medium.
3. Centrifuge.
4. Water bath.

3 Method

All the procedures are to be carried out at room temperature (*see* **Note 7**) using sterile disposable plasticwares unless otherwise specified. Direct manipulation and pipetting of patient material should be performed under sterile conditions in a Class II biological safety cabinet (*see* **Note 2**).

3.1 Isolation of Mononuclear Cells

1. Bring the density gradient media at room temperature and mix it by gently swirling.
2. Transfer the peripheral blood or bone marrow aspirate sample into 50 mL centrifuge tube(s).
3. Dilute the blood or bone marrow aspirate sample with an equal volume of RPMI-1640 medium or PBS (*see* **Note 8**). Mix by inverting the tube.
4. Add 15 mL of density gradient media to a 50 mL conical bottom polystyrene centrifuge tube (*see* **Note 9**).
5. Gently layer the diluted sample onto the density gradient media, while being careful not to disturb the density gradient media/sample interface. It is recommended to hold the centrifuge tube at a 45° angle and slowly pipette the sample down the side of the tube using a 3 mL sterile plastic Pasteur pipette (*see* **Note 10**). Do not layer more than 30 mL of diluted sample in a 50 mL centrifuge tube.
6. Centrifuge at 800× *g* for 30 min at room temperature with the centrifuge brake turned OFF (*see* **Note 11**). Ensure the centrifuge is balanced as vibration will affect cell separation.

7. At the end of the centrifugation step, an upper plasma layer and a lower density gradient media layer should be observed. Erythrocytes and polynucleated granulocytes are pelleted at the bottom of the tube. Mononuclear cells should be visible at the density gradient media/plasma interface.
8. Using a 3 mL plastic Pasteur pipette, remove the mononuclear cells from the plasma/density gradient media interface without disturbing the erythrocyte/granulocyte pellet.
9. Collect the mononuclear cells in a 50 mL conical bottom polystyrene centrifuge tube and wash by adding at least three times the volume of RPMI-1640 medium.
10. Centrifuge at 250 × g for 10 min with the centrifuge brake turned ON. The supernatant will be cloudy due to presence of platelets and can be discarded (*see* **Notes 12** and **13**).
11. Resuspend the mononuclear cells in 10 mL of supplemented RPMI-1640 medium. Count the cells and determine their viability (*see* **Notes 5** and **14**).
12. The mononuclear cells can be analyzed by flow cytometry, further purified by FACS or immunoselection by magnetic beads, and used for a variety of cellular, immunological, molecular, and functional assays. These cells can also be cryopreserved at this point for future use (*see* **Note 15**).

3.2 Cryopreservation of Mononuclear Cells

1. Add 2 mL of DMSO to 8 mL aliquots of FBS.
2. Resuspend the mononuclear cells in RPMI-1640 medium and add an equal volume of DMSO-FBS to achieve a final cell concentration of 5–20 × 10^6 cells/mL.
3. Gently pipette 1 mL of the mononuclear cells' suspension into the appropriate number of cryogenic vials. Immediately place the cryogenic vials in the Freezing Container and leave overnight in a −80 °C freezer (*see* **Notes 16** and **17**) before transferring the cryogenic vials to liquid nitrogen for long-term storage (*see* **Note 18**).

3.3 Thawing of Mononuclear Cells

1. Bring supplemented RPMI-1640 medium to 37 °C.
2. Transfer 10 mL of supplemented RPMI-1640 medium to a centrifuge tube.
3. Remove the cryogenic vial from liquid nitrogen and thaw the cells rapidly in a 37 °C water bath by gently shaking the vial in the water (*see* **Note 19**).
4. As soon as the cells are thawed, wipe the cryogenic vial with tissue pre-wetted with 70% ethanol and add the cells dropwise to the supplemented RPMI-1640 medium in the centrifuge tube.

5. Centrifuge the cells at 300× *g* for 10 min and discard the supernatant.
6. Wash the cells again with 10 mL of supplemented RPMI-1640 media, determine cell concentration and viability (*see* **Note 5**) and resuspend the mononuclear cells at required cell concentration in supplemented RPMI-1640 medium.

4 Notes

1. Collected patient samples will depend on research requirements and Research Ethics Committee approvals. We routinely collect peripheral blood in 9 mL BD Vacutainer® Venous Blood Collection tubes containing Potassium EDTA, as well as 1–3 mL of bone marrow aspirate in heparinized (5000 units/mL) RMPI-1640 medium.
2. Human blood, blood products and subsequent in vitro tissue cultures must be regarded as containing potential blood-borne pathogens and should be handled in accordance with appropriate biosafety practices and standard precautions.
3. The presence of blood clots in samples will affect the quality and yield of mononuclear cells. In particular, bone marrow aspirates should be collected in a manner which prevents clot formation, e.g., appropriate anticoagulant treatment. Samples should be processed as soon as possible, and we recommend to process samples within 4 h of collection. Samples should be kept at room temperature and, if overnight storage is necessary, should be kept at 4 °C. However, delays in the processing of samples can result in loss of cell viability, lower yield, and altered expression of surface markers [15].
4. Up to 20% FBS can be added to the culture media. Some downstream processes may require the use of heat-inactivated FBS [16]. L-glutamine (~2mM) may be added. The addition of antibiotics (e.g. penicillin-streptomycin) are not necessary and should be avoided.
5. We use ethidium bromide (EtBr) and acridine orange (AO) fluorescent nuclear stains (1 mg/mL of EtBr +1 mg/mL AO in PBS) to quantify cell viability/cell death. Non-fluorescent stains, such as trypan blue (0.4% in PBS), can also be used [17]. For both methods, standard use of a Neubauer haematocytometer is required. The advantage of using EtBr/AO staining is the sensitivity of the assay and the ease of quantifying cell viability due to viable cells fluorescing green and non-viable cells fluorescing red/orange. The disadvantages of this staining technique are the requirement of a fluorescent microscope, as well as the toxicity profiles of AO and especially EtBr.

6. DMSO is a dipolar aprotic solvent which has been used as a cryoprotectant capable of minimizing freezing-related cell damage for over 60 years [18]. We use RPMI-1640 medium containing 40% FCS and 10% DMSO as our freezing media, but other formulations and alternatives are available [19].
7. Ideally, room temperature should be between 15 and 25 °C. It is desirable fluctuations in room temperature of the laboratory are minimized. The temperature and, as a result, the density of the density gradient media will affect the quality and yield of mononuclear cells. All centrifugation steps are to be performed at room temperature.
8. We routinely use RPMI-1640 medium to dilute whole blood, but PBS can be used with no effect on the quality and yield of mononuclear cells. In contrast, bone marrow aspirates should only be diluted in RPMI-1640 medium.
9. Smaller blood volumes can be processed in 15 mL conical bottom polystyrene centrifuge tubes without affecting the separation efficiency. Regardless of tube size, a ratio of density gradient media to sample volume of 1:2 should be maintained, i.e., half the volume of density gradient media to diluted sample volume. Centrifugation times are not affected by the tube size.
10. Several blood separation tubes are commercially available that can be used for the fast isolation of PBMC from the blood samples with less dependence on the experience of laboratory personnel [6]. These tubes contain inserts which separate the blood sample from the density gradient media, minimizing the potential of mixing. An additional advantage of using these tubes is that the centrifuge brake can be kept ON for the first centrifugation step, thus reducing the sample processing time. However, such tubes are not suitable for bone marrow aspirate processing.
11. We use this centrifuge speed and time in combination with Lymphoprep™ density gradient media. Consult manufacturer's recommendations if using another density gradient media. Centrifuge brake must be turned OFF as a slow stop is essential to maintain the density gradient.
12. The washing step removes the density gradient media which is hypertonic for the cells, and should be repeated if a large quantify of the density gradient media was recovered in the previous step. The mononuclear yield may be improved by increasing the centrifuge speed, but this will result in increased platelet contamination of the mononuclear cells. More than 90% of the mononuclear cells can be recovered using this procedure; however, other factors may influence mononuclear cell separation efficiency and yield, such as laboratory personnel's experience, room temperature, quality of the density gradient medium, and type of separation tubes.

13. Contamination of the mononuclear cell pellet with red blood cells is easily observed by red color in the pellet. Red cell contamination, which may interfere with flow cytometric analysis, DNA and RNA isolation, and some in vitro assays, can be removed by treating the mononuclear cells with red cell lysis buffer (155 mM NH_4Cl, 12 mM $NaHCO_3$, 0.1 mM EDTA) for 5 min, followed by washing with PBS.
14. Mononuclear cell yields depend on the type of blood cancer and tumor burden. On average, 1 mL of leukemia bone marrow yields $1–3 \times 10^7$ mononuclear cells and, depending on the white blood cell count, 1 mL of peripheral blood yields $1 \times 10^6 - 1 \times 10^8$ mononuclear cells. Mononuclear cell viability in fresh samples should be >95%.
15. Cell separation should be performed before cryopreservation, as immunoselection may not be optimal with cryopreserved samples. For example, isolation of $CD138^+$ multiple myeloma cells must be performed before cryopreservation as the CD138 antigen is lost from the cell surface when the sample is frozen [20].
16. Prolonged exposure to DMSO can induce cellular changes; thus, minimize the length of time the mononuclear cells are kept at room temperature in the freezing media.
17. Freezing containers, such as Nalgene® "Mr. Frosty," must be half filled with 100% isopropyl alcohol and kept at room temperature before transferring to a −80 °C freezer for at least 4 h. With this method, a 1 °C per min cooling rate can be achieved before transferring the cryogenic vials to a long-term liquid nitrogen storage vessel.
18. Mononuclear cells can be cryopreserved indefinitely at −196 ° C in liquid nitrogen, and viable cells recovered when needed. The cryogenic vials can be in the liquid phase or the vapor phase of the liquid nitrogen storage vessel.
19. It is important that the frozen mononuclear cell solution is thawed rapidly. Slow thawing will reduce cell viability due to formation of ice crystals, resulting in cellular damage.

Acknowledgments

This work was funded by Science Foundation Ireland, Irish Blood Transfusion Service, and the Bone Marrow for Leukaemia Trust.

References

1. Wild CP, Weiderpass E, Stewart BW (2020) International Agency for Research on C. World cancer report: cancer research for cancer prevention. International Agency for Research on Cancer, Lyon
2. Greaves M (2016) Leukaemia 'firsts' in cancer research and treatment. Nat Rev Cancer 16(3): 163–172. https://doi.org/10.1038/nrc.2016.3
3. Larson RC, Maus MV (2021) Recent advances and discoveries in the mechanisms and functions of CAR T cells. Nat Rev Cancer 21(3): 145–161. https://doi.org/10.1038/s41568-020-00323-z
4. Quentmeier H, Pommerenke C, Dirks WG, Eberth S, Koeppel M, MacLeod RAF et al (2019) The LL-100 panel: 100 cell lines for blood cancer studies. Sci Rep 9(1):8218. https://doi.org/10.1038/s41598-019-44491-x
5. Böyum A (1968) Isolation of mononuclear cells and granulocytes from human blood. Isolation of monuclear cells by one centrifugation, and of granulocytes by combining centrifugation and sedimentation at 1 g. Scand J Clin Lab Invest Suppl 97:77–89
6. Grievink HW, Luisman T, Kluft C, Moerland M, Malone KE (2016) Comparison of three isolation techniques for human peripheral blood mononuclear cells: cell recovery and viability, population composition, and cell functionality. Biopreserv Biobank 14(5): 410–415. https://doi.org/10.1089/bio.2015.0104
7. Yeo C, Saunders N, Locca D, Flett A, Preston M, Brookman P et al (2009) Ficoll-Paque versus Lymphoprep: a comparative study of two density gradient media for therapeutic bone marrow mononuclear cell preparations. Regen Med 4(5):689–696. https://doi.org/10.2217/rme.09.44
8. Dainiak MB, Kumar A, Galaev IY, Mattiasson B (2007) Methods in cell separations. Adv Biochem Eng Biotechnol 106:1–18. https://doi.org/10.1007/10_2007_069
9. Amet R, Previtali V, Mihigo HB, Sheridan E, Brophy S, Hante NK et al (2022) A novel arylguanidinium derivative, VP79s, targets the signal transducer and activator of transcription 3 signaling pathway, downregulates myeloid cell leukaemia-1 and exhibits preclinical activity against multiple myeloma. Life Sci 290: 120236. https://doi.org/10.1016/j.lfs.2021.120236
10. Bright SA, Byrne AJ, Vandenberghe E, Browne PV, McElligott AM, Meegan MJ et al (2019) Selected nitrostyrene compounds demonstrate potent activity in chronic lymphocytic leukaemia cells, including those with poor prognostic markers. Oncol Rep 41(5):3127–3136. https://doi.org/10.3892/or.2019.7068
11. Bright SA, McElligott AM, O'Connell JW, O'Connor L, Carroll P, Campiani G et al (2010) Novel pyrrolo-1,5-benzoxazepine compounds display significant activity against resistant chronic myeloid leukaemia cells in vitro, in ex vivo patient samples and in vivo. Br J Cancer 102(10):1474–1482. https://doi.org/10.1038/sj.bjc.6605670
12. Ghnewa YG, O'Reilly VP, Vandenberghe E, Browne PV, McElligott AM, Doherty DG (2017) Retinoic acid induction of CD1d expression primes chronic lymphocytic leukemia B cells for killing by CD8(+) invariant natural killer T cells. Clin Immunol 183:91–98. https://doi.org/10.1016/j.clim.2017.08.002
13. Maginn EN, Browne PV, Hayden P, Vandenberghe E, MacDonagh B, Evans P et al (2011) PBOX-15, a novel microtubule targeting agent, induces apoptosis, upregulates death receptors, and potentiates TRAIL-mediated apoptosis in multiple myeloma cells. Br J Cancer 104(2):281–289. https://doi.org/10.1038/sj.bjc.6606035
14. McElligott AM, Maginn EN, Greene LM, McGuckin S, Hayat A, Browne PV et al (2009) The novel tubulin-targeting agent pyrrolo-1,5-benzoxazepine-15 induces apoptosis in poor prognostic subgroups of chronic lymphocytic leukemia. Cancer Res 69(21): 8366–8375. https://doi.org/10.1158/0008-5472.Can-09-0131
15. Kaplan J, Nolan D, Reed A (1982) Altered lymphocyte markers and blastogenic responses associated with 24 hour delay in processing of blood samples. J Immunol Methods 50(2): 187–191. https://doi.org/10.1016/0022-1759(82)90224-1
16. Phelan K, May KM (2017) Mammalian cell tissue culture techniques. Curr Protoc Mol Biol 117(1):A.3F.1–A.3F.23. https://doi.org/10.1002/cpmb.31
17. Atale N, Gupta S, Yadav UCS, Rani V (2014) Cell-death assessment by fluorescent and non-fluorescent cytosolic and nuclear staining techniques. J Microsc 255(1):7–19. https://doi.org/10.1111/jmi.12133

18. Lovelock JE, Bishop MW (1959) Prevention of freezing damage to living cells by dimethyl sulphoxide. Nature 183(4672):1394–1395. https://doi.org/10.1038/1831394a0

19. Awan M, Buriak I, Fleck R, Fuller B, Goltsev A, Kerby J et al (2020) Dimethyl sulfoxide: a central player since the dawn of cryobiology, is efficacy balanced by toxicity? Regen Med 15(3):1463–1491. https://doi.org/10.2217/rme-2019-0145

20. Frigyesi I, Adolfsson J, Ali M, Christophersen MK, Johnsson E, Turesson I et al (2014) Robust isolation of malignant plasma cells in multiple myeloma. Blood 123(9):1336–1340. https://doi.org/10.1182/blood-2013-09-529800

Chapter 11

Serum-Free Production of Human Stem Cell-Derived Liver Spheres for Cancer Metastasis Research

Alvile Kasarinaite, James Drew, Mantas Jonaitis, Elaine Ma, Laura M. Machesky, and David C. Hay

Abstract

Renewable and scalable human liver tissue platforms are a powerful tool to study organ physiology and model diseases, such as cancer. Stem cell-derived models provide an alternative to cell lines, which can display limited relevance to primary cells and tissue. Historically, two-dimensional (2D) cultures have been used to model liver biology as they are easy to scale and deploy. However, 2D liver models lack functional diversity and phenotypic stability in long-term culture. To address those issues, protocols for generating the three-dimensional (3D) tissue aggregates have been developed. Here, we describe a methodology to generate 3D liver spheres from pluripotent stem cells. Liver spheres are composed of three key liver cell types (hepatic progenitor cells, endothelial cells, and hepatic stellate cells) and have been used to study human cancer cell metastasis.

Key words 3D cell culture, Pluripotent stem cells, Liver, Hepatocyte, Endothelial cell, Hepatic stellate cell, Tissue engineering, Cell differentiation, Metabolism, Cancer, Metastasis

1 Introduction

Liver diseases claim two million lives annually [1]. Within the UK, the annual death rate from liver diseases has increased by 4 fold over the last 50 years [2]. An orthotopic transplant is currently the only long-term corrective solution for late-stage liver diseases, yet <10% of transplantation needs are met globally [1].

The shortage of donor organs highlights the need to develop alternative therapies to treat end stage liver disease and failure [3]. Cell based modelling of human liver physiology and disease development are key to those endeavours. Primary human

The original version of this chapter was revised. The correction to this chapter is available at https://doi.org/10.1007/978-1-0716-3056-3_19

Dania Movia and Adriele Prina-Mello (eds.), *Cancer Cell Culture: Methods and Protocols*, Methods in Molecular Biology, vol. 2645, https://doi.org/10.1007/978-1-0716-3056-3_11,
© The Author(s), under exclusive license to Springer Science+Business Media, LLC, part of Springer Nature 2023, Corrected Publication 2023

hepatocytes (PHHs) are regarded as the gold standard model to study human liver biology "in the dish" [4, 5]. However, the sourcing of healthy PHHs can prove challenging and expensive [5]. To bypass those issues, other models have been used to study liver biology, such as cancer-derived cell lines. Despite cancer cells being readily available and relatively cheap to use, they display limited similarity to primary cells [6]. More recently, pluripotent stem cell-based liver cell models have been used to model human liver biology [7, 8]. These cells display self-renewal properties and are capable of differentiating down the three germ layers, promising an unlimited supply of human somatic cells [9].

Many two-dimensional (2D) hepatocyte differentiation protocols from the pluripotent stem cells have been developed [10]. However, the prolonged use of 2D cultures is problematic due to rapid cellular dedifferentiation [10]. To improve liver tissue stability, and to take into account important cell-to-cell interactions within the liver, 3D models were developed. 3D culture systems have been shown to enhance both cell phenotype and stability in vitro, and provide liver support when transplanted in vivo [11].

We have recently automated this process to produce 3D liver tissue which is composed of hepatic progenitor cells, endothelial cells, and hepatic stellate cells [12, 13]. Importantly, in vitro engineered liver spheres were able to model key aspects of liver biology, such as metabolic function and macrovesicular steatosis [12–14].

3D culture systems also have the potential to recapitulate the cancer cell-liver niche in vitro [13]. Recently, liver cancer was identified as the third most common cause of death globally [15]. Approximately 80% of all liver cancer cases are the result of hepatocellular carcinoma (HCC) [16, 17]. However, the liver is also a common site for cancer cell metastasis [18, 19]. Herein, we describe a human 3D liver tissue system designed to study cancer cell metastasis in vitro (Fig. 1).

Using this approach, phenotypically stable hPSC-derived liver tissue can be produced in culture (Fig. 2). The spheroid system exhibits some characteristics of mature liver tissue, such as reduced proliferation, cytochrome P450 activity (Fig. 2), and reduced alpha-fetoprotein (AFP) secretion [13]. The architecture of the liver spheres was confirmed by immunostaining with hepatocytes (HNF4a, E-Cadherin) located on the periphery and endothelial tube structures (VE-Cadherin, Vimentin and CD31) identified inside the spheroid (Fig. 3a) [11, 20].

Cancer cell co-culture with the liver spheres was performed before fixation and whole-mount imaging to study the interaction of cancer cells with the liver spheres (Fig. 3b) [13].

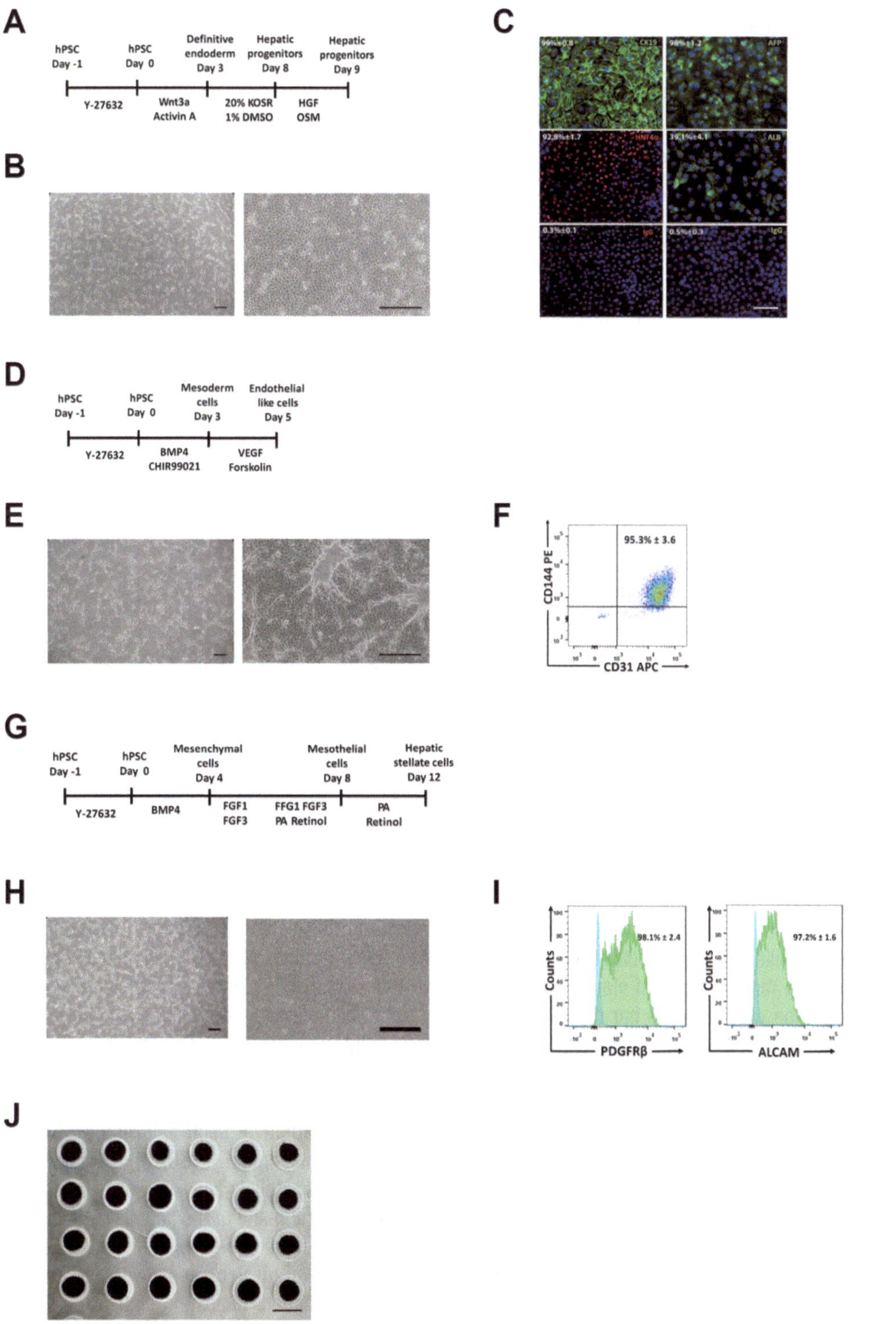

Fig. 1 Pluripotent stem cell differentiation and liver sphere formation. (**a**) Schematic representation of the protocol used to differentiate human pluripotent stem cells (hPSCs) to hepatic progenitor cells. (**b**) Phase contrast images of representative fields of view of hPSC at the start of the differentiation (left) and at the hepatic progenitor stage (right). Images of hPSC were taken at 4× magnification with the scale bar

2 Materials

With exception of CYP P450 and ELISA experiments, all procedures should be performed using sterile techniques and working in a cell culture hood class II. All reconstituted cell culture media should be used withing a month following preparation. If needed, media can be frozen to −20 °C and stored for 6 months, do not freeze again after thaw. Growth factors can be stored at −20 °C for short-term storage up to 6 months, or at −80 °C for long-term storage for at least 12 months. Once thawed, growth factors should be kept at 4 °C and used within 2 weeks. Cell culture media should be supplemented with the growth factors fresh every time before feeding.

2.1 Culture Plates Preparation

1. Phosphate-buffered saline (PBS).
2. Laminin-521 (LN-521).
3. Ca^{2+}/Mg^{2+}DPBS: Dulbecco's phosphate-buffered saline (DPBS) with calcium and magnesium.
4. Low melting temperature agarose.
5. Sterile water.
6. Poly 2-hydroxyethyl methacrylate (poly-HEMA).
7. 95% Ethanol: 95% ethanol, 5% PBS.
8. 6–, 12–, 48- and 96-well plates.
9. Petri-dishes.

Fig. 1 (continued) representing 200 μm. Images of endothelial-like cells were taken at 10× magnification with the scale bar representing 100 μm. (**c**) Hepatic progenitor cell immunostaining for alpha-fetoprotein (AFP), albumin, cytokeratin 19 (CK19) and HNF4α-Hepatocyte Nuclear Factor 4α (HNF4α) on hepatic progenitor cells. Immunoglobulin G (IgG) antibodies were used to demonstrate staining specificity. For each marker, five random fields of view containing at least 400 cells were counted. Images were taken at 20× magnification with the scale bar representing 100 μm. (**d**) Schematic representation of the protocol used to differentiate hPSCs to early endothelial cells. (**e**) Phase contrast images of representative fields of view of hPSC at the start of the differentiation (left) and at the endothelial stage (right). Images of hPSC were taken at 4× magnification with the scale bar representing 200 μm. Images of endothelial-like cells were taken at 10× magnification and the scale bar representing 100 μm. (**f**) Representative result following MACS-mediated enrichment for CD144 expressing cells analyzed by flow cytometry (mean ± SD, $n = 4$). (**g**) Schematic representation of the procedure used to differentiate hPSCs to hepatic stellate cells. (**h**) Phase contrast images of representative fields of view of hPSC at the start of the differentiation and at the hepatic stellate cell stage. Images of hPSC were taken at 4× magnification, the scale bar representing 200 μm; images of derived cells were taken at 10× magnification, the scale bar representing 100 μm. (**i**) Representative histograms of flow cytometry analysis of hepatic stellate cells on day 12 of differentiation, indicating platelet-derived growth factor receptor beta (PDGFRβ) positive cells and activated leukocyte cell adhesion molecule (ALCAM) positive cells from 3 independent experiments. (**j**) Phase contrast images of representative fields of view of agarose microwells containing liver spheres 24 h post-aggregation. Images were taken at 4× magnification with the scale bar representing 400 μm

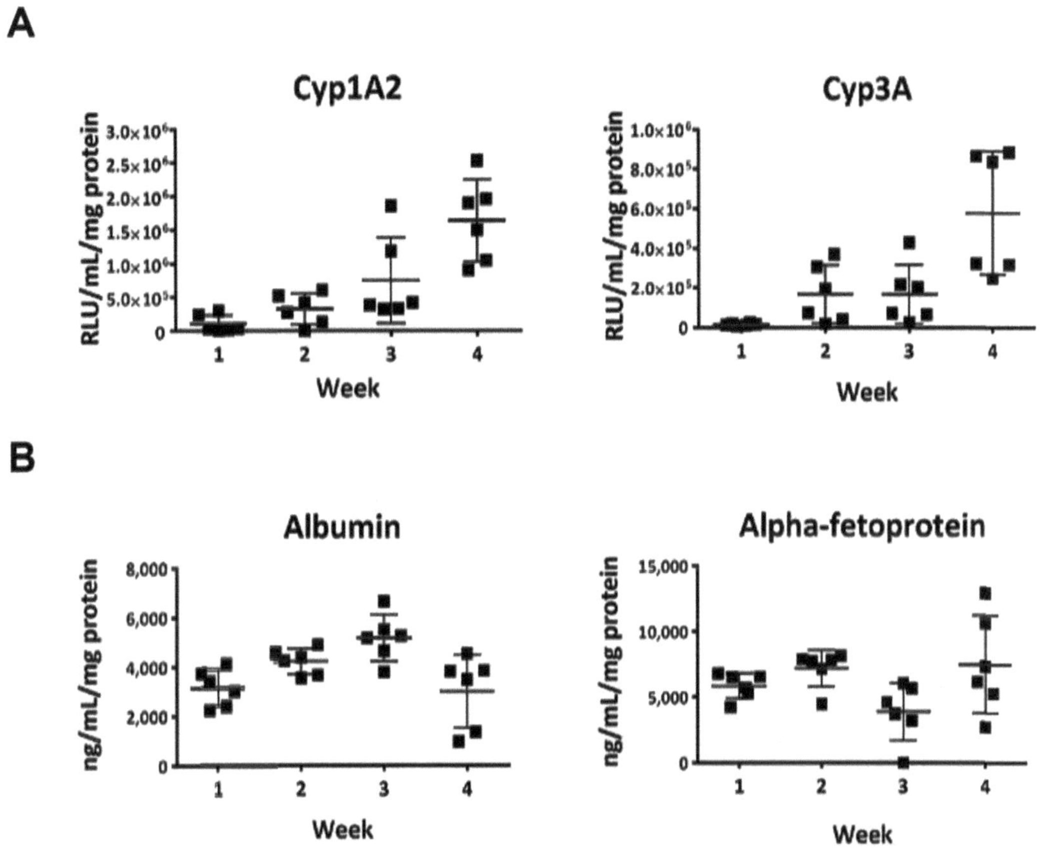

Fig. 2 Liver sphere phenotypic characterization. (**a**) Cytochrome P450 1A2 and 3A activity were analyzed at different time points during liver sphere culture (mean ± SD, $n = 6$). (**b**) Secretion of the serum proteins albumin and alpha-fetoprotein were measured by ELISA at the denoted times (mean ± SD, $n = 6$)

2.2 Stage 1: Human Pluripotent Stem Cell (hPSC) Culture

1. Bovine Serum Albumin (BSA).
2. DPBS.
3. Ca^{2+}/Mg^{2+}DPBS.
4. Cell dissociation reagent.
5. Human Embryonic Stem Cell Line H9.
6. Human iPSC line p106.
7. Human recombinant laminin 521.
8. Complete mTeSR1 medium: 400 m mTeSR1™ basal medium, 100 mL mTeSR1™ 5× supplement.
9. 10 mM Rho-associated Kinase (ROCK) Inhibitor Y27632 (*see* **Note 1**).

2.3 Stage 2a: Hepatic Stellate Cell (HSC) Differentiation

1. BSA.
2. Dexamethasone.
3. DMEM low glucose medium.

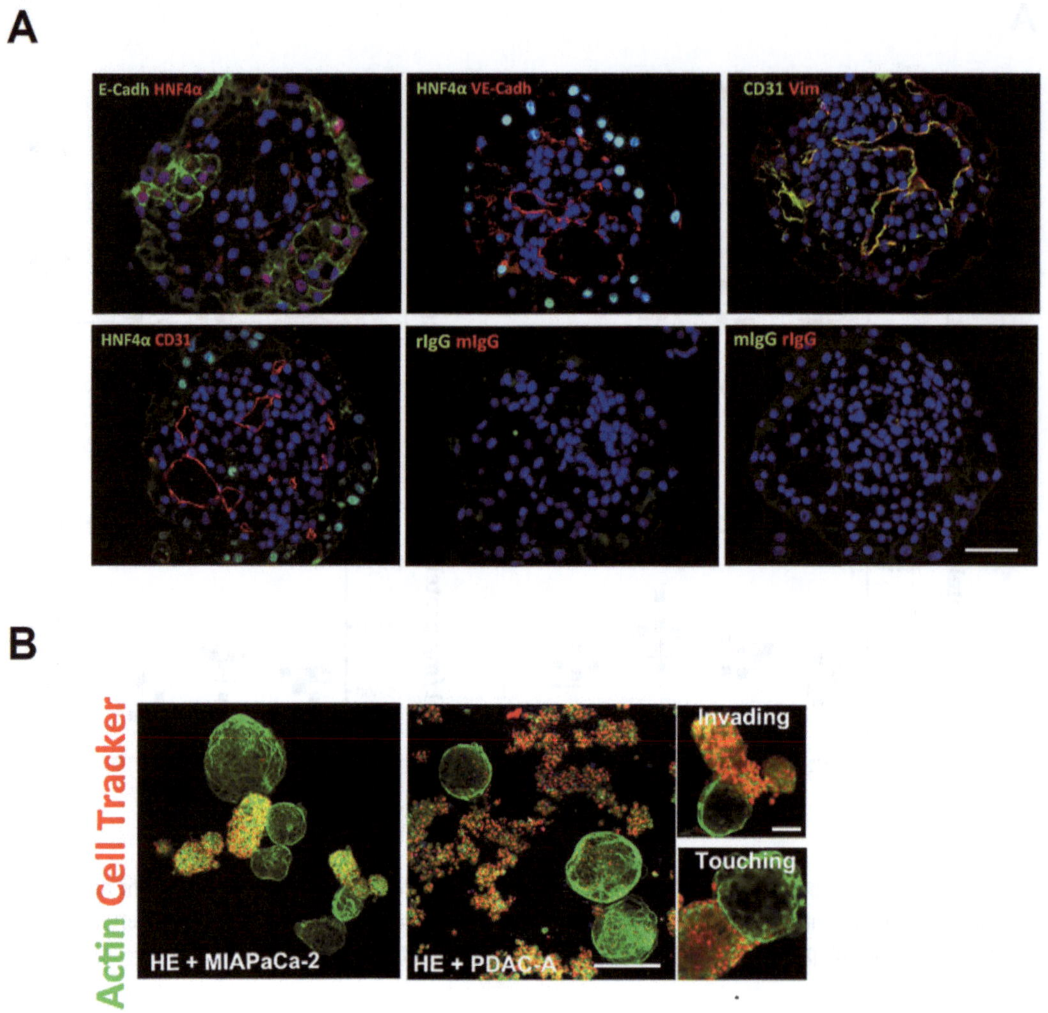

Fig. 3 Liver spheres model in vivo liver morphology and clinically relevant situations (**a**) Immunofluorescent images of liver spheres staining E-Cadherin and HNF4α (hepatocytes) and CD31, VE-Cadherin and Vimentin (endothelial cells). IgG antibodies were used throughout to demonstrate staining specificity. Images were taken at 20× magnification and the scale bar represents 120 μm. (**b**) Liver sphere interaction with two pancreatic cancer cell lines (MIAPaCa-2 and PDAC-A). Representative maximum intensity projections of liver spheres cultured with pancreatic cancer cells. Right panels show different types of interactions observed

4. Human Recombinant BMP4 (*see* **Note 2**).
5. Human Recombinant FGF-1 (*see* **Note 3**).
6. Human Recombinant FGF-3 (*see* **Note 4**).
7. Insulin-transferrin-sodium selenite.
8. L-ascorbic acid.
9. Linoleic Acid.
10. MCDB-201-water.

11. Palmitic Acid (*see* **Note 5**).
12. Penicillin-streptomycin (10,000 U/mL).
13. Retinol (*see* **Note 6**).
14. Cell dissociation reagent.
15. Β-mercaptoethanol.
16. Hepatic stellate cell differentiation medium: 0.5 mL of 50 mM β-mercaptoethanol, 0.5 mL of 2.5 mM dexamethasone, 0.5 mL of 100 mM L-ascorbic acid, 1.25 mL of insulin-transferrin-sodium selenite, 1.25 mL of linoleic acid, 5 mL of penicillin-streptomycin, 200 mL of MCDB-201-water, 285 mL of DMEM Low Glucose.

2.4 Stage 2b: Hepatic Progenitor Cell (HPC) Differentiation

1. Human Activin A (*see* **Note** 7).
2. B27 supplement (50×).
3. Dimethyl sulfoxide (DMSO) (100×).
4. GlutaMAX-I (100×).
5. Human hepatocyte growth factor (HGF) (*see* **Note 8**).
6. HepatoZYME-SFM.
7. Hydrocortisone-21 hemisuccinate sodium salt (HCC) (*see* **Note 9**).
8. Knockout (KO)-DMEM.
9. Knockout serum replacement (KOSR).
10. Minimal essential medium Non-Essential Amino Acids.
11. Oncostatin M (*see* **Note 10**).
12. Penicillin-streptomycin (10,000 U/mL).
13. RPMI 1640.
14. Cell dissociation reagent.
15. Wnt3a (*see* **Note 11**).
16. Β-mercaptoethanol.
17. Endoderm differentiation medium: 5 mL of penicillin-streptomycin, 10 mL of B27 supplement (50×), 500 mL of RPMI 1640 medium.
18. Hepatic progenitor differentiation medium: 1 mL of 50 mM β-mercaptoethanol, 2.5 mL of GlutaMAX-I, 5 mL of DMSO, 5 mL of penicillin-streptomycin, 5 mL of minimal essential medium Non-Essential Amino Acids, 100 mL of KOSR serum replacement, 400 mL of KO-DMEM.
19. Hepatocyte maturation medium: 5 mL of GlutaMAX-I, 5 mL of penicillin-streptomycin, 5 mL of 1 mM HCC, 500 mL of Hepato-ZYME.

2.5 Stage 2c: Endothelial Cell (EC) Differentiation

1. B27 CTS (2%).
2. BSA.
3. Human CD144 (VE-Cadherin) MicroBeads.
4. CHIR-99021 (*see* **Note 12**).
5. CTS Neurobasal media.
6. DMSO.
7. DMEM/F-12 cell medium: DMEM/F-12 cell medium with Glutamax.
8. Forskolin (*see* **Note 13**).
9. GlutaMAX-I (100×).
10. Human Recombinant BMP4.
11. Human Recombinant vascular endothelial growth factor (VEGF) 165 Protein (*see* **Note 14**).
12. LS Columns.
13. MACS MultiStand.
14. CTS N2 (100×).
15. Penicillin-streptomycin (10,000 U/mL).
16. StemPro-34.
17. β-mercaptoethanol.
18. Mesoderm priming medium N2B27: 5 mL of penicillin-streptomycin, 0.5 mL of 50 mM β-mercaptoethanol, 2.5 mL of GlutaMAX-I, 5 mL of CTS N2, 10 mL of CTS B27, 250 mL of DMEM/F-12 cell medium, 250 mL of CTS neurobasal media.
19. Endothelial cell differentiation medium SP34: 5 mL of penicillin-streptomycin, 500 mL of StemPro-34.

2.6 Stage 3a: Liver Sphere Culture

1. Agarose, low gelling temperature.
2. Human hepatocyte growth factor (HGF).
3. Human Basic Fibroblast Growth Factor (bFGF) (*see* **Note 15**).
4. Human Endothelial-SFM.
5. Human Epithelial Growth Factor (EGF) (*see* **Note 16**).
6. Human Recombinant VEGF 165 Protein (*see* **Note 17**).
7. Micro-mold spheroids.
8. Oncostatin M.
9. William's E Medium: William's E Medium, no phenol red.
10. Liver sphere culture medium: 2.5 mL of GlutaMAX-I, 5 mL of penicillin-streptomycin, 27.5 mL of KOSR serum replacement, 250 mL of Human Endothelial-SFM, 250 mL of William's E medium.

2.7 Stage 3b: Formation and Characterization of Liver Cancer Co-culture Spheres

1. MIAPaCa-2 Pancreatic Ductal Adenocarcinoma (PDAC) cell line.
2. PDAC-A (KRas G12D; p53 R172H; Pdx-1::Cre expressing mouse pancreatic cells) cell line.
3. Poly-HEMA coated 96-well plates.
4. CellTracker Red Dye (ThermoFisher, C34552).
5. Alexa Fluor-488 Phalloidin.
6. Fetal bovine serum (FBS).
7. BSA.
8. GlutaMAX-I (100×).
9. Penicillin-Streptomycin (10,000 U/mL).
10. DMEM: DMEM low Glucose medium.
11. Cancer cell medium: 10% FBS, 1% penicillin-streptomycin, 1% L-glutamine, DMEM.
12. Opera PhenixTM High-Content Screening system.

2.8 Quantification of CYP P450 Activity in Liver Spheres

1. GlutaMAX-I.
2. Penicillin-streptomycin (10,000 U/mL).
3. Human Epithelial Growth Factor (EGF) (*see* **Note 16**).
4. Human FGF.
5. Human hepatocyte growth factor (HGF).
6. OSM.
7. Human Recombinant VEGF 165 Protein (*see* **Note 17**).
8. Knockout serum replacement (KOSR).
9. Complete Williams E medium: 1% GlutaMAX-I, 1% penicillin-streptomycin, 10% KOSR, 10 ng/mL EGF, 10 ng/mL FGF, 10 ng/mL HGF, 20 ng/mL OSM, 50 ng/mL VEGF, Williams E medium.
10. 100 μM Luciferin-ME.
11. 50 μM Luciferin-PFBE.
12. P450-Glo assay kit (Promega) (*see* **Note 18**).
13. 96-well white plates.
14. RIPA buffer.
15. Sonicator bath.
16. Bicinchoninic acid (BCA) assay.

2.9 Quantification of Human Serum Albumin or Alpha-Fetoprotein Secretion in Liver Spheres

1. GlutaMAX-I.
2. Penicillin-streptomycin (10,000 U/mL).
3. Human Epithelial Growth Factor (EGF) (*see* **Note 16**).
4. Human FGF.

5. Human hepatocyte growth factor (HGF).
6. OSM.
7. Human Recombinant VEGF 165 Protein (*see* **Note 17**).
8. Knockout serum replacement (KOSR).
9. Complete Williams E medium: 1% GlutaMAX-I, 1% penicillin-streptomycin, 10% KOSR, 10 ng/mL EGF, 10 ng/mL FGF, 10 ng/mL HGF, 20 ng/mL OSM, 50 ng/mL VEGF, Williams E medium.
10. Human serum albumin ELISA kit.
11. Human alpha-fetoprotein ELISA kit.
12. RIPA buffer.
13. Sonicator bath.
14. Bicinchoninic acid (BCA) assay.

3 Methods

3.1 Preparation of Cell Culture Substrates

The method reported here were previously described in [12, 13, 20].

3.1.1 Preparation of Laminin-521 (LN-521)-Coated 6-Well Plates/Petri-Dishes

1. Thaw a stock vial of 100 μg/mL recombinant laminin 521 (LN-521) at 4 °C overnight.
2. Dilute the thawed LN-521 in ice-cold Ca^{2+}/Mg^{2+}DPBS to make a 5 μg/mL solution.
3. Coat the plate substrate with the LN-521 solution (*see* **Note 19**). If using a 6 well-plate, add 1 mL of the 5 μg/mL solution per well. If using Petri dishes with growth area of 55 cm^2, add 5 mL of the 5 μg/mL solution per petri dish. Ensure that the LN-521 solution is covering the whole culture surface.
4. Incubate the plates/Petri dishes in a 37 °C/5% CO_2 cell culture incubator for 2 h, or at 4 °C overnight on a flat surface.
5. Store coated plates at 4 °C on a flat surface. Seal the plates to reduce evaporation and use within 2 weeks.
6. Before use, allow the plate to warm up by placing the plates/Petri dishes in a 37 °C/5% CO_2 cell culture incubator for 30 min or warm them up at room temperature (15–21 °C) for 2 h (*see* **Note 20**).

3.1.2 Preparation of 2% Agarose Molds in 12-Well Plates

The method reported here were previously described in [12, 13, 20].

1. Dissolve 2 g of low melting temperature agarose into 100 mL of sterilized distilled water. Dissolve completely by heating in a microwave.
2. Add 520 μL of melted agarose to a 256-well format mold and leave to solidify for 5–10 min.

3. Transfer each agarose microplate into a single well of a 12-well plate.
4. Add 1.5 mL of DPBS to each well and remove the air bubbles from the microwells by gently pipetting up and down several times using a P1000 pipette tip. This is performed to ensure uniform cell seeding.
5. Store plates at 4 °C for up to 6 months.

3.1.3 Preparation of Poly 2-Hydroxyethyl Methacrylate (Poly-HEMA)-Coated 48- and 96-Well Plates

This protocol was previously described in [20].

1. Dissolve 2 g of poly-HEMA in 100 mL of 95% ethanol.
2. Stir the solution overnight using a hot plate set at 55 °C.
3. Add 250 μL/well of poly-HEMA solution in a 48-well plate and 100 μL/well of poly-HEMA solution in a 96-well plate.
4. Dry overnight at 60 °C using an oven.

3.2 Stage 1: Passaging Human Pluripotent Stem Cells (hPSC) for Differentiation

The method reported here was previously described in [12, 13, 20, 21, 22].

1. Culture hPSCs on LN-521-coated 6-well plates and change mTeSR1 medium every 24 h. A single-cell solution of hPSCs is needed for differentiating the three cell types required for liver spheroid aggregation.
2. Passage once the cells reach 75–85% of confluency.
3. When ready for the differentiation experiment and at 75–85% confluency, aspirate the medium from the wells containing the hPSCs.
4. Wash the cells with 1 mL/well of DPBS at room temperature.
5. Aspirate the DPBS from the wells.
6. Add 1 mL/well of cell dissociation reagent and leave the cells in a 37 °C/5% CO_2 cell culture incubator for 5–8 min (*see* **Note 21**).
7. In the meantime, warm the plates previously coated with LN-521, carefully aspirate the solution, avoiding interfering with the laminin coating, followed by immediate addition of 1 mL/well of complete mTeSR1 medium supplemented with 10 μM Rho-associated kinase (ROCK) inhibitor Y-27632 (*see* **Note 22**). Store the LN-521-coated plates in a 37 °C/5% CO_2 cell culture incubator until **step 13**.
8. Once dissociated, add 1 mL/well of pre-warmed of complete mTeSR1 medium supplemented with 10 μM Rho-associated kinase (ROCK) inhibitor Y-27632 to stop the dissociation reaction.
9. With the help of a cell scraper, collect the cells into a 15 mL centrifuge tube.
10. Centrifuge the cells at 0.2× *g* for 5 min at room temperature.

11. Aspirate the supernatant carefully without disturbing the cell pellet.
12. Add 5–10 mL of pre-warmed complete mTeSR1 medium supplemented with 10 μM Rho-associated kinase (ROCK) inhibitor Y-27632 to the cells and gently pipette up and down to ensure a homogeneous single-cell solution (*see* **Note 23**).
13. Seed the cells into the warm LN-521-coated wells (*see* **Note 24**).
14. Bring the final volume to 2 mL/well by adding pre-warmed complete mTeSR1 medium supplemented with 10 μM ROCKi Y-27632, if needed.
15. Incubate the plate in a 37 °C/5% CO_2 cell culture incubator for 24 h to start the differentiation to either hepatocyte, endothelial, or stellate cells (*see* **Note 25**).

3.3 Stage 2a: Hepatic Stellate Cell (HSC) Differentiation and Preparation for Sphere Formation

This protocol was adapted from [23] and previously described in [12, 13].

1. Seed 1 mL/well of cell suspension at a density of 75,000 cells/ cm^2 following **steps 3–14** of Subheading 3.2.
2. Incubate overnight in a 37 °C/5% CO_2 cell culture incubator.
3. Replace the medium with 2 mL/well of hepatic stellate cell differentiation medium supplemented with 20 ng/mL of BMP4. Incubate at 37 °C/5% CO_2.
4. Replenish the medium every 48 h for 4 days.
5. On day 4, change the medium with hepatic stellate cell differentiation medium supplemented with 20 ng/mL of FGF1 and 20 ng/mL of FGF3. Incubate at 37 °C/5% CO_2.
6. On day 6, change the medium with hepatic stellate cell differentiation medium supplemented with 20 ng/mL of FGF1, 20 ng/mL of FGF3, 5 μM retinol and 100 μM palmitic acid. Incubate at 37 °C/5% CO_2.
7. On day 8, change the medium with hepatic stellate cell differentiation medium supplemented with 5 μM retinol and 100 μM palmitic acid.
8. Replenish the media every 48 h until day 12.
9. At day 12, cells are ready for harvesting and sphere formation.
10. For sphere formation, wash the wells once with DPBS and add 1 mL/well of pre-warmed TrypLE. Incubate for 3–6 min at 37 °C/5% CO_2 to obtain a single-cell suspension.
11. Once cells are detached, collect the cell suspension into a 50 mL tube, wash the wells with 5 mL of cold liver sphere medium supplemented with 10 μM Y-27632 to collect any leftover cells, and transfer the media into the tube containing

the cell suspension. Ensure a single-cell suspension by filtering the cell suspension through a 30 μm filter.

12. Count the cells using an automatic cell counter using trypan blue to exclude dead cells (*see* **Note 26**).
13. Spin down cell suspension by centrifuging the cells at 0.2× *g* for 10 min.
14. Resuspend the cell pellet at 3.84×10^6 cells/mL in liver sphere medium supplemented with 10 μM Y-27632, 10 ng/mL EGF, 10 ng/mL FGF, 10 ng/mL HGF, 20 ng/mL OSM, and 50 ng/mL VEGF (*see* **Note 27**).
15. Store the cell suspension at 4 °C (max up to 4 h), until all cell types are ready for the liver sphere aggregation step (*see* Subheading 3.6) (*see* **Note 28**).

3.4 Stage 2b: Hepatic Progenitor (HB) Cell Differentiation and Preparation for Sphere Formation

This protocol was previously described in [21, 22, 24].

1. Seed 1 mL/well of cell suspension at a density of 40,000 cells/cm^2 following **steps 3–14** of Subheading 3.2.
2. Incubate 24 h in a 37 °C/5% CO_2 cell culture incubator.
3. After 24 h or when cell confluency reaches 40%, replace the media with 2 mL/well of endoderm differentiation medium supplemented with 100 ng/mL of Activin A and 50 ng/mL Wnt3a.
4. Incubate at 37 °C/5% CO_2.
5. Replenish medium every 24 h for 3 days.
6. On day 3, change the medium with hepatic progenitor differentiation medium. Incubate at 37 °C/5% CO_2.
7. Replenish medium every 48 h for 5 days.
8. On day 8, change the medium with hepatocyte maturation medium supplemented with 10 ng/mL of HGF and 20 ng/mL of OSM.
9. Incubate at 37 °C/5% CO_2.
10. On day 9, cells can be harvested for sphere aggregation. Wash once with DPBS and add 1 mL/well of prewarmed TrypLE. Incubate for 6–10 min at 37 °C/5% CO_2 to obtain a single-cell suspension.
11. Once cells are detached, collect the cell suspension into a 50 mL tube, wash the wells with 7 mL of cold liver sphere medium supplemented with 10 μM Y-27632 to collect any leftover cells, and transfer the media into the tube containing the cell suspension. Ensure a single-cell suspension by filtering the cell suspension through a 30 μm filter.
12. Count the cells using an automatic cell counter using trypan blue to exclude dead cells (*see* **Note 26**).

13. Spin down cell suspension by centrifuging the cells at 0.2× *g* for 10 min.
14. Resuspend the cell pellet at 7.68×10^6 cells/mL in liver sphere medium supplemented with 10 μM Y-27632, 10 ng/mL EGF, 10 ng/mL FGF, 10 ng/mL HGF, 20 ng/mL OSM, and 50 ng/mL VEGF (*see* **Note 27**).
15. Store the cell suspension at 4 °C (max up to 4 h), until all cell types are ready for the liver sphere aggregation step (*see* Subheading 3.6) (*see* **Note 28**).

3.5 Stage 2c: Endothelial Cell (ELC) Differentiation and Preparation for Sphere Formation

This protocol was adapted from [25] and previously described in [12, 13].

1. Seed 1 mL/well of cell suspension at a density of 25,000 cells/cm^2 following **steps 3–14** of Subheading 3.2.
2. Incubate 24 h in a 37 °C/5% CO_2 cell culture incubator.
3. After 24 h, replace the medium with 3 mL/well of mesoderm priming medium supplemented with 10 μM CHIR99021 and 25 ng/mL of BMP4.
4. Incubate for 72 h at 37 °C/5% CO_2.
5. Change the medium to StemPro-34 medium supplemented with 200 ng/mL of VEGF and forskolin 2 μM.
6. Replenish the medium after 24 h.
7. This protocol generates a non-homogeneous endothelial cell population ranging from 41% to 64% purity. To ensure a pure >90% population of CD144/CD31 positive endothelial cells, magnetic cell separation (MACS) enrichment for CD144 is required. To collect the cells for CD144 enrichment using MACS, wash once with DPBS and dispense 1 mL/well of pre-warmed TrypLE. Incubate for 3–6 min at 37 °C/5% CO_2 to obtain a single-cell suspension.
8. Collect the cell suspension into a 50 mL tube, wash the wells with 5 mL of cold DPBS with 0.5% BSA to collect any leftover cells, and transfer the wash into the tube containing the cell suspension. Ensure a single-cell suspension by filtering the cell suspension through a 30 μm filter.
9. Count the cells using an automatic cell counter using trypan blue to exclude dead cells (*see* **Note 26**).
10. Spin down cell suspension by centrifuging the cells at 0.2× *g* for 5 min.
11. MACS enrichment is performed as recommended by manufacturers' instructions (*see* **Note 29**).
12. Count the cells to calculate the yield of CD144 positive endothelial cells and spin them at 0.2× *g* for 10 min (*see* **Note 26**).

13. Resuspend the cell pellet at 4.6×10^6 cells/mL in liver sphere medium supplemented with 10 μM Y-27632, 10 ng/mL EGF, 10 ng/mL FGF, 10 ng/mL HGF, 20 ng/mL OSM, and 50 ng/mL VEGF (*see* **Note 27**).
14. Store the cell suspension at 4 °C (max up to 4 h), until all cell types are ready for the liver sphere aggregation step (*see* Subheading 3.6) (*see* **Note 28**).

3.6 Stage 3a: Aggregation and Maintenance of Liver Spheres

This step describes how to prepare the cell suspension with the three cell types differentiated via the protocols described in Subheadings 3.3, 3.4, and 3.5, using the 2% agarose molds formed via the protocol described in Subheading 3.1.2. This protocol was previously described in [12, 13, 20].

1. For each well, prepare a cell suspension containing 100 μL of HB suspension, 50 μL of ELC suspension, 20 μL of HSC suspension, and 20 μL of liver sphere medium supplemented with 10 μM Y-27632, 10 ng/mL EGF, 10 ng/mL FGF, 10 ng/mL HGF, 20 ng/mL OSM, and 50 ng/mL VEGF (*see* **Note 30**).
2. Carefully aspirate the DPBS from the wells of the agarose mold plates prepared in Subheading 3.1.2.
3. Add 190 μL/well of cell suspension.
4. Return the 12-well plates to a 37 °C/5% CO_2 cell incubator for 2 h to allow the cells to settle.
5. After 2 h, add 1.5 mL/well of fresh and warm liver sphere medium supplemented with 10 μM Y-27632, 10 ng/mL EGF, 10 ng/mL FGF, 10 ng/mL HGF, 20 ng/mL OSM, and 50 ng/mL VEGF.
6. Return the plates to the 37 °C/5% CO_2 cell incubator for 24 h to allow the cells to aggregate.
7. Examine the formation of spheres by brightfield microscopy.
8. Replenish the medium every 48 h with 2 mL/well of fresh and warm liver sphere medium supplemented with 10 ng/mL EGF, 10 ng/mL FGF, 10 ng/mL HGF, 20 ng/mL OSM, and 50 ng/mL VEGF.

3.7 Stage 3b: Formation of Liver Cancer Co-culture Spheres

This protocol was previously described in [13].

1. After maintaining the spheres in culture for 4 weeks, transfer the spheres into poly-HEMA coated 96-well plates (*see* Subheading 3.1.3).
2. Label the single cell suspensions of MIAPaCa-2 (a pancreatic ductal adenocarcinoma cell line) and PDAC (Pancreatic ductal adenocarcinoma cell line)-A (KRas G12D; p53 R172H; Pdx-1::Cre expressing mouse PDAC cell line) with CellTracker

Red dye (ThermoFisher, C34552) as per manufacturer's instructions.

3. Add labelled single cell suspensions of MIAPaCa-2 and PDAC at 5×10^4 cells to poly-HEMA coated 96-well plates containing liver spheres in liver sphere medium (*see* **Notes 31** and **32**).
4. After 24 h, fix the samples, stain with Alexa Fluor-488 Phalloidin, and embed in 1% agarose.
5. Image Z-stack tile scans of each well with an Opera PhenixTM High-Content Screening system and score the interactions between cancer cells and liver spheres in ImageJ.

3.8 Quantification of CYP P450 Activity in Liver Spheres

Once a week, CYP P450 metabolic activity can be measured to assess liver sphere maturation. This protocol was previously described in [12, 13, 20].

1. Switch the liver spheres' medium to complete Williams E medium 24 h prior to running the assay.
2. Remove most of the supernatant from the liver spheres and transfer it into 15 mL centrifuge tubes. Store the supernatant at −20 °C for up to 1 month for further analysis (*see* Subheading 3.9).
3. Once the sphere has sedimented by gravity, remove any remaining supernatant.
4. Supplement the complete Williams E medium with 100 μM Luciferin-ME to measure CYP P450 1A2 activity, or with 50 μM Luciferin-PFBE to measure CYP P450 3A activity.
5. Add 350 μL/well of supplemented complete Williams E medium.
6. Transfer the sphere and supplemented complete Williams E medium into a well of a poly-HEMA-coated 48-well plate.
7. Incubate for 24 h in a 37 °C/5% CO_2 cell incubator.
8. Collect all medium from the well and dispense 50 μL/well into 96-well white plates in triplicate (*see* **Note 33**).
9. Add 2 mL/well of fresh liver sphere medium supplemented with 10 ng/mL EGF, 10 ng/mL FGF, 10 ng/mL HGF, 20 ng/mL OSM, and 50 ng/m VEGF, aspirate the medium and spheres, place them back in a 2% agarose mold 12-well plate, and incubate them in a 37 °C/5% CO_2 cell incubator.
10. Add 50 μL/well of the luciferase reagent from the P450-Glo assay kit into the white plates and measure the relative levels of CYP P450 1A2 or CYP P450 3A activity according to manufacturer's instructions.
11. Repeat CYP P450 characterization once a week up to a month. On the last timepoint, collect all the spheres from the well by gently pipetting up and down with P1000 pipette tip into a

1.5 mL Eppendorf tube, and replace the liver medium with RIPA buffer.

12. Freeze the spheres within the RIPA buffer at −20 °C (*see* **Note 34**).
13. After 24 h, thaw the samples at room temperature (15–21 °C) and sonicate the spheres within the tubes (23.3 W) for 15 min to break down the structures of the spheres.
14. Perform the BCA protein quantification assay according to manufacturer's instructions, using 10 μL of sample in triplicate.
15. Normalize the relative levels of CYP P450 1A2 or CYP P450 3A activity over time on the per mg of proteins in the liver sphere as determined by the BCA assay.

3.9 Quantification of Human Serum Albumin or Alpha-Fetoprotein Secretion in Liver Spheres

This protocol was previously described in [12, 13, 20, 22].

1. Thaw the supernatants harvested in **step 2** of Subheading 3.8.
2. Measure the relative levels of human serum albumin or alpha-fetoprotein production with an ELISA kit as per manufacturer's instructions.
3. Use tissue culture media as a negative control.
4. Normalize the protein concentration to per mg of protein of the liver spheres as determined by the BCA assay (**steps 12–14** of Subheading 3.8).

4 Notes

1. Prepare a stock solution of Y-27632 (ROCKi) by dissolving Y-27632 in sterile 0.2% bovine serum albumin (BSA)/DPBS to a final concentration of 10 mM. Store at −80 °C in small aliquots. Use at 1:1000 dilution.
2. Prepare a solution of BMP4 by dissolving BMP4 in sterile 0.2% BSA/DPBS to a final concentration of 50 μg/mL. Store at −80 °C in small aliquots.
3. Prepare a stock solution of FGF1 by dissolving FGF1 in sterile 0.2% BSA/DPBS to a final concentration of 20 μg/mL. Store at −80 °C in small aliquots. Use at 1:1000 dilution.
4. Prepare a stock solution of FGF3 by dissolving FGF3 in sterile 0.2% BSA/DPBS to a final concentration of 20 μg/mL. Store at −80 °C in small aliquots. Use at 1:1000 dilution.
5. Prepare a stock solution of palmitic acid by dissolving palmitic acid in sterile 0.2% BSA/DPBS to a final concentration of 100 μM. Store at −80 °C in small aliquots. Use at 1:1000 dilution.

6. Prepare a stock solution of retinol by dissolving retinol in sterile EtOH to a final concentration of 500 mM. Store at −80 °C in small aliquots. Dilute 1:100 in EtOH and then use at 1:1000 dilution.
7. Prepare a 1000× stock solution of human activin A by dissolving human activin A lyophilized protein in sterile 0.2% BSA/DPBS to a final concentration of 100 μg/mL. Store at −80 °C in small aliquots. Use at 1:1000 dilution.
8. Prepare a 1000× stock solution of human hepatocyte growth factor (HGF) by dissolving human HGF lyophilized protein in sterile 0.2% BSA/DPBS to a final concentration of 10 μg/mL. Store at −80 °C in small aliquots. Use at 1:1000 dilution.
9. Prepare a solution of hydrocortisone 21-hemisuccinate sodium salt (HCC) by dissolving HCC in PBS to a final concentration of 10 μM. Filter the solution with the sterile 0.22 μm filters and store in −80 °C in 5 mL aliquots. Use at 1:100 dilution.
10. Prepare a 1000× stock solution of oncostatin M (OSM) by dissolving OSM in sterile 0.2% BSA/DPBS to a final concentration of 20 μg/mL. Store at −80 °C in small aliquots. Use at 1:1000 dilution.
11. Prepare a 1000× stock solution of Wnt3A by dissolving mouse Wnt3A lyophilized protein in sterile 0.2% BSA/DPBS to a final concentration of 10 μg/mL. Store at −80 °C in small aliquots. Use at 1:200 dilution.
12. Prepare a solution of CHIR99021 by dissolving CHIR99021 in sterile DMSO to a final concentration of 10 μM. Store at −80 °C in small aliquots.
13. Prepare a solution of forskolin by dissolving forskolin in sterile 0.2% BSA/DPBS to a final concentration of 10 mM. Store at −80 °C in small aliquots.
14. Prepare a 1000× stock solution of VEGF by dissolving the lyophilized protein in 0.2% BSA/DPBS to a final concentration of 10 μg/mL. Store at −80 °C in small aliquots. Use at 1:1000 dilution.
15. Prepare a 1000× stock solution of bFGF by dissolving the lyophilized protein in sterile 0.2% BSA/DPBS to a final concentration of 10 μg/mL. Store at −80 °C in small aliquots. Use at 1:1000 dilution.
16. Prepare a 1000× stock solution of EGF by dissolving the lyophilized protein in sterile 0.2% BSA/DPBS to a final concentration of 10 μg/mL. Store at −80 °C in small aliquots. Use at 1:1000 dilution.
17. Prepare a 1000× stock solution of VEGF by dissolving the lyophilized protein in 0.2% BSA/DPBS to a final concentration of 10 μg/mL. Store at −80 °C in small aliquots. Use at 1:1000 dilution.

18. Our protocol has been optimized on this kit. If using a different commercial kit, consider running some trials.
19. The use of LN-521 is recommended to obtain reproducible results. For other matrices, further optimization in cell seeding densities will be required.
20. Do not let the LN-521 evaporate. Seal the plates with parafilm to avoid evaporation and to reduce the risk of contamination. Addition of 1 mL of Ca^{2+}/Mg^{2+}DPBS can reduce the risk of evaporation.
21. Incubation time depends on cell type and confluency, therefore inspection under a brightfield microscope is needed to determine the optimal incubation time. Do not exceed 15 min incubation.
22. The use of ROCKi Y-27632 enhances cell attachment and survival.
23. Tap the cell pellet to ensure the single-cell suspension and avoid cell clumps.
24. To ensure an even distribution of the cell suspension, move gently the plate ten times back and forth and left to right.
25. Maintenance and the scaling-up of a human pluripotent stem cells (hPSC) stock population accordingly to meet the three differentiations is key to ensure sufficient cell numbers.
26. We recommend using automated cell count, as manual cell count can introduce user-to-user variability.
27. A second cell count is recommended after cell re-suspension, to ensure the right cell concentration is achieved.
28. Cells can also be fixed for FACS characterization at this stage. For negative and positive endothelial cells, populations can be collected after MACS enrichment for FACS characterization.
29. Our protocol for MACS enrichment was as follows:
    - Resuspend cell pellet in 80 μL per 10^7 cells and add 20 μL of CD144-magnetic beads.
    - Mix well and incubate at 4 °C for 15 min.
    - Wash cells with 1–2 mL of cold DPBS supplemented with 0.5% BSA and centrifuge the cell suspension at 0.2× *g* for 5 min.
    - Resuspend up to 10^8 cells in 500 μL of DPBS supplemented with 0.5% BSA.
    - Place the MACS column in the magnetic field and prepare the column by rinsing it with 3 mL of DPBS supplemented with 0.5% BSA.
    - Apply the cell suspension onto the column, flow-through contains the un-labeled cells.

- Wash the column three times with 3 mL of DPBS supplemented with 0.5% BSA.
- Add 5 mL of DPBS supplemented with 0.5% BSA and quickly collect the magnetically labeled cells by pushing the plunger into the column.

30. For liver spheroid aggregation, we recommend a HB:ELC:HSC cell ratio of 10:3:1. Each agarose mold contains 256 micro-molds; thus, each sphere contains approximately 3000 HB, 900 ELC, and 300 HSC at the time of seeding.
31. We recommend optimization of cancer cell lines: liver cells ratios. As different cancer cell lines grow at different rates, a range of ratios should be trialed.
32. Our protocol describes the addition of cancer cells 4 weeks after the sphere aggregation. However, cancer cells can be added in the HB:ELC:HSC cell suspension before sphere aggregation or during medium change after the liver sphere formation. If introducing cancer cells before liver spheroid formation, for each well, prepare a cell suspension containing 100 μL of HB suspension, 50 μL of ELC suspension, 20 μL of HSC suspension, and 20 μL of cancer cell suspension at 3.84×10^6 cells/mL (*see* **Note 31**) in liver sphere medium, supplemented with 10 μM Y-27632, 10 ng/mL EGF, 10 ng/mL FGF, 10 ng/mL HGF, 20 ng/mL OSM, and 50 ng/mL VEGF. In addition, cancer cells can be added to the liver spheroids within the agarose molds. Carefully aspirate most of the media from each well and mold before gently pipetting 150 μL of single cancer cell suspension into the mold. Incubate for 2 h at 37 °C to allow cells to settle, before adding 1.5 mL/well of fresh and warm liver sphere medium supplemented with 10 μM Y-27632, 10 ng/mL EGF, 10 ng/mL FGF, 10 ng/mL HGF, 20 ng/mL OSM, and 50 ng/mL VEGF.
33. Measurements can be performed at a later timepoint. For this, cover the plates with lids and seal with parafilm. Store the plates at −20 °C for up to 1 month.
34. Liver spheres can be stored in RIPA buffer at −20 °C for up to 1 month, to perform the measurements at a later time.

References

1. Asrani SK, Devarbhavi H, Eaton J, Kamath PS (2019) Burden of liver diseases in the world. J Hepatol 70(1):151–171
2. Statistics. British Liver Trust. Available from: https://britishlivertrust.org.uk/about-us/media-centre/statistics/. Accessed 17 Jan 2022
3. Romano M, Elgueta R, McCluskey D, Ortega-Prieto AM, Stolarczyk E, Dazzi F et al (2021) Pluripotent stem cell-derived hepatocytes inhibit T cell proliferation in vitro through tryptophan starvation. Cell 11(1):24
4. LeCluyse EL, Alexandre E, Hamilton GA, Viollon-Abadie C, Coon DJ, Jolley S et al (2005) Isolation and culture of primary

human hepatocytes. In: Helgason CD, Miller CL (eds) Basic cell culture protocols. Humana Press, Totowa, pp 207–229

5. Soldatow VY, LeCluyse EL, Griffith LG, Rusyn I (2013) In vitro models for liver toxicity testing. Toxicol Res 2(1):23–39
6. Human primary cells and immortal cell lines: differences and advantages (2019) PromoCell. Available at: https://promocell.com/in-the-lab/human-primary-cells-and-immortal-cell-lines/. Accessed 13 Sept 2021
7. Sinton MC, Meseguer-Ripolles J, Lucendo-Villarin B, Wernig-Zorc S, Thomson JP, Carter RN et al (2021) A human pluripotent stem cell model for the analysis of metabolic dysfunction in hepatic steatosis. iScience 24:101931
8. Meseguer-Ripolles J, Lucendo-Villarin B, Tucker C, Ferreira-Gonzalez S, Homer N, Wang Y et al (2021) Dimethyl fumarate reduces hepatocyte senescence following paracetamol exposure. iScience 24(6):102552
9. Medvedev SP, Shevchenko AI, Zakian SM (2010) Induced pluripotent stem cells: problems and advantages when applying them in regenerative medicine. Acta Nat 2(2):18–28
10. Szkolnicka D, Hay D et al (2020) Liver stem cells principles of tissue engineering. In: Lanza R, Langer R, Vacanti J, Atala A (eds), 5th edn ch 40. Elsevier, Amsterdam
11. Rashidi H, Luu N-T, Alwahsh SM, Ginai M, Alhaque S, Dong H et al (2018) 3D human liver tissue from pluripotent stem cells displays stable phenotype in vitro and supports compromised liver function in vivo. Arch Toxicol 92(10):3117–3129
12. Meseguer-Ripolles J, Kasarinaite A, Lucendo-Villarin B, Hay DC (2021) Protocol for automated production of human stem cell derived liver spheres. STAR Protoc 2(2):100502
13. Lucendo-Villarin B, Meseguer-Ripolles J, Drew J, Fischer L, Ma E, Flint O, Simpson KJ, Machesky LM, Mountford JC, Hay DC (2020) Development of a cost-effective automated platform to produce human liver spheroids for basic and applied research. Biofabrication 13:015009
14. Lyall MJ, Cartier J, Thomson JP, Cameron K, Meseguer-Ripolles J, O'Duibhir E et al (2018) Modelling non-alcoholic fatty liver disease in human hepatocyte-like cells. Philos Trans R Soc B Biol Sci 373(1750):20170362
15. Cancer. World Health Organisation (WHO). Available from: https://www.who.int/news-room/fact-sheets/detail/cancer. Accessed 17 Jan 2022
16. Types of liver cancer. Cancer Research UK. Available from: https://www.cancerresearchuk.org/about-cancer/liver-cancer/types. Accessed 24 Jan 2022
17. Marengo A, Rosso C, Bugianesi E (2016) Liver cancer: connections with obesity, fatty liver, and cirrhosis. Annu Rev Med 67(1):103–117
18. Valderrama-Treviño AI, Barrera-Mera B, Ceballos-Villalva JC, Montalvo-Javé EE (2017) Hepatic metastasis from colorectal cancer. Euroasian J Hepatol Gastroenterol 7(2): 166–175
19. Miarka L, Hauser C, Helm O, Holdhof D, Beckinger S, Egberts J-H et al (2019) The hepatic microenvironment and TRAIL-R2 impact outgrowth of liver metastases in pancreatic cancer after surgical resection. Cancers 11(6):745
20. Lucendo-Villarin B, Rashidi H, Alhaque S, Fischer L, Meseguer-Ripolles J, Wang Y, O'Farrelly C, Themis M, Hay DC (2019) Serum free production of three-dimensional human hepatospheres from pluripotent stem cells. J Vis Exp 149:e59965
21. Meseguer-Ripolles J, Lucendo-Villarin B, Wang Y, Hay DC (2018) Semi-automated production of hepatocyte like cells from pluripotent stem cells. J Vis Exp JoVE 137:57995
22. Wang Y, Tatham MH, Schmidt-Heck W, Swann C, Singh-Dolt K, Meseguer-Ripolles J et al (2019) Multiomics analyses of HNF4α protein domain function during human pluripotent stem cell differentiation. iScience 16: 206–217
23. Coll M, Perea L, Boon R, Leite SB, Vallverdú J, Mannaerts I et al (2018) Generation of hepatic stellate cells from human pluripotent stem cells enables in vitro modeling of liver fibrosis. Cell Stem Cell 23(1):101–113.e7
24. Wang Y, Alhaque S, Cameron K, Meseguer-Ripolles J, Lucendo-Villarin B, Rashidi H et al (2017) Defined and scalable generation of hepatocyte-like cells from human pluripotent stem cells. J Vis Exp JoVE 121:55355
25. MacAskill MG, Saif J, Condie A, Jansen MA, MacGillivray TJ, Tavares AAS et al (2018) Robust revascularization in models of limb ischemia using a clinically translatable human stem cell-derived endothelial cell product. Mol Ther 26(7):1669–1684

Chapter 12

A Self-Assembly Method for Creating Vascularized Tumor Explants Using Biomaterials for 3D Culture

Lance L. Munn and Despina Bazou

Abstract

Validation of potential therapeutic targets in cancer requires functional live assays that recapitulate the biology, anatomy, and physiology of human tumors. We present a methodology for maintaining mouse and patient tumor samples ex vivo for in vitro drug-screening as well as for the guidance of patient-specific chemotherapies. The harvested tumor biopsy, excised from mice or patients, is integrated into a support tissue that includes extended stroma and vasculature. The methodology is more representative than tissue culture assays, faster than patient-derived xenograft models, easy to implement, amenable to high-throughput assays and does not carry the ethical issues or expense associated with animal studies. Our physiologically relevant model can be successfully used for high-throughput drug screening.

Key words Cancer, Chemotherapy, High-throughput assay, Tumor explant, Vascular bed, Vascularization

1 Introduction

Methodologies for testing and applying existing and emerging treatments on a patient-by-patient basis would help to prioritize treatment strategies based on the potential for response or the likelihood of toxicities. By testing patient tumor samples at the time of resection, it should be possible to formulate patient-specific treatment strategies in case of progression through initial therapies. In some cancers, such as pancreatic cancer, treatment decisions must be made in the time between surgery and progression—typically weeks or months.

Many approaches have been developed to recreate tumors in culture to study cancer biology or identify promising drugs. Cancer cell lines or circulating cancer cells (CTCs) grown in 2-*D* cultures or as 3D spheroids are valuable tools for identifying targetable pathways, but inconsistently predict clinical response—likely due to the omission of important antagonistic or synergistic

Dania Movia and Adriele Prina-Mello (eds.), *Cancer Cell Culture: Methods and Protocols*,
Methods in Molecular Biology, vol. 2645, https://doi.org/10.1007/978-1-0716-3056-3_12,
© The Author(s), under exclusive license to Springer Science+Business Media, LLC, part of Springer Nature 2023

interactions between tumor components. Recent efforts to grow patient-derived xenograft (PDX) tumors in mice for subsequent drug testing are promising, but this approach necessitates that the original human tumor be passaged through mice. In addition, in our 10-year experience in an ongoing project at the Massachussetts General Hospital (MGH), the PDX take rate is less than 60%, and it requires ~145 days to produce a primary PDX from a pancreatic ductal adenocarcinoma (PDAC) patient sample. For personalized medicine, many treatment decisions need to be made more rapidly than this.

Currently available organoid cultures have similar limitations in speed and reliability and, like PDX systems, can only provide useful information in specific applications [1]. Tissue slice [2] cultures retain cell subpopulations, histoarchitecture, and phenotypic heterogeneity of the primary tumor. However, they still lack an extended, supporting vasculature and often use nonrepresentative matrices such as matrigel for support. The inconsistent ability of organoid and PDX assay systems to predict clinical success has generated substantial interest in human-based tissue-mimetic constructs for disease modeling and drug testing, also referred to as "organs-on-a-chip" [3, 4]. Clinical oncologists need such tools to improve the reliability of drug screening and implementation of personalized medicine [5, 6].

The vasculature and its associated stroma control much of tumor biology, including metabolism, invasion, and metastasis. To date, there are a few vascularized models of human tumors that use microfluidic models [7] and bioreactors [8]. Although the microfluidic model achieved tumor vascularization and early tumor intravasation steps were identified, the tumor spheroid consisted of a single cancer cell population and lacked other components of the tumor microenvironment. Bioreactors have been successful in the culture of tumor explants under dynamic flow, but are difficult to adapt for high-throughput drug screening and are not amenable for longitudinal microscopic imaging; furthermore, the explant blood vessels are isolated, not integrated into a surrounding network formed in vitro.

Tumor growth and response to treatment are shaped by the interactions with host tissue-derived stroma. To recreate the correct micro-anatomy—and thus, a truly biomimetic tumor—the cancer and the host tissue have to co-evolve naturally. Only in this way will the correct mixture and organization of host macrophages, fibroblasts, stromal matrix, blood vessels, and cancer cells be produced. Therefore, it may be more appropriate to use intact ex vivo tumor tissue rather than synthetic tumors created from isolated cancer cells. These considerations are of particular relevance to pancreatic cancer, a tumor known to have a very substantial stromal component that mediates tumor progression.

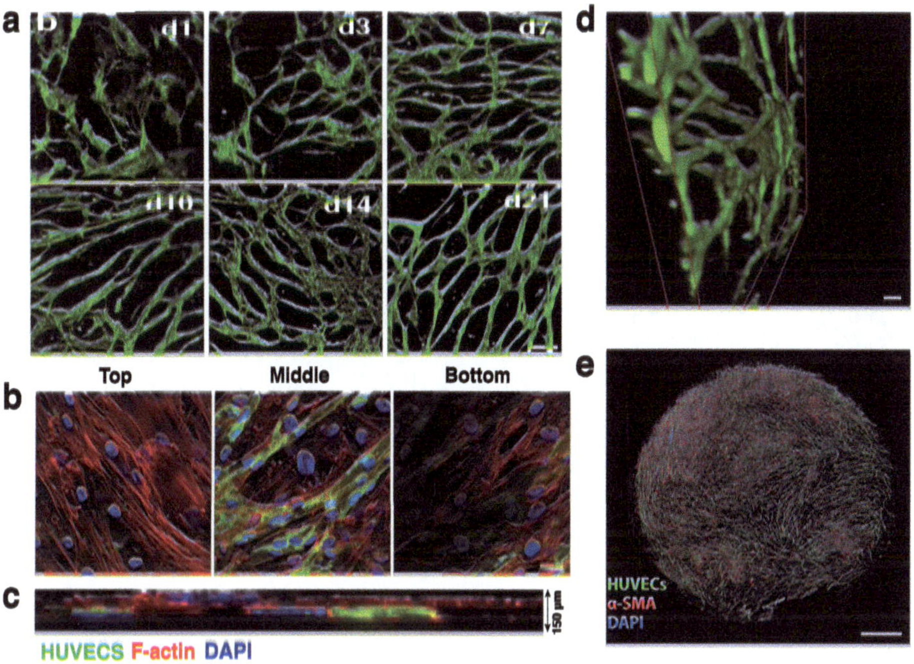

Fig. 1 In vitro vascularization. (**a**) Co-culturing HUVECs (green) with SMCs (not shown here for clarity) leads to the formation of an interconnected vessel network by day 21. Scale bar, 200 μm. (**b**) Co-cultures (shown at Day 7 in culture) form a multi-layered tissue with SMCs surrounding the HUVECs (green) are seen underneath as well as above the vessel network. F-actin of both cell populations is stained with Phalloidin-568 (red). Blue depicts nuclei stained with DAPI. Scale bar, 20 μm. (**c**) The thickness of the multi-layered tissue (HUVECs, SMCs, and matrix) is approximately 150 μm (a side view of a confocal stack is shown). (**d**) 3-D reconstruction of part of an in vitro network. Scale bar, 50 μm. (**e**) The complete network fully covers a well of a 96-well plate (Scale bar, 800 μm). (Reproduced with permission from Ref. [9])

We have established such a method by implanting tumor explants into a supporting self-assembled tissue bed that contains stromal cells, matrix, and vasculature [9]. The resulting vascularized tumor explants (VTEs) maintain their anatomy and viability and can be used for studying tumor biology, for drug screening, or for testing personalized therapies. We will describe the following steps for implementing this methodology: generation of a vascularized network (Subheadings 2.1 and 3.1) (Fig. 1), preparation of mouse- and patient-derived tumor explants (Subheadings 2.2, 3.2, and 3.3) (Fig. 3), imaging and analysis (Subheadings 2.3, 3.4, 3.5, and 3.6) (Fig. 2). Using this methodology, the harvested patient biopsy [10] or mouse tumor explant integrates with the stromal bed and vasculature, providing the correct extracellular matrix (collagen I, IV, fibronectin), associated stromal cells, and a lumenized vessel network.

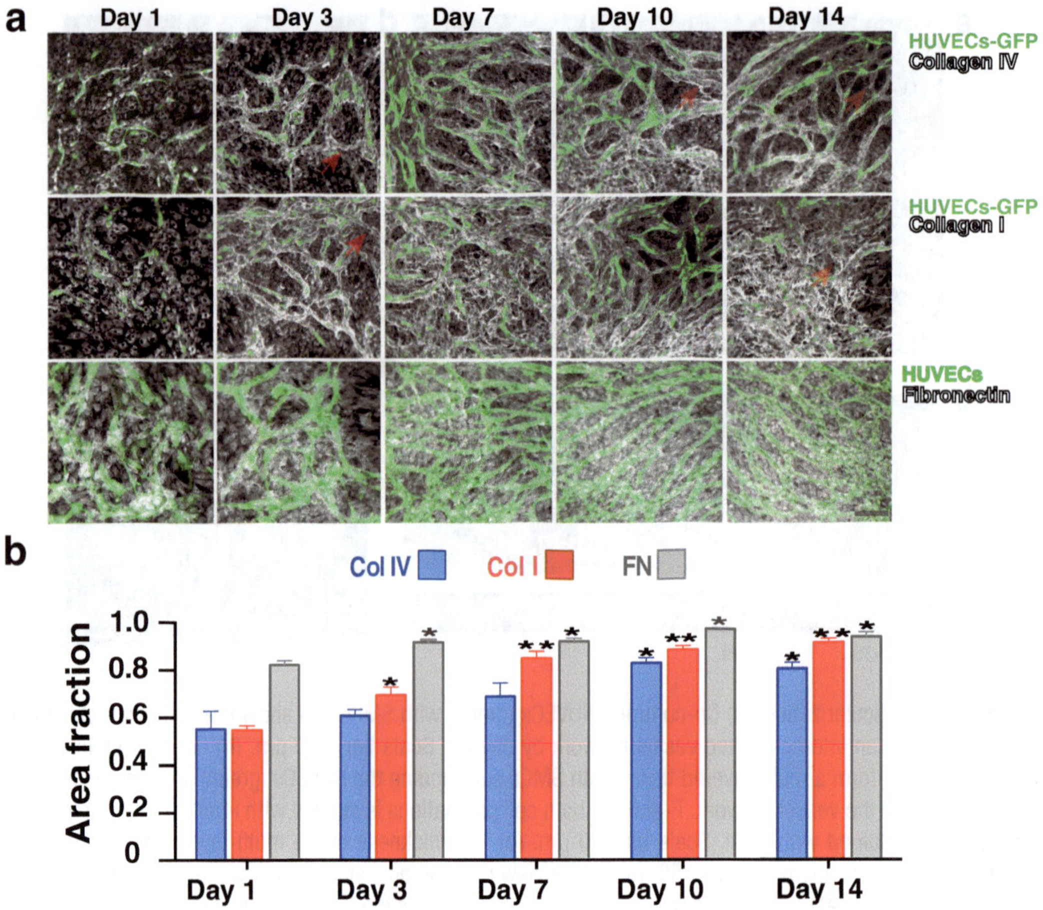

Fig. 2 Co-cultures define their own microenvironment. (**a**) *Top row*: Collagen IV (grey) is strongly present around the vessel network (green) on day 3, and by day 10, it has completely covered the vessel network. *Middle row*: Collagen I (grey) also shows a similar peri-endothelial localization pattern in co-culture. *Bottom row*: Fibronectin (grey) is significantly produced by the vessel network. Scale bar, 50 μm. (**b**) Collagen I, IV, and fibronectin abundance levels increase over time in the developing vessel network (assessed by the fractional area of staining in the immunofluorescent images). Error bars are standard errors from $n = 3$. (Reproduced with permission from Ref. [9])

2 Materials

2.1 Cellular Components

1. Human umbilical vein endothelial cells (HUVECs) (*see* **Note 1**).
2. Endothelial growth medium (EGM).
3. Gelatin solution: 0.1% tissue-grade gelatin solution (*see* **Note 2**).
4. 0.025% (wt/vol) Trypsin-EDTA.
5. Human vascular endothelial growth factor (VEGF).

6. Pulmonary artery smooth muscle cells (SMCs).
7. Smooth muscle growth medium (SmGM-2).
8. Endothelial cell growth medium-2 (EGM-2).

2.2 Mouse Tumor Explants

1. Human melanoma (Mu89) tumors grown orthotopically in 8-week-old severe combined immunodeficient (SCID) females (*see* **Note 3**).
2. Human breast (BT474) tumors grown in the mammary fat pad of female nude mice (*see* **Note 3**) [11].
3. Human pancreatic (PANC-1) tumors grown orthotopically in 8-week-old severe combined immunodeficient mice (SCID) females (*see* **Note 3**) [12].
4. Patient-derived tumor explants prepared from freshly resected human PDAC that had not been treated (chemotherapy or chemoradiation) prior to surgery (*see* **Note 4**).

2.3 In Situ Immunofluorescence Staining and Confocal Imaging Analysis

2.3.1 Reagents

1. Washing solution: 1× phosphate buffered saline (PBS).
2. Fixation solution: 3% w/w paraformaldehyde, 1× PBS.
3. Permeabilization solution: 0.5% TritonX, 1× PBS.
4. 0.1 M Glycine in PBS (*see* **Note 5**).
5. Blocking solution: 5% donkey serum, 0.1% TritonX, 1× PBS.
6. Primary antibodies (e.g., anti-CD31, anti-α-SMA, anti-collagen I, anti-collagen IV, anti-fibronectin, and anti-Tenascin-C primary antibodies).
7. Fluorescently conjugated secondary antibodies.
8. 1:200 DAPI nuclear stain.
9. 1:200 AlexaFluor568 phalloidin.

2.3.2 Equipment

1. Phase contrast/epi-fluorescence microscope (Olympus IX70, 4× and 10× air lens, PRIOR automated stage, OpenLab software).
2. Confocal microscope (Olympus IX81 inverted confocal microscope, 20× air lens and 60× oil lens).
3. Image acquisition software (Fluoview).
4. Image analysis software (ImageJ, Adobe Photoshop).

3 Methods

3.1 Formation of Co-cultures of Endothelial Cells (ECs) and SMCs

1. Seed ECs and SMCs together in EGM-2 on a 96-well plate at concentrations of 10^4 cells/mL (ECs) and 5×10^4 cells/mL (SMCs).
2. Mix briefly to achieve an even cell distribution on the well substrate (*see* **Note 6**).

3. Replace the culture media every 3 days.
4. Monitor co-cultures for vasculature network formation under phase contrast/epi-fluorescence microscope (Fig. 1a–e).

3.2 Formation of Mu89, BT474, PANC-1 Vascularized Tumor Explant (VTE) Culture

1. When tumors reach 4–5 mm diameter, resect them and cut them into explants of size between 0.2 and 0.5 mm for addition into the co-culture wells.
2. Incorporate the mice tumor explants at day 3 of vasculature network formation (one explant per well) (*see* **Note 7**) (Fig. 3a–e).
3. Monitor the VTEs (i.e., tumor explants with the vascular network) for up to 10 days under phase contrast/epi-fluorescence microscope, with the culture media being changed every third day.

3.3 Formation of Patient-Derived PDAC Vascularized Tumor Explant (VTE) Culture

1. Following surgery, collect tumor pieces from the Pathology Department and cut them into explants of size between 0.2 and 0.5 mm for addition into the co-culture wells.
2. Incorporate the patients' tumor explants at day 3 of vasculature network formation (one explant per well) (*see* **Note 7**) (Fig. 3f, g).
3. Monitor the patient-derived VTEs (i.e., tumor explants with the vascular network) for up to 10 days under phase contrast/epi-fluorescence microscope, with the culture media being changed every third day.

3.4 Immuno-fluorescence Staining and Imaging of ECs-SMCs Co-cultures

1. Wash the co-cultures with PBS to remove any cell media.
2. Incubate with the fixation solution for 20 min at room temperature.
3. Wash 3× with PBS.
4. Incubate with the permeabilization solution for 10 min at 4 °C.
5. Rinse the co-cultures 3× with PBS-Glycine.
6. Incubate with blocking solution for 60 min at room temperature.
7. Remove the blocking solution and incubate the co-cultures with primary antibodies (targeting the protein of interest) overnight at 4 °C. As an example, in Fig. 2a, we report the staining of CD31, α-SMA, Collagen I, Collagen IV, and Fibronectin.
8. Wash 3× with PBS.
9. Remove the primary antibody solution and incubate with the appropriate fluorescently labelled secondary antibodies for 60 min at room temperature.
10. Wash 3× with PBS.

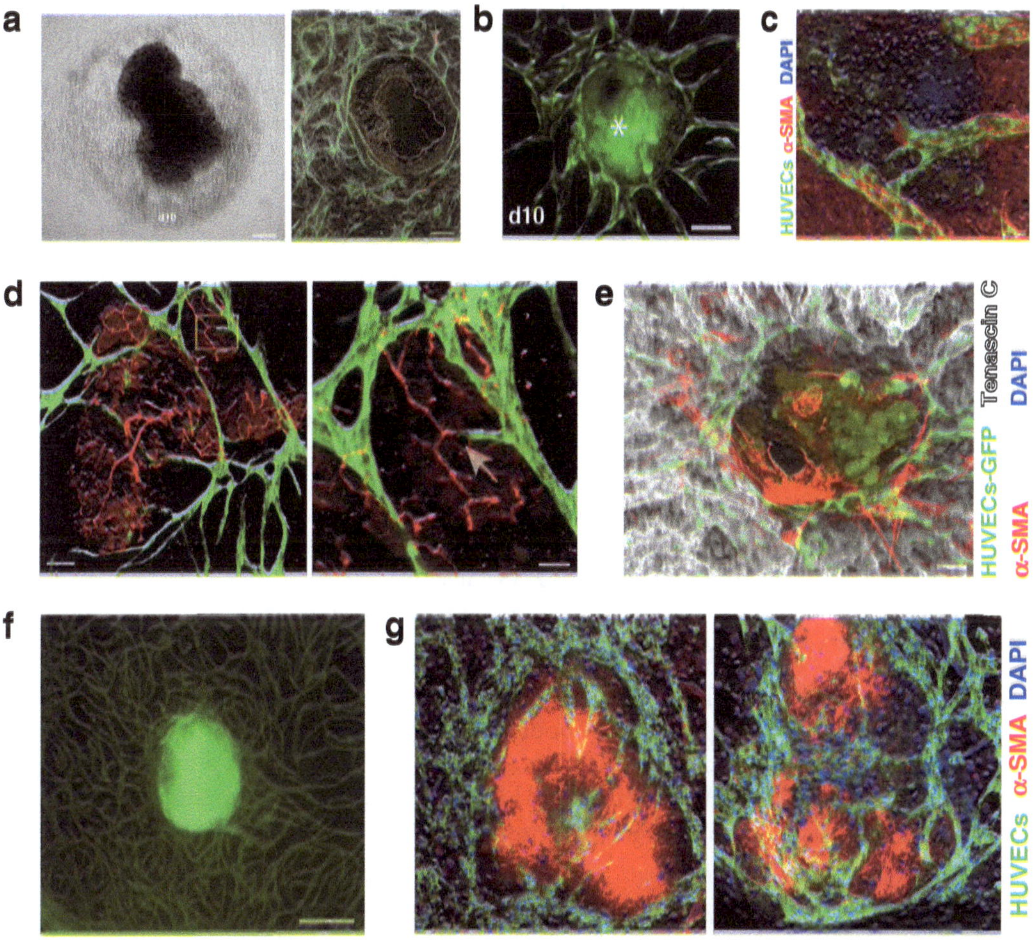

Fig. 3 VTE cultures. (**a**) Brightfield image showing the disruption of Mu89 tumors when placed on the EC-SMC co-culture at Day 0. The corresponding mosaic confocal image of the same tumor (purple outline) shows that the formed vessels (green) create a "ring" around the Mu89 tumor explant. (**b**) Vascularization of PANC-1 VTEs (asterisk), showing the vessels (green) surrounding and penetrating the tumor explants (due to auto-fluorescence, the tumor explants are also shown in green) over 10 days. Scale bar, 150 μm. (**c**) High-resolution confocal image showing a vessel (green) with a clearly formed lumen extending into a tumor explant. Scale bar, 50 μm. Nuclei stained with DAPI (blue). (**d**) Vessels (green) connect with the mouse explant vessels (red). Scale bar, 50 μm. Zoomed-in yellow area shows two vessels connecting (arrow); scale bar, 25 μm. (**e**) Tenascin-C (grey) was strongly expressed by Mu89 VTEs ex-vivo. Scale bar, 100 μm. (**f**) A patient-derived pancreatic VTE on Day 7 in culture was highly vascularized in vitro. Vessels (green); Scale bar, 150 μm. (**g**) Two different patient-derived VTEs (which appear red due to high levels of αSMA) show significant vessel infiltration within 7 days in culture. Vessels (green), nuclei (blue, DAPI). Scale bar, 150 μm. (Reproduced with permission from Ref. [9, 10])

11. To label actin filaments, incubate with AlexaFluor 568 Phalloidin for 30 min at room temperature.
12. To label cell nuclei, incubate with DAPI nuclear stain for ~5 min at room temperature.
13. Wash 3× with PBS.
14. Acquire images with confocal microscopy with 1–5 μm slice thickness.
15. Process the confocal projections with the ImageJ analysis software.

3.5 Immunofluorescence Staining and Imaging of VTEs

1. Wash co-cultures with PBS to remove any cell media.
2. Incubate with the fixation solution for 20 min at room temperature.
3. Wash 3× with PBS.
4. Incubate with the permeabilization solution for 10 min at 4 °C.
5. Rinse 3× with PBS-Glycine.
6. Incubate with blocking solution for 60 min at room temperature.
7. Remove the blocking solution and incubate the VTEs with primary antibodies (targeting the protein of interest) overnight at 4 °C. As an example, in Fig. 3a–g, we report the staining of VTEs with CD31, Tenascin-C (*see* **Note 8**) and α-SMA.
8. Wash 3× with PBS.
9. Remove the primary antibody and incubate with appropriate fluorescently labelled secondary antibodies for 60 min at room temperature.
10. Wash 3× with PBS.
11. Stain cell nuclei with DAPI nuclear stain for 5 min at room temperature.
12. Wash 3× with saline prior to confocal microscopy (image slice of 1 μm).

3.6 Extracellular Matrix Quantification in EC-SMC Co-cultures

1. For all analyses, use at least triplicate co-cultures with two images randomly selected from each repeat, i.e., six images in total to ensure good statistical power.
2. Use the ImageJ software to manually threshold the corresponding confocal images to include the collagen I, IV, and fibronectin-positive signals.
3. Convert the images to binary and then quantify the stained area as a fraction of the total fluorescence (Fig. 2b).

4 Notes

1. In our studies, human umbilical vein endothelial cells (HUVECs) were acquired from the Center for Excellence in Vascular Biology, Brigham & Women's Hospital, Harvard Medical School, Boston, MA. These can also be obtained commercially from several suppliers (e.g., Lonza, Sigma, Thermofisher, etc).
2. Dilute 2% gelatin solution in PBS to coat tissue culture flasks or dishes.
3. The Massachusetts General Hospital Subcommittee on Research Animal Care approved all mouse experiments.
4. The use of human tumor tissue was approved by the Partners human research committee.
5. We use glycine at a concentration of 0.1 M in PBS to reduce background fluorescence.
6. To ensure efficient mixing of ECs and SMCs in a 96-well plate, rotate the plate in a figure-of-eight motion.
7. When adding the tumor explants to the culture together with the ECs and SMCs, the explants contact the plastic and tend to disperse, as cells from the tumor migrate away on the plastic. When this happens in the co-cultures, cells from the explant are able to displace the vascular bed (Fig. 3a). For this reason, the tumor explants are added on day 3; the explants retain their morphology and the vasculature migrates to surround the tumors over 10 days in culture (Fig. 3b–g).
8. Tenascin-C is an extracellular matrix molecule that is highly associated with melanoma and in fact was expressed in Mu89 explants (Fig. 3e).

Acknowledgments

We acknowledge support from the National Institutes of Health (R01HL106584). We also thank Julia Kahn for her help with tumor implantations.

References

1. Huang L, Holtzinger A, Jagan I, BeGora M, Lohse I, Ngai N et al (2015) Ductal pancreatic cancer modeling and drug screening using human pluripotent stem cell- and patient-derived tumor organoids. Nat Med 21(11): 1364–1371. https://doi.org/10.1038/nm.3973
2. Merz F, Gaunitz F, Dehghani F, Renner C, Meixensberger J, Gutenberg A et al (2013) Organotypic slice cultures of human glioblastoma reveal different susceptibilities to treatments. Neuro-Oncology 15(6):670–681. https://doi.org/10.1093/neuonc/not003

3. Neuzi P, Giselbrecht S, Lange K, Huang TJ, Manz A (2012) Revisiting lab-on-a-chip technology for drug discovery. Nat Rev Drug Discov 11(8):620–632. https://doi.org/10.1038/nrd3799
4. van der Meer AD, van den Berg A (2012) Organs-on-chips: breaking the in vitro impasse. Integr Biol (Camb) 4(5):461–470. https://doi.org/10.1039/c2ib00176d
5. Bersini S, Jeon JS, Dubini G, Arrigoni C, Chung S, Charest JL et al (2014) A microfluidic 3D in vitro model for specificity of breast cancer metastasis to bone. Biomaterials 35(8): 2454–2461. https://doi.org/10.1016/j.biomaterials.2013.11.050
6. Jeon JS, Bersini S, Gilardi M, Dubini G, Charest JL, Moretti M et al (2015) Human 3D vascularized organotypic microfluidic assays to study breast cancer cell extravasation. Proc Natl Acad Sci U S A 112(1):214–219. https://doi.org/10.1073/pnas.1417115112
7. Ehsan SM, Welch-Reardon KM, Waterman ML, Hughes CC, George SC (2014) A three-dimensional in vitro model of tumor cell intravasation. Integr Biol (Camb) 6(6):603–610. https://doi.org/10.1039/c3ib40170g
8. Ferrarini M, Steimberg N, Ponzoni M, Belloni D, Berenzi A, Girlanda S et al (2013) Ex-vivo dynamic 3-D culture of human tissues in the RCCS bioreactor allows the study of Multiple Myeloma biology and response to therapy. PLoS One 8(8):e71613. https://doi.org/10.1371/journal.pone.0071613
9. Bazou D, Maimon N, Gruionu G, Munn LL (2016) Self-assembly of vascularized tissue to support tumor explants in vitro. Integr Biol (Camb) 8(12):1301–1311. https://doi.org/10.1039/c6ib00108d
10. Bazou D, Maimon N, Gruionu G, Grahovac J, Seano G, Liu H et al (2018) Vascular beds maintain pancreatic tumour explants for ex vivo drug screening. J Tissue Eng Regen Med 12(1):e318–ee22. https://doi.org/10.1002/term.2481
11. Jain RK, Munn LL, Fukumura D (2012) Mammary fat pad tumor preparation in mice. Cold Spring Harb Protoc 2012(10):1115–1116. https://doi.org/10.1101/pdb.prot071514
12. Jain RK, Munn LL, Fukumura D (2012) Pancreatic tumor preparation in mice. Cold Spring Harb Protoc 2012(12). https://doi.org/10.1101/pdb.prot072363

Chapter 13

A Step-by-Step Methodological Guide for Developing Zonal Multicellular Scaffold-Based Pancreatic Cancer Models

Priyanka Gupta and Eirini G. Velliou

Abstract

The tumor microenvironment (TME), a complex heterogeneous mixture of various cellular, physical, and biochemical components and signals, is a major player in the process of tumor growth and its response to therapeutic methods. In vitro 2D monocellular cancer models are unable to mimic the complex in vivo characteristics of cancer TME involving cellular heterogeneity, presence of extracellular matrix (ECM) proteins, as well as spatial orientation and organization of different cell types forming the TME. In vivo animal-based studies have ethical concerns, are expensive and time-consuming, and involve models of non-human species. In vitro 3D models are capable of tiding over several issues associated with both 2D in vitro and in vivo animal models. We have recently developed a novel zonal multicellular 3D in vitro model for pancreatic cancer involving cancer cells, endothelial cells, and pancreatic stellate cells. Our model (i) can provide long-term culture (up to 4 weeks), (ii) can control the ECM biochemical configuration in a cell specific manner, (iii) shows large amounts of collagen secretion by the stellate cells mimicking desmoplasia, and (iv) expresses cell-specific markers throughout the whole culture period. This chapter describes the experimental methodology to form our hybrid multicellular 3D model for pancreatic ductal adenocarcinoma, including the immunofluorescence staining on the cell culture.

Key words Pancreatic cancer, 3D in vitro model, Multicellular, Desmoplasia, Fibrosis

1 Introduction

The development and assessment of cancer therapeutic methods in vitro require appropriate niche-mimicking models that can recapitulate as accurately as possible various aspects of the tissue environment. More specifically, ideal models must be capable of providing cell–cell and cell–matrix cross talk, spatial architecture, mechanical stiffness, as well as realistic biochemical gradients. Unlike 2D cell cultures/models, in vitro 3D models can provide many of these essential characteristics while circumventing issues associated with in vivo animal models like high cost, crossing over different species and ethical concerns related to use of animals in research [1–5].

Dania Movia and Adriele Prina-Mello (eds.), *Cancer Cell Culture: Methods and Protocols*,
Methods in Molecular Biology, vol. 2645, https://doi.org/10.1007/978-1-0716-3056-3_13,
© The Author(s), under exclusive license to Springer Science+Business Media, LLC, part of Springer Nature 2023

To date, there are three main platforms for the development of 3D in vitro cancer models: (1) spheroids, i.e., cell aggregates, (2) hydrogel-based systems, and (3) polymeric scaffold-based systems [6–9]. The selection of an appropriate 3D in vitro platform depends on the cell type and origin, the cell quantity, the experimental timeline, and the experimental question to be answered [10]. Polymeric scaffold-based in vitro models allow for a robust control of internal structure (pore shape, size, and interconnectivity), provide high mechanical strength, and allow for the introduction of fluid flow within the system [5, 8, 11, 12]. Additionally, surface modification of the scaffolds with proteins enables extracellular matrix (ECM) mimicry, which is crucial in tissue and/or tumor development. These characteristics make polymeric scaffolds a promising platform for in vitro tumor niche studies.

Pancreatic Ductal Adenocarcinoma (PDAC) is a disease of the pancreatic duct cells and has a high mortality rate with the current 5-year survival rate being at only 10% [13, 14]. Such dismal statistics of PDAC are attributed to its late-stage diagnosis and resistance to currently available treatment methods. The high treatment resistance is partly related to the very complex tumor microenvironment (TME) of PDAC. PDAC's TME contains a plethora of cellular, biochemical, biomechanical, and structural components which all interact in a complex pattern, leading to the progression of the disease. For example, the stellate cells of the pancreas, when activated during the diseased state, secrete an extensive amount of Collagen I ECM protein, resulting in a dense fibrotic reaction also known as desmoplasia. This increase of collagen and other ECM proteins, including fibronectin, results in the formation of a dense mass leading to increased stiffness, the formation of aberrant blood vessels and/or their collapse, and extreme hypoxic tumors. All those changes promote treatment resistance [5, 15, 16]. Therefore, in vitro 3D PDAC models that can capture various TME aspects are essential for in-depth studies of the disease, such as studies focused on the TME link to the disease progression, its resistance to therapies and its effect on the efficiency of novel therapeutic methods. Various groups have developed 3D in vitro models for PDAC using spheroids [17–20], hydrogels [21, 22], and polymeric scaffolds [5, 23, 24].

In this methodology chapter, we describe a recently developed novel hybrid multicellular model of PDAC, using polyurethane-based polymeric scaffolds, pancreatic cancer cells, pancreatic (activated) stellate cells, and endothelial cells. This model allows for long-term culture (up to 4 weeks), mimicks of in vivo cell and TME behavior, and can be used for therapeutic assessment [25]. All three cell types used for the development of this model are proliferative throughout the culture period (Fig. 1a) and express cell-specific markers (Fig. 1b). Additionally, the model is also able to mimic in its stromal compartment the desmoplastic reaction, which is a

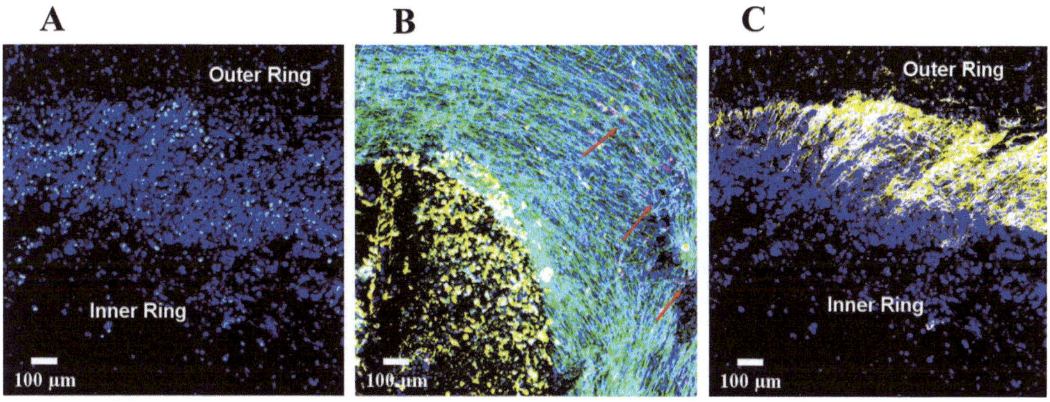

Fig. 1 Representative immunofluorescence CLSM images of a zonal multicellular model of PDAC. (**a**) Ki-67 positive (green) proliferative cells. (**b**) cellular distribution of PANC-1 (pan-Cytokeratin, yellow), HMEC (CD-31, red), and PS-1 (αSMA, green). (**c**) collagen I (human) ECM protein secretion. All cell nuclei were stained with DAPI (blue). (Image reproduced from Ref. [25])

hallmark for PDAC (Fig. 1c). A schematic of the hybrid multicellular (tri-culture) PDAC model can be seen in Fig. 2.

Figure 2 schematically outlines the methodology for the development of the zonal hybrid model for PDAC.

2 Materials

1. Clinical grade polyurethane (PU) beads.
2. 99.8% anhydrous pure 1, 4 Dioxane
3. PANC-1 pancreatic cancer cell line.
4. PS-1 pancreatic stellate cells [20].
5. Human microvascular endothelial cell (HMEC) line.
6. DMEM: DMEM high glucose medium.
7. MCDB131: MCDB131 medium without L-glutamine.
8. L-glutamine.
9. Fetal bovine serum (FBS).
10. Antibiotic-antimycotic (e.g., penicillin-streptavidin).
11. Epidermal Growth Factor (EGF).
12. Hydrocortisone.
13. DMEM complete medium (for PANC-1 and PS-1 cell lines): DMEM; 10% FBS; 1% antibiotic-antimycotic; 10 mM L-glutamine.
14. MCDB131 complete medium (for HMEC cell line): MCDB131; 10% FBS; 1% antibiotic-antimycotic; 10 ng/ml EGF); 1 μg/mL hydrocortisone; 10 mM L-glutamine (*see* **Note 1**).

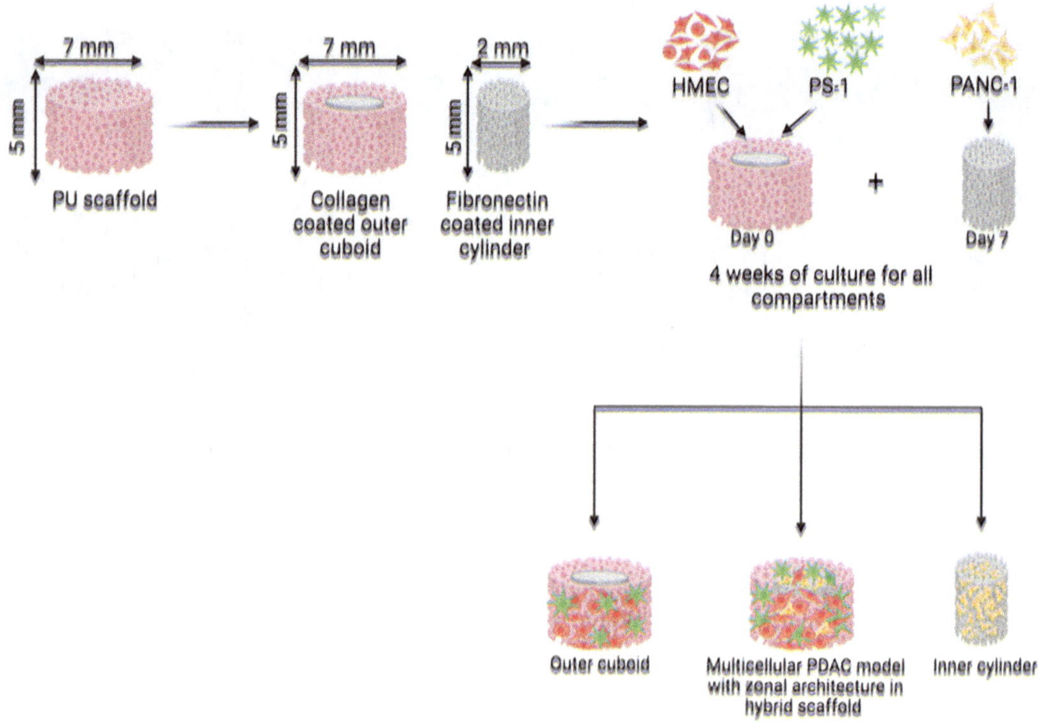

Fig. 2 Schematic presentation of the development of 3D in vitro hybrid PDAC model. The model contains PANC-1 cancer cells (yellow), HMEC endothelial cells (red), and PS-1 stellate cells (green) on ECM protein-coated PU scaffolds. Figure created using BioRender

15. Cell dissociation reagent.
16. Phosphate Buffered Saline (PBS).
17. Fibronectin: 25 μg/ml Bovine Plasma Fibronectin, PBS.
18. Collagen: 25 μg/ml Calf Skin Collagen I, PBS.
19. Trypan Blue.
20. Live-dead staining solution: 2 μM calcein, 4 μM ethidium homodimer, DMEM or MCDB131 complete medium.
21. 70% Ethanol: 70% ethanol, 30% sterile water
22. T75, T175 flasks.
23. 24, 48 well multi-well cell culture plates
24. Sterile tweezers.
25. Scaffold Puncher (3–4 mm diameter).
26. Non-sterile razor blades or scalpel with blades.
27. 15 ml and 50 ml conical tubes
28. Cryovials.
29. Liquid Nitrogen.

3 Method

3.1 Development of Scaffolds for Cancer and Stromal Compartment of Hybrid Model

1. Prepare PU scaffolds via Thermal-Induced Phase Separation (TIPS) as per the published protocol [5, 9, 11, 12, 23, 25, 26].
2. Cut scaffolds into 7 × 5 mm cuboids using razor blades or scalpels.
3. Use the scaffold puncher at the center of each cuboid to punch out the cylindrical inner cancer compartment. The remaining hollow cuboid is the outer stromal compartment.
4. Sterilize the scaffolds by exposing them to 70% ethanol (3 h) and UV ray (1 h).
5. Incubate the two compartments with the appropriate DMEM or MCDB131 complete medium in the cell incubator (37 °C, 5% CO_2) for up to 48 h to confirm the sterility of the scaffolds. Thereafter, handle the scaffolds under aseptic conditions only (under biosafety cabinet, handling with sterile tweezers).
6. Transfer the scaffolds into sterile conical tubes using sterile tweezers and add sterile PBS just enough to completely submerge all scaffolds. Use separate tubes for the inner and outer compartments. Centrifuge at 3630× *g* for 10 min.
7. Remove PBS from the tubes and add 25 μg/ml of Fibronectin and 25 μg/ml Collagen I solution to the cancer and stromal compartment scaffolds, respectively. Centrifuge at 1420× *g* for 20 min for coating the scaffolds with appropriate ECM protein (*see* **Note 2**).
8. Remove the ECM protein solution and add PBS, enough to completely submerge the scaffolds.
9. Centrifuge at 910× *g* for 10 min to complete the ECM coating process.
10. Store the scaffolds at 4 °C till use (up to 1 month).

3.2 Cell Seeding on PU Scaffolds and Formation of Hybrid PDAC Model

1. Grow PANC-1, HMEC, and PS-1 cells in appropriate media using T flask till required cell density is reached.
2. At time point Day 0, dissociate the HMEC and stellate cells, centrifuge and count cells using trypan blue (*see* **Note 3**).
3. Re-suspend in complete medium (*see* **Note 4**) and mix the two cell types at appropriate seeding density (*see* **Note 5**) and ratio using the required amount of medium (about 30 μl/scaffold of media volume should be used).

4. Transfer the Collagen I-coated outer compartments to a 24-well plate using sterile tweezers. Gently squeeze out excess PBS from the scaffolds prior to transferring them to the wells.
5. Slowly add the cell suspension mixture on to the scaffolds in a drop wise manner at appropriate seeding density. About 30 μl of cell suspension should be added to each scaffold (*see* **Note 6**).
6. Transfer the plate to a humidified cell incubator (37 °C/ 5% CO_2) and incubate the scaffolds for about 30 min.
7. After 30 min, add slowly from the side, 1.5 ml of complete medium (*see* **Note 4**) to each well, and return the plate to the cell incubator (37 °C/ 5% CO_2) for 24 h.
8. At time point Day 6, dissociate PANC-1 cells, centrifuge at 250× *g* for 5 mins, and count cells using trypan blue.
9. Re-suspend cells in complete medium (*see* **Note 4**) at appropriate seeding density (*see* **Note 5**) using the required amount of medium (about 10 μl/ scaffold of media volume should be used).
10. Transfer the Fibronectin-coated inner compartments to a 48-well plate using sterile tweezers. Gently squeeze out excess PBS from the scaffolds prior to transferring them to the wells.
11. Slowly seed PANC-1 cells on to the scaffolds in a drop-wise manner at appropriate seeding density. About 10 μl of cell suspension should be added to each scaffold.
12. Transfer the plate to a humidified cell incubator (37 °C/ 5% CO_2) and incubate the scaffolds for about 30 min.
13. Add slowly from the side, 1 ml of complete medium (*see* **Note 4**) to each well, and return the plate to the cell incubator (37 °C/ 5% CO_2) and grow the cells for 24 h.
14. After 24 h incubation (Day 7), transfer the inner cancer compartment within the hollow of the stromal compartment very gently with sterile tweezers. Add 1.5 ml of fresh appropriate complete medium (*see* **Note 4**) and culture the hybrid model for up to 3 weeks.
15. Change medium every alternate day and change culture plate every 7 days.
16. At appropriate time points, depending on the experimental design, transfer scaffolds to cryovials, and snap freeze by immersing them in liquid nitrogen for about 10 min.
17. Store scaffolds at −80 °C till further use for analysis (up to 6 months).

3.3 Live–Dead Analysis of Hybrid Scaffolds

1. Prepare a solution of live-dead stain using complete medium (*see* **Note 4**). Transfer the solution to the required number of wells in a 24-well plate (about 500 μl of solution per well). Keep away from light, cover the vial or tube with aluminum foil.
2. Take out samples to be analyzed from −80 °C freezer and keep them in liquid nitrogen.
3. Separate the inner and outer compartment very gently using thin tweezers. Follow steps 4 and 5 individually for the two compartments.
4. Transfer a single scaffold on to a sterile flat surface and cut thin sections using a razor blade (*see* **Note 7**).
5. Transfer the slices immediately to the wells containing Live–Dead staining solution using a tweezer. 1–3 slices can be added to each well.
6. Once all samples are added, cover the well plate with aluminum foil and incubate for 1 h at 37 °C and 5% CO_2.
7. Remove the staining solution and wash the scaffold slices 3× with PBS.
8. Add complete medium to the samples and image immediately using a Confocal Laser Scanning Microscope (CLSM).
9. Scaffolds can be prepared similarly for immunofluorescence staining to detect cell-specific markers, cell-secreted ECM proteins, proliferative markers, apoptotic markers, etc. (*see* **Note 8**) [25].

4 Notes

1. When preparing a co-culture (HMEC and stellate cells) or the hybrid model (PANC-1, HMEC, and stellate cells), use 2× the concentration for EGF and hydrocortisone to prepare the MCDB131 complete medium.
2. Do not reuse the ECM protein working solution once it has been used for scaffold coating.
3. The time points for cell seeding and seeding density are model- and cell-dependent. They can be optimized as per requirements.
4. For the co-culture (HMEC and stellate cells) and the hybrid model (PANC-1, HMEC, and stellate cells) use complete DMEM and MCDB131 medium in a 1:1 ratio.
5. Seeding density for each cell line should be decided based on the ratio of the three cell types needed within the model. For our work, we have used three different cell PANC-1:HMEC:

PS-1 ratios: 1:1:1, 1:2:2, and 1:2:9. At least 0.25×10^6 cells/scaffold should be used for the inner cancer compartment. Cell density can be increased depending on the rate of cell proliferation and doubling time of the cell types used.

6. While adding the cells on to the scaffolds, mix the cell suspension regularly to maintain a homogenous cell suspension. Add the cells in a drop-wise manner and make sure that the media is soaked in by the scaffolds.
7. Wear double gloves while cutting scaffolds to avoid injury to fingers. Make sure sections are cut uniformly and relatively thin. Thick samples will not be visible under confocal microscopy.
8. Pan cytokeratin, CD-31, and α-smooth muscle actin (SMA) can be used as cell-specific markers for PANC-1, HMEC, and PS-1 cell lines, respectively.

Acknowledgments

E.V. is grateful to the Royal Academy of Engineering for an Industrial Fellowship and to the Medical Research Council UK for a New Investigator Research Grant (MR/V028553/1), which also financially supports P.G.

The authors would like to dedicate this chapter in memory of Dr. Alan Millington of the University of Surrey UK, who, sadly, passed away after battling pancreatic cancer in March 2022.

References

1. Chim LK, Mikos AG (2018) Biomechanical forces in tissue engineered tumor models. Curr Opin Biomed Eng 6:42–50. https://doi.org/10.1016/j.cobme.2018.03.004
2. Adcock AF, Trivedi G, Edmondson R, Spearman C, Yang L (2015) Three-dimensional (3D) cell cultures in cell-based assays for in-vitro evaluation of anticancer drugs. J Anal Bioanal Tech 6(3):1
3. Bermejo-Rodríguez C, Pérez-Mancera PA (2015) Use of DNA transposons for functional genetic screens in mouse models of cancer. Curr Opin Biotechnol 35:103–110
4. Pérez-Mancera PA, Guerra C, Barbacid M, Tuveson DA (2012) What we have learned about pancreatic cancer from mouse models. Gastroenterology 142(5):1079–1092
5. Totti S, Allenby MC, Dos Santos SB, Mantalaris A, Velliou EG (2018) A 3D bioinspired highly porous polymeric scaffolding system for in vitro simulation of pancreatic ductal adenocarcinoma. RSC Adv 8(37):20928–20940. https://doi.org/10.1039/C8RA02633E
6. Nyga A, Cheema U, Loizidou M (2011) 3D tumour models: novel in vitro approaches to cancer studies. J Cell Commun Signal 5(3):239–248
7. Rodrigues J, Heinrich MA, Teixeira LM, Prakash J (2021) 3D in vitro model (r)evolution: unveiling tumor–stroma interactions. Trends Cancer 7(3):249–264. https://doi.org/10.1016/j.trecan.2020.10.009
8. Totti S, Vernardis SI, Meira L, Pérez-Mancera PA, Costello E, Greenhalf W et al (2017) Designing a bio-inspired biomimetic in vitro system for the optimization of ex vivo studies of pancreatic cancer. Drug Discov Today 22(4):690–701. https://doi.org/10.1016/j.drudis.2017.01.012
9. Wishart G, Gupta P, Schettino G, Nisbet A, Velliou E (2021) 3d tissue models as tools for

radiotherapy screening for pancreatic cancer. Br J Radiol 94(1120):20201397. https://doi.org/10.1259/bjr.20201397

10. Gupta P, Miller A, Olayanju A, Madhuri TK, Velliou E (2022) A systematic comparative assessment of the response of ovarian cancer cells to the chemotherapeutic cisplatin in 3D models of various structural and biochemical configurations-does one model type fit all? Cancers 14(5):1274
11. Blanco TM, Mantalaris A, Bismarck A, Panoskaltsis N (2010) The development of a three-dimensional scaffold for ex vivo biomimicry of human acute myeloid leukaemia. Biomaterials 31(8):2243–2251. https://doi.org/10.1016/j.biomaterials.2009.11.094
12. Velliou EG, Dos Santos SB, Papathanasiou MM, Fuentes-Gari M, Misener R, Panoskaltsis N et al (2015) Towards unravelling the kinetics of an acute myeloid leukaemia model system under oxidative and starvation stress: a comparison between two-and three-dimensional cultures. Bioprocess Biosyst Eng 38(8): 1589–1600
13. O'reilly D, Fou L, Hasler E, Hawkins J, O'connell S, Pelone F et al (2018) Diagnosis and management of pancreatic cancer in adults: a summary of guidelines from the UK National Institute for Health and Care Excellence. Pancreatology 18(8):962–970
14. Society A. American Cancer Society. Cancer Facts & Figures 2020. Am Cancer Soc 2020: 1–52
15. Ansari D, Friess H, Bauden M, Samnegård J, Andersson R (2018) Pancreatic cancer: disease dynamics, tumor biology and the role of the microenvironment. Oncotarget 9(5):6644
16. Longo V, Brunetti O, Gnoni A, Cascinu S, Gasparini G, Lorusso V et al (2016) Angiogenesis in pancreatic ductal adenocarcinoma: a controversial issue. Oncotarget 7(36):58649
17. Boj SF, Hwang C-I, Baker LA, Chio IIC, Engle DD, Corbo V et al (2015) Organoid models of human and mouse ductal pancreatic cancer. Cell 160(1-2):324–338
18. Brancato V, Comunanza V, Imparato G, Corà D, Urciuolo F, Noghero A et al (2017) Bioengineered tumoral microtissues recapitulate desmoplastic reaction of pancreatic cancer. Acta Biomater 49:152–166
19. Di Maggio F, Arumugam P, Delvecchio FR, Batista S, Lechertier T, Hodivala-Dilke K et al (2016) Pancreatic stellate cells regulate blood vessel density in the stroma of pancreatic ductal adenocarcinoma. Pancreatology 16(6): 995–1004
20. Froeling FE, Mirza TA, Feakins RM, Seedhar A, Elia G, Hart IR et al (2009) Organotypic culture model of pancreatic cancer demonstrates that stromal cells modulate E-cadherin, β-catenin, and ezrin expression in tumor cells. Am J Pathol 175(2):636–648
21. Chiellini F, Puppi D, Piras AM, Morelli A, Bartoli C, Migone C (2016) Modelling of pancreatic ductal adenocarcinoma in vitro with three-dimensional microstructured hydrogels. RSC Adv 6(59):54226–54235
22. Chang C-Y, Lin C-C (2021) Hydrogel models with stiffness gradients for interrogating pancreatic cancer cell fate. Bioengineering 8(3):37
23. Gupta P, Totti S, Pérez-Mancera PA, Dyke E, Nisbet A, Schettino G et al (2019) Chemoradiotherapy screening in a novel biomimetic polymer based pancreatic cancer model. RSC Adv 9(71):41649–41663. https://doi.org/10.1039/C9RA09123H
24. Ricci C, Mota C, Moscato S, D'Alessandro D, Ugel S, Sartoris S et al (2014) Interfacing polymeric scaffolds with primary pancreatic ductal adenocarcinoma cells to develop 3D cancer models. Biomatter 4(1):e955386
25. Gupta P, Pérez-Mancera PA, Kocher H, Nisbet A, Schettino G, Velliou EG (2020) A novel scaffold-based hybrid multicellular model for pancreatic ductal adenocarcinoma—toward a better mimicry of the in vivo tumor microenvironment. Front Bioeng Biotechnol 8(290). https://doi.org/10.3389/fbioe.2020.00290
26. Wishart G, Gupta P, Nisbet A, Schettino G, Velliou E (2021) On the evaluation of a novel hypoxic 3D pancreatic cancer model as a tool for radiotherapy treatment screening. Cancers (Basel) 13(23). https://doi.org/10.3390/cancers13236080

Chapter 14

Measuring Immune Cell Movement Toward the Soluble Microenvironment of Human Tissues Using a Boyden Chamber-Based Migration Assay

Eimear Mylod, Joanne Lysaght, and Melissa J. Conroy

Abstract

Migration assays are used to measure cell movement toward a variety of chemoattractants in a controlled environment. Here we describe a method for a Boyden chamber-based migration assay using conditioned media generated from the tumor, liver, and visceral adipose tissue of cancer patients.

Key words Chemotaxis, Chemokines, Immune cells, Boyden chamber, Cell migration

1 Introduction

The migration of cells is an integral process required from embryonic development through to maintenance of homeostasis in adulthood. Cell migration is central to the immune response, wound healing, and angiogenesis. It is also a common feature of pathological processes, such as cancer cell metastasis or chronic inflammation [1].

Cell movement can take many forms. Chemotaxis is the directional movement of cells on the basis of a gradient of a chemoattractant [2, 3]. Haptotaxis is the directional movement of cells in response to adhesive substrates such as membrane-bound chemoattractants expressed on the extracellular matrix [2, 3]. Chemokinesis is random movement following the application of an agent [2]. Chemokines are a family of small proteins which drive the migration of cells and play a crucial role in the positioning of immune cells [4].

Boyden chambers, also referred to as transwells, were first introduced by Boyden for the analysis of leukocyte migration [5]. Using Boyden chambers is a widely accepted method to analyze cell migration. Boyden chambers contain a porous membrane

Dania Movia and Adriele Prina-Mello (eds.), *Cancer Cell Culture: Methods and Protocols*,
Methods in Molecular Biology, vol. 2645, https://doi.org/10.1007/978-1-0716-3056-3_14,
© The Author(s), under exclusive license to Springer Science+Business Media, LLC, part of Springer Nature 2023

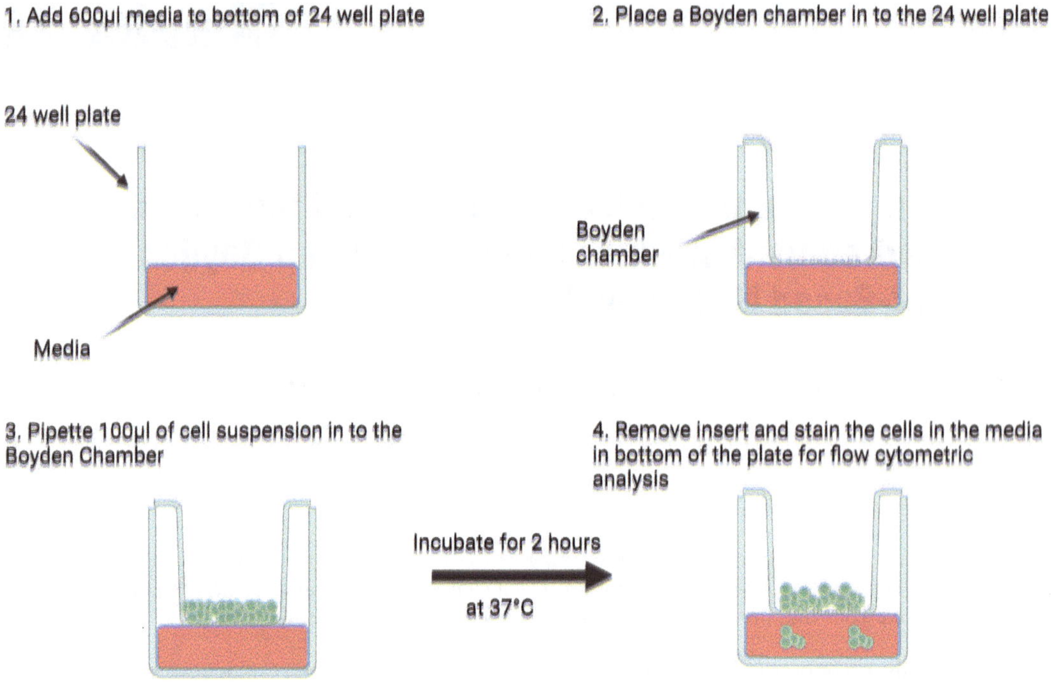

Fig. 1 Schematic depicting the protocol for a Boyden chamber chemotaxis assay. Created with Biorender.com

of a defined size, which is sealed on either end. To study chemotaxis, the chamber is then placed in a larger well, containing media with or without a chemoattractant. After an appropriate incubation time, the frequency and type of cells which have migrated can be quantified (Fig. 1).

The study of cell movement is of great interest in a range of different conditions, including cancer, obesity, multiple sclerosis, arthritis, asthma, viral infection, and COPD [6–8]. Chemokine receptor antagonists are commonly studied for their therapeutic utility to stunt inflammatory immune cell migration and attenuate the pathology of inflammatory diseases. As such, unearthing the key chemokines which drive immune cell movement around the body in these diseases is imperative [6–11]. Boyden Chambers are often employed prior to in vivo studies as a crude means to identify key chemokine pathways driving immune cell migration in inflammatory diseases and in the early development of chemokine-targeted therapies [9, 10, 12, 13].

Here we will describe how our group has generated and utilized conditioned media derived from the visceral adipose tissue (VAT), liver, and tumor of esophageal adenocarcinoma patients to (1) recapitulate the soluble microenvironment of these tissues, (2) measure immune cell migration toward these soluble microenvironments, and (3) identify the key chemokine pathways involved in such immune cell migration, using Boyden chambers and chemokine receptor antagonists [9, 10, 13–16]. In this protocol, we describe the specific example of the migration of lymphocytes toward tissue-conditioned media following pre-treatment with antagonists. This protocol can be easily adapted to different experimental setups and cell types (*see* **Note 1**). Conditioned media can be generated from a number of tissues, including the muscle, subcutaneous adipose tissue, and the lymph node.

2 Materials

2.1 Reagents

1. Phosphate buffered saline (PBS)
2. 10 mg/mL Gentamycin
3. 5.5 mM glucose: 99 mg glucose, 100 mL distilled sterile water
4. 0.1% Sodium azide: 1 g sodium azide, 100 mL distilled sterile water
5. Ficoll Paque

2.2 Beads and Antibodies

1. CD3/CD28 Dynabeads (*see* **Note 2**)
2. Flow Cytometry cell counting beads
3. Flow Cytometry antibodies

2.3 Plastics

1. Petri Dishes
2. Flow Tubes
3. 24-well plates
4. Sterile Pasteur pipettes
5. 5 μm pore Transwell® inserts (*see* **Note 3**)
6. Cryovials
7. 70 μm mesh filters
8. 50 mL centrifuge tubes 9. 30 mL Sterilin Universal Container

2.4 Equipment

1. Tweezers
2. Scalpel
3. Scissors
4. Weighing scales
5. Laminar flow hood
6. Cell culture incubator
7. Flow Cytometer
8. DynaMag™-2 Magnet (*see* **Note 2**)

2.5 Media and Supplements

1. Supplemented M199 media: M199 cell media, 50 μg/mL Gentamycin
2. Supplemented RPMI media: RPMI cell media, 10% fetal bovine serum (FBS), 1% penicillin/streptomycin

3. Serum-free RPMI: RPMI cell media, as purchased
4. FBS positive control: 20% FBS, 80% M199 cell media

2.6 Buffers

1. Adipose transfer buffer: 5.5 mM glucose, 50 μg/mL gentamycin, PBS
2. Flow cytometry buffer: 2% FBS, 0.1% sodium azide, PBS
3. Flow cytometry blocking buffer: 50% FBS, 50% flow cytometry buffer

3 Methods

All procedures must be performed in a sterile laminar flow hood, unless otherwise stated.

3.1 Peripheral Blood Mononuclear Cell (PBMC) Isolation

1. Dilute whole blood from consented healthy individuals 1: 2 with PBS.
2. Using a Pasteur pipette, layer 35 mL of diluted blood on top of 15 mL of ficoll paque in a 50 mL tube.
3. Centrifuge at 400× *g* for 25 min.
4. Carefully transfer the layer of PBMC (which is below the plasma layer) to a new 50 mL tube using a sterile Pasteur pipette.
5. Resuspend the PBMC in 50 mL of PBS and centrifuge at 800× *g* for 5 min.
6. Discard the supernatant by inverting the tube, and vortex the pellet.
7. Resuspend the pellet in 50 mL of PBS and centrifuge at 400× *g* for 10 min.
8. Discard the supernatant by inverting the tube, and vortex the pellet.
9. Resuspend the pellet in 2 mL RPMI, and proceed to counting the PBMC.

3.2 Activation and Treatment of Lymphocytes for Boyden Chamber Assay

1. To analyze T cell migration, incubate PBMC for 5 days at a 1:1 ratio with CD3/CD28 T-cell expander Dynabeads to activate the T cells.
2. Following activation, remove the beads as per manufacturers' protocol, and resuspend the activated T cells at 200,000 cells/ 100 μL with serum-free RPMI. Seed cells in a 96 well plate at a volume of 100 μL per well.
3. To analyze natural killer (NK) cell migration, no prior activation is needed, and pre-treatment with antagonists can begin immediately after PBMC are isolated.

4. Resuspend PBMC at a density of 200,000 cells/100 μL with serum-free RPMI. Seed cells in a 96-well plate at a volume of 100 μL per well (*see* **Note 4**).
5. To antagonise a chemokine receptor, pre-treat PBMC in the appropriate wells with the specific antagonist for 1 h prior to the migration assay (*see* **Note 5**).

3.3 Visceral Adipose Tissue-Conditioned Media Generation

1. Immediately transfer visceral adipose tissue (VAT) sample collected from patients under anesthetic at the start of surgery, into a universal container of adipose transfer buffer.
2. Place the VAT sample under a laminar flow hood and wash with PBS to remove blood contamination.
3. Weigh 5 g of VAT and chop it into small pieces of approximately 0.5 g on a petri dish using sterile scissors (*see* **Note 6**).
4. Transfer the chopped VAT to a sterile petri dish and add 10 mL of supplemented M199 media.
5. Place in an incubator at 37 °C/5% CO_2 for 72 h.
6. Discard VAT, and sterile filter the conditioned media through a 70 μm mesh filter to ensure no pieces of VAT remain in the conditioned media.
7. Aliquot the adipose tissue-conditioned media (ACM) into labelled cryovials and cryopreserve the vials of ACM at −80 °C for long-term storage.

3.4 Liver Tissue-Conditioned Media Generation

1. Immediately transfer liver tissue samples collected from patients under anesthetic at the commencement of surgery, into a universal container of adipose transfer buffer.
2. Place the liver sample under a laminar flow hood and wash with PBS to remove blood contamination.
3. Weigh the liver sample and chop into small pieces using a scissors or scalpel.
4. Transfer the chopped liver to a 24 well-plate, and add 350 μL of supplemented M199 media every 0.1 g of the liver.
5. Place in an incubator at 37 °C/5% CO_2 for 72 h.
6. Discard the liver, and sterile filter the conditioned media through a 70 μm mesh filter to ensure no pieces of the liver remain in the conditioned media.
7. Aliquot the liver tissue-conditioned media (LCM) into labelled cryovials and cryopreserve the vials of LCM at −80 °C for long-term storage.

3.5 Tumor-Conditioned Media Generation

1. Place tumor biopsies, sampled from the esophageal tumor at the time of surgical resection, under a laminar flow hood and weigh them using a fine scale.
2. Place a tumor biopsy in a well of a 24-well plate.

3. Add 1 ml of supplemented M199 media every 100 mg of biopsy.
4. Place in an incubator at 37 °C/5% CO_2 for 24 h.
5. Gently remove the biopsy from the plate using sterile tweezers. Snap freeze in a cryovial in liquid nitrogen and place at −80 °C for long-term storage.
6. Aliquot the tumor tissue-conditioned media (TCM) into labelled cryovials and cryopreserve the TCM at −80 °C for long term storage.

3.6 Boyden Chamber Immune Cell Migration Assay

1. Defrost and warm ACM, LCM, TCM, M199, and FBS to 37 °C.
2. Prepare the 20% FBS positive control.
3. Under a laminar flow hood, add 600 μL of ACM, LCM, TCM, M199, or M199 + 20% FBS to their respective wells of a 24-well plate, as seen in the experimental layout in Table 1 (*see* **Note 7**).
4. Gently place each transwell insert into each well of the 24-well plate using forceps.
5. Gently and slowly transfer the immune cells resuspended in 100 μL of serum-free RPMI into each transwell, using a pipette (*see* **Note 8**).
6. Close the lid and incubate the plate at 37 °C/5% CO_2 (*see* **Note 9**) for 2 h.

Table 1
Experimental layout of the Chemotaxis Assay

Well	Top chamber (lymphocytes)	Bottom chamber (control or conditioned media)	Purpose
1	Untreated cells	M199	Negative control
2	Untreated cells	M199 + 20% FBS	Positive control
3	Untreated cells	ACM	Baseline migration toward ACM
4	Cells pre-treated with antagonist	ACM	Experimental
5	Untreated cells	LCM	Baseline migration toward LCM
6	Cells pre-treated with antagonist	LCM	Experimental
7	Untreated cells	TCM	Baseline migration toward TCM
8	Cells pre-treated with antagonist	TCM	Experimental

3.7 Flow Cytometric Staining

1. Gently remove the plate from the incubator and transfer to laminar flow hood.
2. Carefully remove the transwells using forceps and discard them.
3. Remove the media in the bottom well (which should contain the migrated cells) and place in a tube suitable for flow cytometric analysis (referred to here as a flow tube).
4. Add 1 mL of flow cytometry buffer to each tube and centrifuge at 120× *g* for 5 min.
5. Invert flow tube to discard supernatant.
6. Resuspend the cell pellet in 1 mL of flow cytometry-blocking buffer and incubate at room temperature for 5 min.
7. Add 1 mL of flow cytometry buffer to each tube and centrifuge at 120× *g* for 5 min.
8. Invert flow tube to discard supernatant.
9. Repeat **steps** 7 and **8**.
10. Invert flow tube to discard supernatant and resuspend in 100 μL of flow cytometry buffer.
11. If using a mixed cell population such as PBMC, add appropriate antibodies and incubate at 4 °C for 30 min in the dark (*see* **Note 10**).
12. Add 1 mL of flow cytometry buffer to each tube, and centrifuge at 120× *g* for 5 min.
13. Invert flow tube to discard supernatant.
14. Repeat **steps 12** and **13**.
15. Invert flow tube to discard supernatant and resuspend in 200 μL of flow cytometry buffer (*see* **Note 11**).

3.8 Flow Cytometric Acquisition and Analysis

1. Set up flow cytometer as appropriate for the fluorescent staining selected.
2. Immediately before acquisition, add 10 μL of counting beads to each tube and vortex.
3. Ensure all liquid within the flow tube is acquired.
4. Gating for the count beads and lymphocytes can be seen in Fig. 2.
5. The absolute number of cells which have migrated can be calculated using the equation below (*see* **Note 12**).

$$\frac{\text{No.of beads per } 10\ \mu\text{L} \times \text{No.of cells acquired}}{\text{Volume of cells acquired} \times \text{no.of beads acquired}}$$

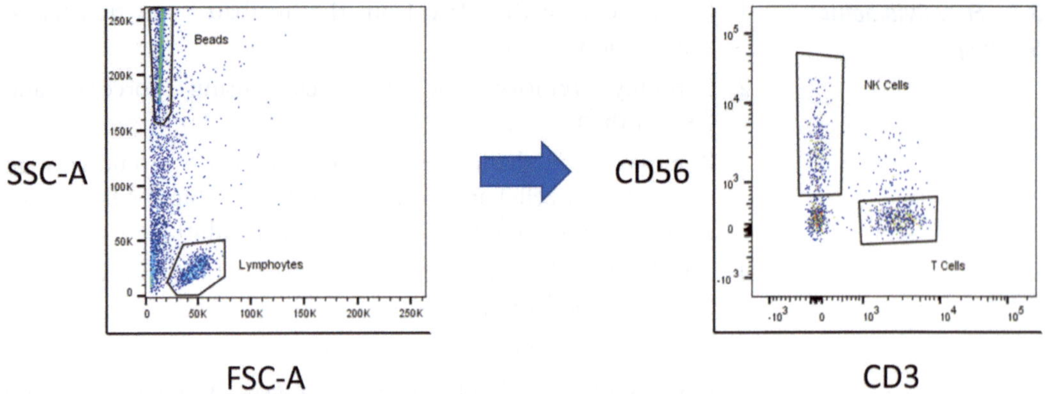

Fig. 2 Representative dot plots. (Left) Representative dot plot showing the locations of the count beads and lymphocytes in an SSC-A vs. FSC-A density plot. (Right) Representative dot plot showing the gating of NK Cells (CD56+CD3-) and CD56- T cells (CD56-CD3+) in a CD56 and CD3 density plot

4 Notes

1. A number of different cell types can be used for migration assays, including cancer cells.
2. Gibco Dynabeads™ Human T-Activator CD3/CD28 for T Cell Expansion and Activation are used for this protocol. To analyze T cell chemotaxis, T cells must first be activated. To remove the Dynabeads, the DynaMag™-2 Magnet is recommended.
3. Boyden chamber transwells are available in a range of pore sizes. For this method, 5 μm pore inserts are recommended. The smaller the pore size, the more challenging it is for cells to migrate through. As such, it is imperative to choose the correct pore size for the cell type of interest. Lymphocyte- and monocyte-based assays generally use filters with 5 μm pore size. The filters with larger 8 μm pore size are typically used for epithelial cells, fibroblasts, and endothelial cells. Filters with smaller 3 μm pore size can be used for neutrophils and eosinophils.
4. It is imperative that the migrating cells are placed in media free from sera such as FBS. If serum-supplemented media is used, then non-specific migration of cells will occur and accurate measurement of chemotaxis will be compromised.
5. The time required to pre-treat the cells will vary depending on the antagonist used. This will need to be optimized for each antagonist.

6. Five grams is used as standard for generating ACM. However, if less is available, a 2:1 ratio of M199 media to weight of VAT can be used. Also, a smaller size plate such as a 6-well plate can be used in place of a petri dish.
7. Ensuring the correct controls are run alongside the experimental wells is integral to ensure that the protocol is working effectively. M199 alone acts as a negative control to account for random migration and gravity. The FBS positive control acts as a positive control because FBS is enriched with chemoattractants and should induce a significant level of cell migration.
8. When adding the cells to the transwells, it is best to slowly pipette down the inside of the transwell where there is a groove, which the pipette tip can be placed into.
9. When transferring the plate to the incubator, do so as gently as possible to avoid disturbing the inserts and the cells. Avoid sudden movements and gently close the incubator door.
10. Depending on the cell type of interest, a number of flow cytometry antibodies can be used. CD3 is a pan T cell marker. CD4 and CD8 can further categorize T cell phenotypes. NK cells can be identified using CD56 and CD3 antibodies.
11. It is vital to be as accurate as possible when resuspending the cells in flow cytometry buffer prior to flow cytometric analysis.
12. The number of beads per μL is noted on the information sheet provided with the count beads, and it is supplier- and lot-specific.

Acknowledgments

The authors would like to thank the patients and staff at the Oesophageal and Gastric Centre at St. James's Hospital for the provision of visceral adipose tissue, liver, and tumor biopsies. The work was performed in accordance with the Code of Ethics of the World Medical Association (Declaration of Helsinki) for experiments involving humans. Patients provided informed consent for sample and data acquisition, and the study received full ethical approval from the St. James's Hospital Ethics Review Board.

References

1. Horwitz R, Webb D (2003) Cell migration. Curr Biol 13:R756–R759
2. Petrie RJ, Doyle AD, Yamada KM (2009) Random versus directionally persistent cell migration. Nat Rev Mol Cell Biol 10:538–549
3. Rikitake Y, Takai Y (2011) Directional cell migration, pp 97–143
4. Proudfoot AEI (2002) Chemokine receptors: multifaceted therapeutic targets. Nat Rev Immunol 2:106–115
5. Boyden S (1962) The chemotactic effect of mixtures of antibody and antigen on polymorphonuclear leucocytes. J Exp Med 115:453–466

6. Jin T, Xu X, Hereld D (2008) Chemotaxis, chemokine receptors and human disease. Cytokine 44:1–8
7. Raman D, Sobolik-Delmaire T, Richmond A (2011) Chemokines in health and disease. Exp Cell Res 317:575–589
8. Gerard C, Rollins BJ (2001) Chemokines and disease. Nat Immunol 2:108–115
9. Mylod E, Melo AM, Donlon NE, Davern M, Bhardwaj A, Reynolds JV, Lysaght J, Conroy MJ (2021) Fractalkine elicits chemotactic, phenotypic, and functional effects on CX3CR1 + CD27 – NK cells in obesity-associated cancer. J Immunol:ji2000987
10. Conroy MJ, Galvin KC, Kavanagh ME, Mongan AM, Doyle SL, Gilmartin N, O'Farrelly C, Reynolds JV, Lysaght J (2016) CCR1 antagonism attenuates T cell trafficking to omentum and liver in obesity-associated cancer. Immunol Cell Biol 94:531–537
11. Horuk R (2009) Chemokine receptor antagonists: overcoming developmental hurdles. Nat Rev Drug Discov 8:23–33
12. O'Boyle G, Fox CRJ, Walden HR et al (2012) Chemokine receptor CXCR3 agonist prevents human T-cell migration in a humanized model of arthritic inflammation. Proc Natl Acad Sci 109:4598–4603
13. Mylod E, O'Connell F, Donlon NE, Butler C, Reynolds JV, Lysaght J, Conroy MJ (2021) The omentum in obesity-associated cancer: a hindrance to effective natural killer cell migration towards tumour which can be overcome by CX3CR1 antagonism. Cancers (Basel) 14: 64
14. Conroy MJ, Fitzgerald V, Doyle SL, Channon S, Useckaite Z, Gilmartin N, O'Farrelly C, Ravi N, Reynolds JV, Lysaght J (2016) The microenvironment of visceral adipose tissue and liver alter natural killer cell viability and function. J Leukoc Biol 100: 1435–1442
15. Conroy MJ, Maher SG, Melo AM, Doyle SL, Foley E, Reynolds JV, Long A, Lysaght J (2018) Identifying a novel role for fractalkine (CX3CL1) in memory CD8+ T cell accumulation in the omentum of obesity-associated cancer patients. Front Immunol. https://doi.org/10.3389/fimmu.2018.01867
16. Kavanagh ME, Conroy MJ, Clarke NE et al (2019) Altered T cell migratory capacity in the progression from barrett oesophagus to oesophageal adenocarcinoma. Cancer Microenviron. https://doi.org/10.1007/s12307-019-00220-6

Chapter 15

A Method for the In Vitro Cytotoxicity Assessment of Anti-cancer Compounds and Materials Using High Content Screening Analysis

Melissa Anne Tutty and Adriele Prina-Mello

Abstract

To date, there is a large bottleneck associated with cancer drug design and development: a lack of appropriate methodologies for screening their potential toxicity. This issue not only causes a high attrition rate for these compounds but also slows down the drug discovery process in general. To overcome this problem, robust, accurate, and reproducible methodologies for assessing anti-cancer compounds are essential. Multiparametric technique and high-throughput analysis, in particular, are favored due to the time- and cost-effective way in which they assess large panels of materials, and due to their large informational output. Following extensive work within our group, we have developed a protocol for assessing the toxicity of anti-cancer compounds using a high-content screening and analysis (HCSA) platform, which is both time-effective and reproducible.

Key words High content screening, In vitro cytotoxicity, A549, HepG2, Fluorescent imaging

1 Introduction

In recent years, the high attrition rate observed in cancer drug development has caused considerable debate around accurate methodologies for the pre-clinical assessment of anti-cancer drugs and delivery systems. To overcome this bottleneck, our group has successfully addressed the methodological gaps currently seen with regard to the accurate and time-effective assessment of cytotoxicity, by establishing a standardized and robust protocol to assess the same in cancer cell lines, using high content screening (HCS) analysis [1]. High Throughput (HTP) assays such as high-content screening analysis, or HCSA (also often referred to as high-content analysis, or HCA), are now becoming more widely used approaches in cell and cancer biology. HCSA has become a robust and established protocol for cytotoxicity assessment [2, 3], due to its multiparametric capabilities and large informational output, and fast

Dania Movia and Adriele Prina-Mello (eds.), *Cancer Cell Culture: Methods and Protocols*,
Methods in Molecular Biology, vol. 2645, https://doi.org/10.1007/978-1-0716-3056-3_15,
© The Author(s), under exclusive license to Springer Science+Business Media, LLC, part of Springer Nature 2023

becoming a gold-standard methodology for the screening of treatments, often anti-cancer treatments [4–8]. The move towards automated microscope-based platforms, which utilize fluorescent-based dyes or assays, in conjunction with automated imaging technology and data analysis in a high-throughput format, can dramatically reduce extracellular interference through direct visualization of subcellular organelles, with this methodology well-suited for large-scale applications such as cancer drug discovery [9]. This methodology provides the flexibility which is required within image-based analysis to allow subcellular data mining, without the issues normally observed with regard to interference in conventional colorimetric cytotoxicity tests. In saying this, HCSA is a convergence between high-resolution fluorescence-based imaging and conventional cell-based assays, providing advanced imaging capabilities via the use of specialist software, combined with data output, which is based on a cell-by-cell basis [10, 11]. This powerful technique also has the capacity to screen hundreds to thousands of compounds due to its fully automated platform, making it an extremely robust and time-effective methodology. With the recent surge in the development of drug compounds and indeed nanomedicines to treat cancer, is it critically important to have well-established and unified screening and analysis platforms that have the capacity to process the influx of these new materials in the drug discovery pipeline, which will ultimately speed up and increase the success rates of safer formulations [10, 12–14].

In our group, we have demonstrated the suitability of HSCA for monitoring the cytotoxic effects of many materials across several cancer cell lines. Currently, there is a broad spectrum of fluorescent dyes and kits to choose from, available from many reputable suppliers (e.g., ThermoFisher, Enzo Life Science, and others). Based on our two decades-long experience, the most appropriate parameters to use as an initial screen of new materials are those that assess membrane permeability (e.g., YO-PRO®-1 iodide), total cell count (e.g., Hoechst 33342), and or subcellular organelles lysosomal mass and/or changes to pH (e.g., LysoTracker®). Hoechst 33342 rapidly and selectively binds to the minor groove of dsDNA, and also preferentially to the adenine-thymine (A-T) rich regions, YO-Pre3O®-1 iodide is a cell membrane impermeable dye which binds to DNA when the membrane integrity of a cell is compromised, and the LysoTracker® dye is a weak base which has free movement into the cell. This dye is protonated and produces a fluorescence upon entering acidic organelles, where it is then retained. Whilst these dyes are the most popular and commonly used markers for HCSA, it is worth noting that many more exist for the same functions. In saying this, our protocol has been optimized for the dyes mentioned above.

In this chapter, we are providing insight into the HCS methodology developed in Trinity College, Dublin, in the past decade and brought forward across several large-scale and network projects for drug compound identification, including the European Commission (EC) MULTIFUN (Multifunctional Nanotechnology for selective detection and Treatment of Cancer) project (project reference ID 262943), and more recently, for interlaboratory comparison under the recently completed EC Horizon 2020 REFINE project (project reference ID 761104).

While data comparison and outcomes assessment has been published separately [8], here below we are reporting on the main protocol for cytotoxicity assessment by means of HCS technology, which has been tested and verified across different nanomaterials, active compounds, medicinal or theranostics product, and cancer treatment.

2 Materials

2.1 Cell Culture Formation

2.1.1 Cell Components

1. Human lung adenocarcinoma epithelial cells (A549 cell line) (*see* **Note 1**)
2. Human hepatocellular carcinoma cells (HepG2 cell line) (*see* **Note 1**)
3. Peripheral blood monocyte cells from acute monocytic leukemia (THP-1 cell line) (*see* **Note 1)**

2.1.2 Reagents

1. Fetal Bovine Serum (FBS) (Merck Life Sciences Ltd., Ireland)
2. Complete DMEM: DMEM medium, 1% penicillin/streptavidin, 10% FBS
3. Agent for the dissociation of adherent mammalian cells from plastic surfaces (e.g., trypsin, TryplE™ etc.) (BioSciences Ltd., Ireland)
4. Agent for washing cell monolayers (e.g., phosphate buffered saline [PBS])

2.1.3 Equipment

1. Chosen plate format, e.g., 96-well plates (*see* **Note 2**) (BioSciences Ltd., Ireland)
2. High content screening plate reader, e.g., Lionheart microscope, (BioTek, Mason Technology, Ireland), or CellInsight™ CX5 High Content Screening (HCS) Platform (Thermo Fisher Scientific, Ireland)
3. A user-friendly software for HCSA and data mining, e.g., Gen5™ Software (BioTek Mason Technology, Ireland) or InCell Investigator Software (GE Healthcare, Buckinghamshire, UK)

2.2 Immunofluorescence Staining for High Content Screening Imaging Analysis

1. Washing solution: 1× PBS (e.g., Gibco™, Thermo Fisher, Ireland)
2. Fixation solution: 3.7% w/w paraformaldehyde, 1× PBS
3. Permeabilization solution: 0.25% Triton-X, 1× PBS
4. Blocking solution: 5% bovine serum albumin (BSA), 0.1% Triton-X, 1× PBS
5. Fluorescent organelle stains: organelle stain (e.g., Lysotracker® red, YO-PRO®1); 1× PBS

2.3 Quantitative Evaluation of Cell Viability by High Content Screening Imaging Analysis

1. Fluorescent nuclear stains: nuclear stain (e.g., Hoechst), 1× PBS
2. Agent for the dissociation of adherent mammalian cells from plastic (e.g., trypsin, TryplE, etc.)
3. Washing solution: 1× phosphate buffered saline (PBS)
4. Fixation solution: 3.7% w/w paraformaldehyde, 1× PBS

3 Methods

Suggested workflow for in vitro cytotoxicity assessment of anticancer compounds and materials using HCSA is shown in Fig. 1 below.

3.1 Cell Culture Formation and Treatment with Compounds

3.1.1 Cell Plating

1. Detach cells from the culturing flask using an agent for the dissociation of adherent mammalian cells from plastic surfaces.
2. Resuspend in appropriate media at a concentration of 1.5×10^6 cells/mL (*see* **Note 1**).
3. Plate cells on chosen plate format, e.g., 96-well plates (*see* **Note 2**) at a concentration of 1.0×10^4 cells/well.
4. Incubate plates overnight in humidified atmosphere at 37 °C and 5% CO_2 to allow cell attachment.

3.1.2 Treatment (or Exposure) of Cultures with Nanomaterials/Compounds/Controls

1. Wash cells with 200 μL/well of 1× PBS before treatment (or exposure) with chosen compounds, delivery system, and/or control samples.
2. Add chosen compounds, delivery system, and/or control samples to the wells in triplicate for the specific time points chosen ($n_{\text{replicates}} = 3$). Compound treatment time points may include 24, 48, and 72 h. Negative control wells contain cell culture medium alone. Positive control wells contain cytotoxic reagents, of known cellular activity, diluted in cell culture medium to provide time-dependent established response. For instance, Tacrine, a positive control for organelle damage, is used at a concentration of 100 μM, and Valinomycin, a positive control for membrane damage, is used at a concentration of 120 μM, for 24 h treatments (Fig. 2).

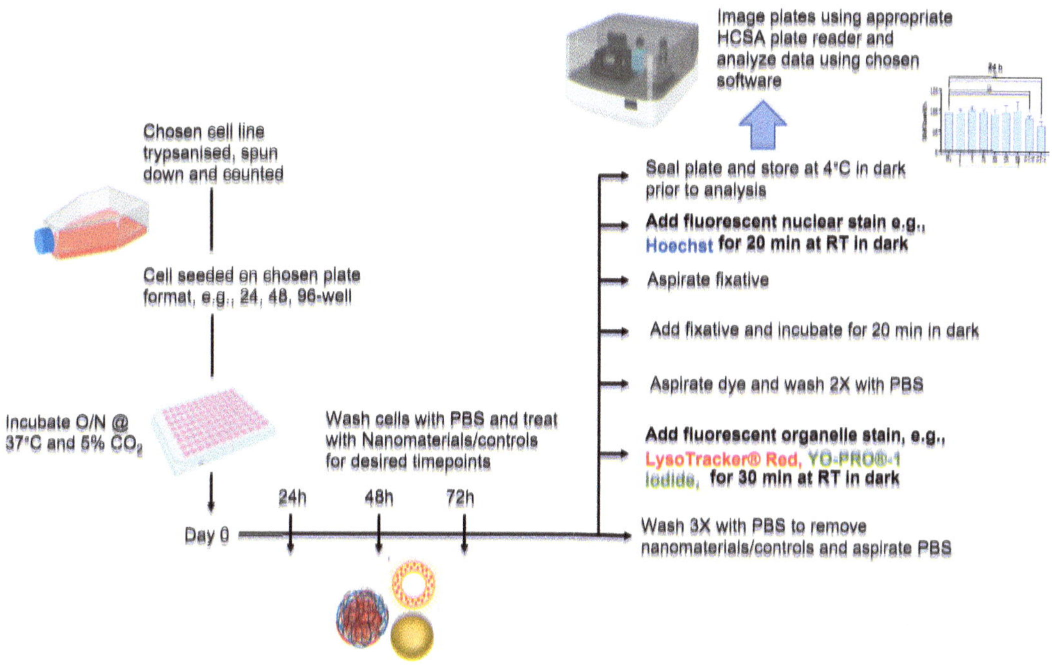

Fig. 1 Suggested workflow for the in vitro cytotoxicity assessment of anti-cancer compounds (and other materials) using HCSA. Briefly, chosen cell line is plated on chosen plate format, i.e., 24, 48, 96 etc., and incubated overnight at 37 °C and 5% CO_2, before cells are incubated with test material. Cells are then washed, stained with appropriate dyes, before being washed and fixed, prior to imaging analysis using HCSA plate reader

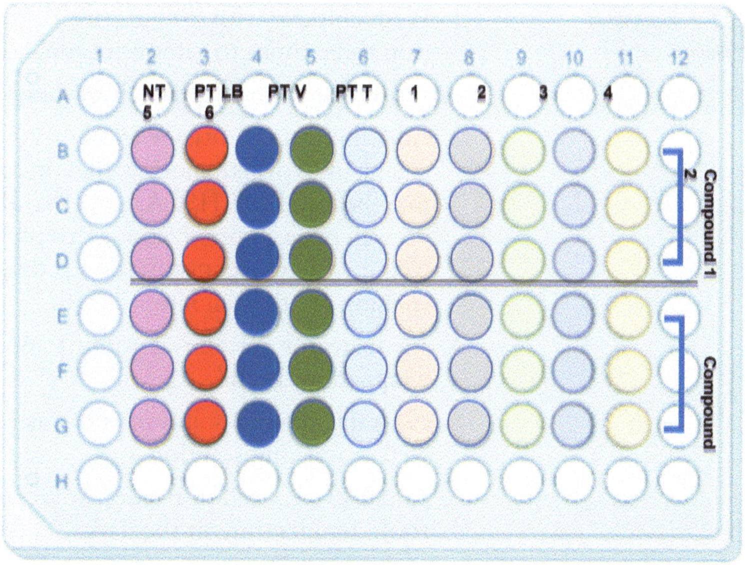

Fig. 2 Sample plate layout for HCSA experiment. Plate layout includes experimental samples (1–6), assay negative control (NT indicating cell culture medium), assay internal positive control (LB indicating lysis buffer), and experimental positive controls (PT–V, where V is Valinomycin, and PT–T, where T is Tacrine)

Table 1
Immunofluorescent targets and suggested dyes for HCSA

Targets	Florescent dyes
Cell membrane permeability	YO-PRO®-1 iodide
Total cell count	Hoechst 33342
Subcellular organelle lysosomal mass/changes in pH	LysoTracker® Red

3. Incubate at 37 °C and 5% CO_2 for the chosen time points, as detailed above
4. Repeat experiment three times ($n_{\text{tests}} = 3$) in order to determine experimental error.

3.2 Immunofluorescence Staining of Cultures for Microscopy Analysis

Immunofluorescence (IF) is an immunochemical technique that allows the detection and localization of various antigens in different types of tissues of various cell culture preparations. It has excellent sensitivity and much greater signal amplification in comparison to immunohistochemistry and employs various microscopy techniques. Two methods are available for IF, direct, or primary staining, i.e., the method used in this protocol, and indirect or secondary staining [15]. As detailed in the introduction, this protocol is not limited to specific dyes and many different ones exist and may be used; however, this protocol has been optimized for the dyes detailed in Table 1.

3.2.1 Fluorescent Staining

1. Wash cultures with 200 μL/well of pre-warmed washing solution three times to remove residual/extracellular materials.
2. In the last wash, aspirate all washing solution.
3. Add fluorescent organelle stain at calculated concentration. Again, it is worth noting that many different plate formats may be used for HSCA, i.e., 24, 48, or 96 well plates. For 96 well plates, optimized for this protocol, it is usually 50 μL/well for the dye concentration provided by the supplier of the staining for the cell line selected.
4. Incubate for 30 min at room temperature in the dark, e.g., wrapped in tin foil.
5. Aspirate staining solution and wash two times with 200 μL/well of washing solution.
6. Add 100 μL/well of fixation solution. Incubate for 20 min at room temperature in the dark.
7. Aspirate the fixation solution.
8. Add 100 μL/well of fluorescent nuclear stain and incubate for 20 min at room temperature in the dark.

9. Aspirate the fluorescent nuclear stain and wash two times with 200 μL/well of washing solution.
10. Seal plate using adhesive plate sealers (Thermo Fisher, Ireland) and store at 4 °C in the dark prior to high-content image analysis.

3.2.2 Quantitative Evaluation of Cell Viability Using High Content Screening Analysis (See Note 3)

1. The protocol defined for the acquisition of images is dependent on the numbers of fluorophores to be measured, the objective used, and filters required.
2. In this protocol, the study design includes the definition of three fluorescent channels, bright field channel, and merged pseudo-colored images.
3. Prior to automated image acquisition using a HCSA plate reader (e.g., Lionheart, as detailed previously), culture plates are autofocused using the autofocus feature.
4. Image contrast is set for each channel using untreated negative control well as baseline.
5. Images are acquired from 6 defined fields, per well, using a 10× 0.45 Plan Apo dry objective.
6. HCSA acquisition protocol is set up according to peak excitation/emission parameters of each chosen dye.
7. Peak excitation and emission wavelengths of suggested fluorescent dyes are as follows:
   (a) Hoechst, excitation/emission 350/461 nm—nuclear stain (nuclear intensity/cell viability)—DAPI filter
   (b) YO-PRO®-1, excitation/emission 491/590 nm—YO-PRO®-1 cell permeability probe (membrane permeability/cell viability)—FITC filter
   (c) LysoTracker® Red, excitation/emission 577/590 nm—Lysotracker®red probe (lysosomal mass and/or pH changes)—TRITC filter
8. Images analyzed by an appropriate software, e.g., Gen 5 software (BioTek Mason Technology, Ireland), or others as previously detailed, for cell viability and fluorescent intensity quantification.
9. Results of compound and positive control-treated wells for each experiment are calculated as a percentage of normalized untreated (culture medium) control wells. Data are inputted and processed using an appropriate software for statistical analysis, e.g., GraphPad Prism 7, Version 7.1.0 (GraphPad Software Inc, San Diego, USA) and appropriate statistical tests, e.g., 2-way ANOVA with Bonferroni post-test analysis.

4 Notes

1. This protocol has been optimized for the A549, THP-1, and HepG2 cell lines; however, it is suitable for many other cell lines. If using different cell lines, optimization of cell seeding density and positive controls will be needed. In our studies, all cell lines were acquired from the American Type Culture Collection (ATCC). These can also be obtained commercially from many other suppliers (e.g., Sigma-Aldrich).
2. Many different plate formats may be used for HSCA, depending on the analysis required, i.e., 24, 48, or 96 well plates.
3. The results obtained from HCSA is both *qualitative* and *quantitative.* Qualitative analysis is concerned with interpreting and understanding what is observed in the fluorescence microscopy output of the experiments, e.g., visible reduction in cell count and/or lysosomal mass, and/or visible increase in nuclear permeability, and quantitative analysis is concerned with actually measuring and counting the events. This combination allows the acquisition of unbiased multiparametric data, at the single cell level. Results from HCSA may also be verified further, if desired, using other quantitative cytotoxicity or viability assays, e.g., CellTiter-Glo® viability assay or LDH/MTT cytotoxicity assays.

Acknowledgments

We acknowledge the partial financial support from the European Union under H2020 REFINE (ref 761104), Safe-N-MedTech (ref. 814607), and EUNCL (ref 654190).

The authors would also like to thank several colleagues: Drs C. Maguire and G. Hannon from the Nanomedicine Group and LBCAM at Trinity College Dublin and also Drs M. Roesslin and G. Klinkenberg as members of the EUNCL, REFINE, and Safe-N-Medtech European projects for the technical discussions and insight on aspects related to High Content and Throughput Screening.

References

1. Williams Y, Prina-Mello A, Volkov Y (2014) 4.24 – high content screening and analysis with nanotechnologies. In: Brahme A (ed) Comprehensive biomedical physics. Elsevier, Oxford, pp 379–389
2. Collins AR, Annangi B, Rubio L, Marcos R, Dorn M, Merker C et al (2017) High throughput toxicity screening and intracellular detection of nanomaterials. WIREs Nanomed Nanobiotechnol 9(1):e1413. https://doi.org/10.1002/wnan.1413
3. Movia D, Prina-Mello A, Volkov Y, Giordani S (2010) Determination of spiropyran cytotoxicity by high content screening and analysis for safe application in bionanosensing. Chem Res

Toxicol 23(9):1459–1466. https://doi.org/10.1021/tx100123g

4. Movia D, Gerard V, Maguire CM, Jain N, Bell AP, Nicolosi V et al (2014) A safe-by-design approach to the development of gold nanoboxes as carriers for internalization into cancer cells. Biomaterials 35(9):2543–2557. https://doi.org/10.1016/j.biomaterials.2013.12.057
5. Verma NK, Crosbie-Staunton K, Satti A, Gallagher S, Ryan KB, Doody T et al (2013) Magnetic core-shell nanoparticles for drug delivery by nebulization. J Nanobiotechnol 11(1):1. https://doi.org/10.1186/1477-3155-11-1
6. Shan F, Close DA, Camarco DP, Johnston PA (2018) High-content screening comparison of cancer drug accumulation and distribution in two-dimensional and three-dimensional culture models of head and neck cancer. Assay Drug Dev Technol 16(1):27–50. https://doi.org/10.1089/adt.2017.812
7. Esquer H, Zhou Q, Abraham AD, LaBarbera DV (2020) Advanced high-content-screening applications of clonogenicity in cancer. SLAS Discov 25(7):734–743. https://doi.org/10.1177/2472555220926921
8. Kossatz S, Grandke J, Couleaud P, Latorre A, Aires A, Crosbie-Staunton K et al (2015) Efficient treatment of breast cancer xenografts with multifunctionalized iron oxide nanoparticles combining magnetic hyperthermia and anti-cancer drug delivery. Breast Cancer Res 17(1):66. https://doi.org/10.1186/s13058-015-0576-1
9. O'Brien PJ (2014) High-content analysis in toxicology: screening substances for human toxicity potential, elucidating subcellular mechanisms and in vivo use as translational safety biomarkers. Basic Clin Pharmacol Toxicol 115(1):4–17. https://doi.org/10.1111/bcpt.12227
10. Prina-Mello A, Crosbie-Staunton K, Salas G, Morales MP, Volkov Y (2013) Multiparametric toxicity evaluation of SPIONs by high content screening technique: identification of biocompatible multifunctional nanoparticles for nanomedicine. IEEE Trans Magn 49(1):377–382. https://doi.org/10.1109/TMAG.2012.2225024
11. Biswas S (2021) High content analysis across signaling modulation treatments for subcellular target identification reveals heterogeneity in cellular response. Front Cell Dev Biol:8. https://doi.org/10.3389/fcell.2020.594750
12. Jan E, Byrne SJ, Cuddihy M, Davies AM, Volkov Y, Gun'ko YK et al (2008) High-content screening as a universal tool for fingerprinting of cytotoxicity of nanoparticles. ACS Nano 2(5):928–938. https://doi.org/10.1021/nn7004393
13. Collins AR, Annangi B, Rubio L, Marcos R, Dorn M, Merker C et al (2017) High throughput toxicity screening and intracellular detection of nanomaterials. Wiley Interdiscip Rev Nanomed Nanobiotechnol 9(1). https://doi.org/10.1002/wnan.1413
14. Prina-Mello A (2018) Chapter 17 – Multiparametric preclinical assessment of theranostics materials. In: Conde J (ed) Handbook of nanomaterials for cancer theranostics. Elsevier, pp 517–535
15. Im K, Mareninov S, Diaz MFP, Yong WH (2019) An introduction to performing immunofluorescence staining. Methods Mol Biol 1897:299–311. https://doi.org/10.1007/978-1-4939-8935-5_26

Chapter 16

Testing the Effects of Magnetic Hyperthermia in 2D Cell Culture

Gary Hannon and Adriele Prina-Mello

Abstract

Magnetic hyperthermia is an innovative thermal therapy for the treatment of solid malignancies. This treatment approach utilizes magnetic nanoparticles that are stimulated by alternating magnetic fields to induce temperature elevations in tumor tissue, resulting in cell death. Magnetic hyperthermia is clinically approved for treating glioblastoma in Europe and is undergoing clinical evaluation for prostate cancer in the United States. Numerous studies have also demonstrated efficacy in other cancers, however, and its potential utility extends far beyond its current clinical indications. Despite this great promise, assessing the initial efficacy of magnetic hyperthermia in vitro is a complicated endeavor, with multiple hurdles worth considering, such as accurate thermal monitoring, accounting for nanoparticle interference, and a myriad of treatment controls that make robust experimental planning essential to evaluate treatment outcome. Presented here is an optimized magnetic hyperthermia treatment protocol to test the primary mechanism of cell death in vitro. This protocol can be applied to any cell line and ensures accurate temperature measurements, minimal nanoparticle interference, and controls for multiple factors that can influence experimental outcome.

Key words Hyperthermia, Magnetic hyperthermia, Iron oxide nanoparticles, Magnetic nanoparticles, Heat, Cancer, Cell culture, Cell death, Apoptosis, Necrosis

1 Introduction

Magnetic hyperthermia (MH) is a thermal therapy for cancer that utilizes tumor-residing magnetic nanoparticles (MNP) and externally-applied alternating magnetic fields (AMF) to achieve therapeutic temperature elevations in tumors. Since its first reported use in 1957, MH has proven a hugely promising treatment for many cancers and achieved clinical approval to treat glioblastoma in Europe in 2011 [1, 2]. Currently, MH is also undergoing clinical evaluation in prostate cancer and pancreatic ductal adenocarcinoma, as the therapy looks certain to expand to multiple indications [3, 4]. While great strides have been made over the last two decades with this therapy, the scientific and economic

Dania Movia and Adriele Prina-Mello (eds.), *Cancer Cell Culture: Methods and Protocols*,
Methods in Molecular Biology, vol. 2645, https://doi.org/10.1007/978-1-0716-3056-3_16,

© The Author(s), under exclusive license to Springer Science+Business Media, LLC, part of Springer Nature 2023

gap between basic research and clinical translation remains extensive due to the complexity in applying the treatment [3, 5, 6]. Notably, protocols for preliminary efficacy testing of MH in vitro vary considerably in the literature with regards to controls, thermal monitoring and dosage, and endpoints measured [7]. Factors such as treatment controls, nanoparticle interference, and accurate thermal monitoring are essential considerations when optimizing an efficacy protocol for MH in vitro. It is important to ensure the AMF and MNP are evaluated for their cytotoxic effects both alone and in combination, as both alone have previously been shown to be inherently cytotoxic [8–12]. Thermal monitoring must be performed using accurate thermal probes for cells exposed to both MNP and AMF and AMF alone to ensure the therapeutic temperature elevations are exclusively due to the MH treatment, as AMF can induce detectable temperature elevations via Eddy currents [13]. Additionally, inorganic nanomaterials have been widely reported to interfere with assays where optical absorbance is employed as an endpoint; hence, any cell viability assay that relies on these parameters is vulnerable to interference and, therefore, false positive/negative results [14–17]. Described here is an in vitro MH protocol optimized to control for the effects of MNP and AMF alone, sham treatment, nanomaterial interference, intracellular and extracellular MH, and temperature elevations achieved through AMF alone. This protocol is based on Hannon G et al. [18], but can be tailored for any cell line and any research question related to applying MH in vitro, as it controls for the multitude of factors that may influence experimental outcomes with this complex therapy.

2 Materials

2.1 Cell Culture

1. BxPC-3 cell line (*see* **Note 1**)
2. Cell culturing media: RPMI Media, 10% fetal bovine serum (FBS), 1% penicillin (10,000 U/mL)-streptomycin (10,000 μg/mL)
3. Phosphate buffer saline (PBS)
4. Dissociation reagent: 1× TrypLE™ Express Enzyme no phenol red
5. 1.5 mL microcentrifuge tubes

2.2 Magnetic Hyperthermia

1. Magnetic nanoparticles (MNP) (*see* **Note 2**)
2. Alternating magnetic fields (AMF) generator (*see* **Note 3**)
3. Fiber optic temperature sensors (*see* **Note 4**)
4. Wash solution: 70% ethanol, 30% sterile deionized water
5. Parafilm

2.3 Cell Viability Assay

1. Annexin binding buffer: 10 mM HEPES, 140 mM NaCl, 2.5 mM $CaCl_2$
2. Staining buffer: 1% bovine serum albumin (BSA), 0.1% sodium azide, PBS
3. Annexin V: 2.5 μL Annexin V, 1 mL annexin binding buffer (*see* **Note 5**)
4. 7-AAD: 2.5 μL 7-AAD, 1 mL staining buffer (*see* **Note 5**)
5. 50 μM Cisplatin: 20 μL of 2.5 mM cisplatin, 980 μL cell culturing media
6. 10% DMSO: 100μL pure DMSO, 900 μL cell culturing media
7. Flow cytometer (*see* **Note 6**)

2.4 Data Analysis

1. FlowJo software or equivalent
2. GraphPad Prism software or equivalent

3 Methods

3.1 In Vitro Magnetic Hyperthermia (MH)

1. Culture cells until reaching their exponential growth phase.
2. Once 80% confluency is reached, dissociate cells with the dissociation reagent and seed cells at 100,000 cells/well in a 12-well plate (*see* **Note 7**).
3. Incubate for 24 h at 37 °C and 5% CO_2.
4. Aspirate the media from each well and expose cells to either cell culturing media (for negative control, AMF control, and stain controls) or MNP (for nanoparticle control and MH treatment) for 24 h. Each exposure condition must be tested in duplicate (*see* **Note 8**).
5. After 24 h, detach cells from two MNP-treated wells and one media-treated well with 300 μL of dissociation reagent. Centrifuge for 5 min at 845× *g*. Disperse in 1 mL PBS and centrifuge again at 845× *g*. Remove supernatant and redisperse in 100 μL of media (with or without MNP) in sterile 1.5 mL tubes (*see* **Note 9**).
6. Store one of the MNP-treated samples at 37 °C and 5% CO_2 for the duration of the MH treatment to act as a sham control. Wrap the tubes (without lid) containing the second MNP-treated sample and the media-treated sample twice in parafilm. Place both tubes in the center of the coil, within an external heating source (*see* **Note 10**). Wash thermal probes in the wash solution, and immediately pierce them through the double layer of parafilm. Position the probes into the media at the bottom of the tubes (*see* **Note 11** and Fig. 1).

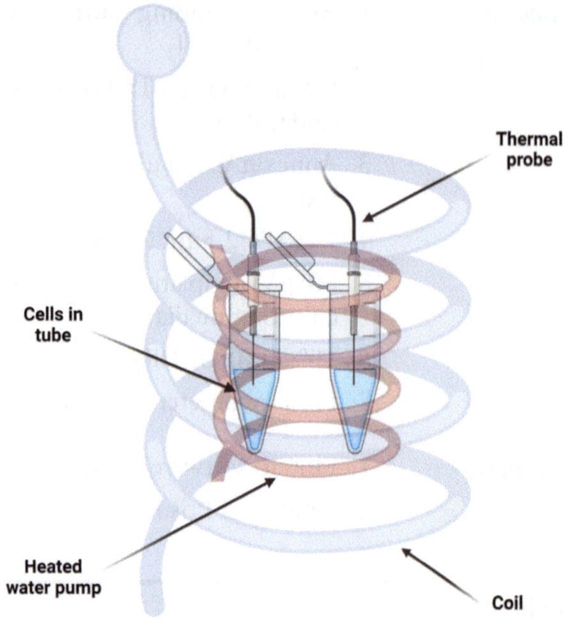

Fig. 1 Experimental setup for in vitro magnetic hyperthermia experiments. Cells are placed in a 1.5 mL tube and wrapped twice in parafilm. The tubes are then placed in the center of the coil within tubing containing heated water derived from a water bath and circulated by a peristaltic pump. Thermal probes are wiped with ethanol and pierced through the parafilm and into the media to monitor temperature. Once the temperature in the tubes reaches 36–37 °C, the AMF is turned on for magnetic hyperthermia treatment

7. Once the temperature read by the probes in both tubes reaches 36–37 °C, switch on the AMF and record temperatures in both tubes for the predetermined duration of treatment (*see* **Notes 12** and **13**).
8. After AMF exposure, remove the thermal probes and immediately wrap the tubes in an additional layer of parafilm to prevent environmental contamination.
9. Centrifuge samples at 845× *g* for 5 min. Remove 100 μL of supernatant and add 10 mL of PBS. Repeat the step three times.
10. Resuspend cells in cell culturing media and reseed them in a 12-well plate. Incubate at 37 °C and 5% CO_2 for 24 h for cell viability assessment.
11. Perform **steps 4–10** on the experimental duplicates.
12. Expose two media-treated wells with cisplatin for 24 h. This serves as a positive control for the cell viability assessment and a positive stain control for Annexin V (*see* **Note 14**) (Fig. 2).

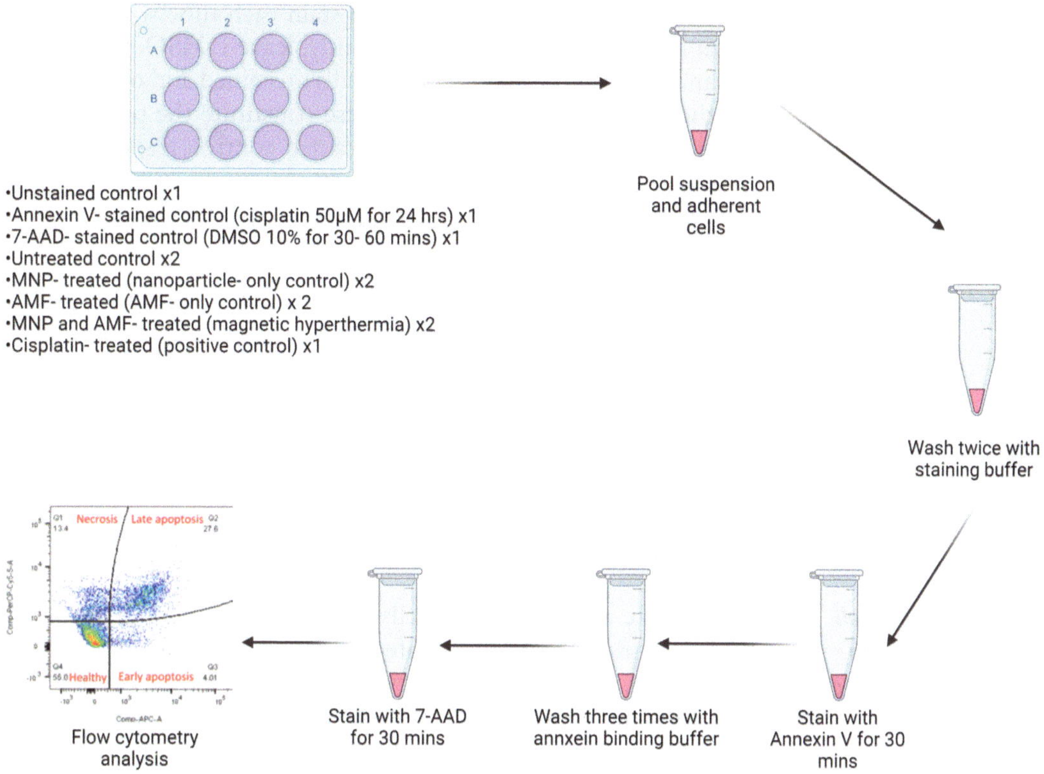

Fig. 2 Staining protocol for cell viability analysis by flow cytometry. After cells are treated, they can be stained with annexin V and 7-AAD for cell viability analysis. First, wells containing adherent cells and cells suspended in media are pooled to ensure no loss of dead/live populations of cells. They are then washed with staining buffer, stained with annexin V, washed with annexin binding buffer, and finally stained with 7-AAD. The cells can then be analyzed by flow cytometry

3.2 Cell Viability Assessment

1. After 24 h, expose one well with 10% DMSO for 30–60 min to induce necrosis. This is the positive control for 7-AAD.
2. To stain the cells, transfer the media from each well to sterile 1.5 mL tubes and centrifuge at 845× *g* for 5 min to capture cells in suspension. While the centrifuge step is occurring, add 300 μL of dissociation reagent to all wells to detach adherent cells.
3. The centrifuged cells initially suspended in media will now have accumulated at the bottom of the tube; remove the supernatant and ensure to leave approximately 100 μL at the bottom of the tube to ensure the cell pellet is not aspirated or damaged.
4. Pool the detached cells from the wells, and add them to their corresponding tubes. Centrifuge the cells at 845× *g* for 5 min, remove supernatant, redisperse in 500 μL of cold (≈4 °C) staining buffer, and spin down again at 845× *g* for 5 min.

5. Aspirate the staining buffer and stain each tube with 50 μL of annexin V for 30 min. Importantly, leave DMSO-treated cells and one tube of media-treated cells unstained and only disperse them in buffer for this step; these constitute the positive stain control for 7-AAD and the unstained control, respectively.
6. Make all tubes up to 500 μL in annexin binding buffer. Centrifuge at 845× *g* for 5 min, aspirate the buffer, and repeat this washing step two more times.
7. Add 50 μL of 7-AAD to each tube. Leave two tubes of cells unstained: unstained control cells and cisplatin-treated cells, which is a positive stain control for annexin V. All tubes are now ready for flow cytometry analysis.
8. Prepare an MNP-only control that will be used for measuring interference. For this, disperse MNP at the concentration used for MH, in staining buffer.
9. Analyze samples by flow cytometry.

3.3 Analyzing Cell Viability Data with Flow Cytometry and Controlling for Interference

1. Analyze the unstained cells and positive stain controls for Annexin V (cisplatin-treated cells) and 7- AAD (DMSO-treated cells) to establish which regions of the scatter plot represent a positive or a negative cell population for each stain. Compensation should be performed to limit spectral overlap.
2. Analyze the MNP-only control to confirm that the MNP does not interfere with the fluorescent readout (*see* **Note 15**) (Fig. 3).

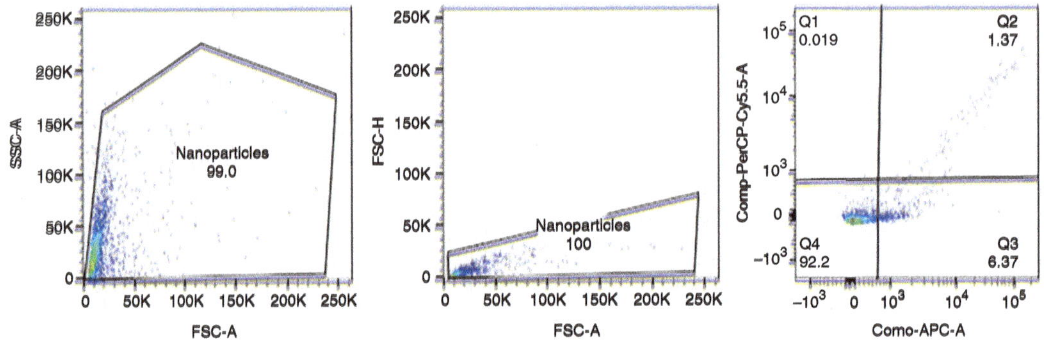

Fig. 3 Testing nanoparticle interference with channels. To confirm the nanoparticles are not contributing false positives for apoptosis and necrosis, the entire population of nanoparticles was gated to determine if they interefered with APC and PerCP-Cy5.5 channels used for this experiment. As <10% of the events were registered as APC and PerCP-Cy5.5 positive, the nanoparticles were deemed not to interfere with this experiment. Additionally, care was taken to avoid the nanoparticle population when gating the cells for this experiment. (Figure reproduced with permission from Ref. [17])

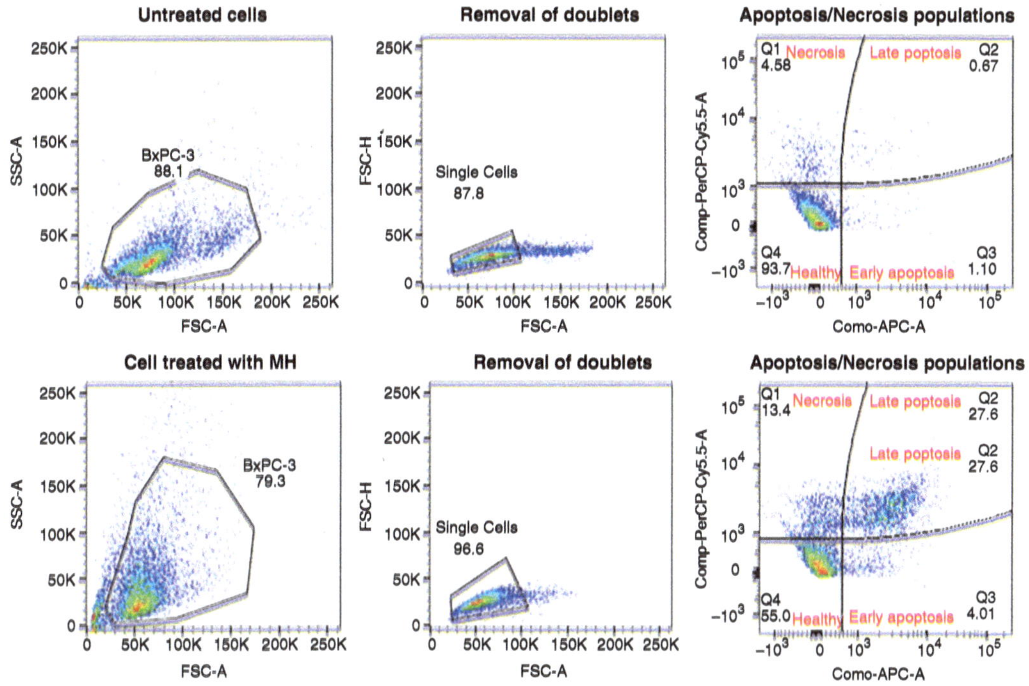

Fig. 4 Gating strategy for BxPC-3 cells treated with magnetic hyperthermia. BxPC-3 cells are identified via FSC-A and SSC-A, doublets are removed using FSC-H and FSC-A, and populations positive for APC were deemed apoptotic, while PerCP-Cy5.5 positive were deemed necrotic. Cells staining positive for both APC and PerCP-Cy5.5 were considered late apoptotic, an intermediate between apoptosis and necrosis. Notably, when gating for cells, the MNP signaling (FSC-A^{low}) is avoided to limit interference

3. Analyze the rest of the samples and establish the rates of apoptosis and necrosis between treatment groups.
4. Extrapolate the efficacy of MH in inducing apoptosis and necrosis (Fig. 4 and **Note 16**).

4 Notes

1. Although BxPC-3 cells are used here, this protocol is applicable to any cell line as long as positive control concentrations are optimized for the specific cell line.
2. This protocol can be adapted for any MNP that can generate therapeutic temperatures effectively under the influence of AMF, as long as the concentration of MNP required is low enough to not carry inherent cytotoxic effects to the cell line tested. For more information, please refer to two excellent recent reviews on the topic [5, 19].
3. It is important to use clinically relevant AMF fields and frequencies for these studies; safety thresholds in this case have been defined as fields from 4.85–50 kA/m at frequencies of 100 kHz [2].

4. Highly accurate thermal probes are essential for this protocol. We use Optocon fiber optic temperature sensors (Weidmann-Optocon, Germany) that have accuracy to ±0.2 °C. Any temperature sensors with equivalent accuracy are suitable, however.
5. Alternative stains and fluorophores for apoptosis and necrosis detection by flow cytometry can be used. For example, 7-AAD can be substituted for PI to achieve similar results. In our case, we found APC-Annexin V and 7-AAD suitable as there is minimal spectral overlap and the MNP we tested did not interfere with these channels. Stock concentrations of these stains are not provided by the manufacturer we used (BioLegend, UK) and so cannot be reported here. Alternative manufacturers can be used and may have alternative recommendations for working concentrations of these stains. We recommend following their guidelines.
6. Any flow cytometer will suffice, as only two-color channel detection is required for this protocol.
7. A minimum of 100,000 cells/well are necessary to successfully capture 10,000 events by flow cytometry at the end of the experiment, as samples are subjected to multiple washing steps and treatments, causing cell loss.
8. The experiment is carried out in duplicate as MH treatment is only carried out on two cell samples at a time (MNP-treated cells and AMF control). The MNP should be dispersed in an aqueous stock solution (which is the case with most commercially available MNP) and added to cell culture media directly before treatment, at a concentration that is non-toxic to the cells. 50 μM of cisplatin for 24 h is used as a positive control for Annexin V staining, which is a stain for apoptosis, while 10% DMSO for 30–60 min is used as a positive control for 7-AAD, which stains for necrosis [20, 21].
9. The small volume mimics a thermal "hot-spot" in a tumor subjected to MH. The "hot-spot" is a small area in the tumor containing a high concentration of MNP which will be exposed to considerable temperature elevations when the AMF is applied [22]. In this protocol, excess MNP is washed away. Typically, if the MNP is washed away, the concentration of intracellular MNP is not high enough for detecting temperature elevations in the media. Therefore, an extracellular MNP treatment group is necessary to monitor MH. The extracellular and intracellular MNP treatments concentrations can be the same or different, as long as they are both non-cytotoxic and the extracellular MNP concentration induces therapeutic temperature elevations under AMF.

10. We used a peristaltic pump connected to a water bath.
11. The double layer of parafilm around the tube serves two purposes: it reduces the risk of film breakage and cell contamination from the local environment, and it provides a secure hold for the thermal probe preventing movement during the hyperthermia treatment.
12. To increase the temperature in the tubes prior to AMF exposure, either the pump speed or water bath temperature can be increased. Likewise, a reduction in pump speed or water bath temperature will reduce the temperature in the tubes. The parameters for the AMF will vary depending on the MNP used. In our case, a field of 35 mT at a frequency of 92 kHz is sufficient to generate therapeutic temperatures for many MNP tested in media that are at non-cytotoxic concentrations. At a minimum, record temperatures every 10 s for 2 min initially once the AMF is switched on, and every minute thereafter. This ensures that both the immediate and long-term temperature elevations observed in the MNP-treated cells are truly from MH, and not an artifact of eddy currents or environmental heating. Importantly, only begin measuring time intervals when your desired therapeutic temperature is reached. This usually takes $\approx$ 1 min from the start of AMF exposure.
13. Consider two critical factors during the thermal treatment: the temperature of the cells exposed to AMF only should not increase >1.5 °C (in our experience, no significant increase in cell death is observed when temperatures reached $\approx$38 °C), and the AMF may need to be continually adjusted during treatment to ensure your predetermined temperature window is maintained.
14. For every cell line we have used in this protocol, 50 μM cisplatin for 24 h was enough to induce significant apoptosis. However, tested time points or cisplatin concentrations may need optimization to suit different cell types.
15. While gating is optimized to avoid MNP interference, it is still worth gating this nanoparticle population to determine if they can introduce positive signals in the channels used. Notably, aggregated MNP or MNP located on cells may introduce false positive signaling if they interfere with the channels used.
16. Acceptance criteria: the experiment can be deemed successful if the following conditions are satisfied: the untreated cell population has viability ≥85%; no significant difference between AMF-treated and untreated cells exist; MNP demonstrate no cytotoxicity on their own; the positive controls elicit significant cell death against the untreated control.

Acknowledgments

This work was supported by the Irish Research Council Postgraduate Scholarship (project identifier: GOIPG/2016/1629) and the European Commission's NoCanTher project (Grant Agreement No. 685795). We would also like to thank Chemicell GmbH for providing the MNP used for developing this protocol and acknowledge Biorender software for the creation of Figs. 1 and 2.

References

1. Gilchrist RK, Medal R, Shorey WD, Hanselman RC, Parrott JC, Taylor CB (1957) Selective inductive heating of lymph nodes. Ann Surg 146(4):596–606. https://doi.org/10.1097/00000658-195710000-00007
2. Hannon G, Tansi FL, Hilger I, Prina-Mello A (2021) The effects of localized heat on the hallmarks of cancer. Adv Ther 4(7):2000267. https://doi.org/10.1002/adtp.202000267
3. Rubia-Rodríguez I, Santana-Otero A, Spassov S, Tombácz E, Johansson C, De La Presa P et al (2021) Whither magnetic hyperthermia? A tentative roadmap. Materials (Basel) 14(4):706. https://doi.org/10.3390/ma14040706
4. Tansi FL, Maduabuchi WO, Hirsch M, Southern P, Hattersley S, Quaas R et al (2021) Deep-tissue localization of magnetic field hyperthermia using pulse sequencing. Int J Hyperth 38(1):743–754. https://doi.org/10.1080/02656736.2021.1912412
5. Liu X, Zhang Y, Wang Y, Zhu W, Li G, Ma X et al (2020) Comprehensive understanding of magnetic hyperthermia for improving antitumor therapeutic efficacy. Theranostics 10(8):3793–3815. https://doi.org/10.7150/thno.40805
6. Kozissnik B, Bohorquez AC, Dobson J, Rinaldi C (2013) Magnetic fluid hyperthermia: advances, challenges, and opportunity. Int J Hyperth 29(8):706–714. https://doi.org/10.3109/02656736.2013.837200
7. Vilas-Boas V, Carvalho F, Espiña B (2020) Magnetic hyperthermia for cancer treatment: main parameters affecting the outcome of in vitro and in vivo studies. Molecules 25(12):2874. https://doi.org/10.3390/molecules25122874
8. Asín L, Ibarra MR, Tres A, Goya GF (2012) Controlled cell death by magnetic hyperthermia: effects of exposure time, field amplitude, and nanoparticle concentration. Pharm Res 29(5):1319–1327. https://doi.org/10.1007/s11095-012-0710-z
9. Ludwig R, Teran FJ, Teichgraeber U, Hilger I (2017) Nanoparticle-based hyperthermia distinctly impacts production of ROS, expression of Ki-67, TOP2A, and TPX2, and induction of apoptosis in pancreatic cancer. Int J Nanomed 12:1009–1018. https://doi.org/10.2147/ijn.S108577
10. Afzalipour R, Khoei S, Khoee S, Shirvalilou S, Raoufi NJ, Motevalian M et al (2021) Thermosensitive magnetic nanoparticles exposed to alternating magnetic field and heat-mediated chemotherapy for an effective dual therapy in rat glioma model. Nanomedicine 31:102319. https://doi.org/10.1016/j.nano.2020.102319
11. Valdiglesias V, Kiliç G, Costa C, Fernández-Bertólez N, Pásaro E, Teixeira JP et al (2015) Effects of iron oxide nanoparticles: cytotoxicity, genotoxicity, developmental toxicity, and neurotoxicity. Environ Mol Mutagen 56(2):125–148. https://doi.org/10.1002/em.21909
12. Hannon G, Tutty M, Prina-Mello A (2018) Immunotoxicity and safety considerations for iron oxide nanoparticles. Clin Appl Magn Nanopart:273–288
13. Stigliano RV, Shubitidze F, Petryk JD, Shoshiashvili L, Petryk AA, Hoopes PJ (2016) Mitigation of eddy current heating during magnetic nanoparticle hyperthermia therapy. Int J Hyperth 32(7):735–748. https://doi.org/10.1080/02656736.2016.1195018
14. Kroll A, Pillukat MH, Hahn D, Schnekenburger J (2012) Interference of engineered nanoparticles with in vitro toxicity assays. Arch Toxicol 86(7):1123–1136. https://doi.org/10.1007/s00204-012-0837-z
15. Ong KJ, MacCormack TJ, Clark RJ, Ede JD, Ortega VA, Felix LC et al (2014) Widespread nanoparticle-assay interference: implications for nanotoxicity testing. PLoS One 9(3):e90650-e. https://doi.org/10.1371/journal.pone.0090650

16. Hannon G, Prina-Mello A (2021) Endotoxin contamination of engineered nanomaterials: overcoming the hurdles associated with endotoxin testing. WIREs Nanomed Nanobiotechnol 13(6):e1738. https://doi.org/10.1002/wnan.1738
17. Andraos C, Yu IJ, Gulumian M (2020) Interference: a much-neglected aspect in high-throughput screening of nanoparticles. Int J Toxicol 39(5):397–421. https://doi.org/10.1177/1091581820938335
18. Hannon G, Bogdanska A, Volkov Y, Prina-Mello A (2020) Comparing the effects of intracellular and extracellular magnetic hyperthermia on the viability of BxPC-3 cells. Nanomaterials 10(3):593
19. Egea-Benavente D, Ovejero JG, Morales MDP, Barber DF (2021) Understanding MNPs behaviour in response to AMF in biological milieus and the effects at the cellular level: implications for a rational design that drives magnetic hyperthermia therapy toward clinical implementation. Cancers (Basel) 13(18):4583. https://doi.org/10.3390/cancers13184583
20. Crowley LC, Marfell BJ, Scott AP, Waterhouse NJ. Quantitation of apoptosis and necrosis by annexin V binding, propidium iodide uptake, and flow cytometry. Cold Spring Harb Protoc. 2016;2016(11). doi: https://doi.org/10.1101/pdb.prot087288.
21. Zimmermann M, Meyer N (2011) Annexin V/7-AAD staining in keratinocytes. Methods Mol Biol 740:57–63. https://doi.org/10.1007/978-1-61779-108-6_8
22. Cazares-Cortes E, Cabana S, Boitard C, Nehlig E, Griffete N, Fresnais J et al (2019) Recent insights in magnetic hyperthermia: from the "hot-spot" effect for local delivery to combined magneto-photo-thermia using magneto-plasmonic hybrids. Adv Drug Deliv Rev 138:233–246. https://doi.org/10.1016/j.addr.2018.10.016

Chapter 17

Cell Viability Assay with 3D Prostate Tumor Spheroids

Ezgi Oner, Steven G. Gray, and Stephen P. Finn

Abstract

WST-8 (Cell Counting Kit 8; CCK-8) is the last generation tetrazolium-based cell viability assay and has recently been accepted as a validated method for measuring the cell viability of 3D in vitro models. Here, we describe how to form 3D prostate tumor spheroids using the polyHEMA technique, apply drug treatments and WST-8 assay to these spheroids, and calculate their cell viability. The advantages of our protocol are the formation of spheroids without adding extracellular matrix components, and the elimination of the critique handling process needed for transferring spheroids. Although this protocol exemplifies the determination of percentage cell viability in PC-3 prostate tumor spheroids, it can be adapted and optimized for other prostate cell lines and other types of cancers.

Key words Three-dimensional cell culture, 3D in vitro models, Tumor spheroids, Prostate cancer, Prostaspheres, PolyHEMA method, Ultra-low attachment surface, WST-8 (Cell Counting Kit-8, CCK-8) cell viability assay, Cytotoxicity

1 Introduction

In vitro cell culture methods used for preclinical experiments have a key role in making "terminate" or "go" decisions with the candidate drug molecule at the beginning of the drug research and development process. The next step in the preclinical phase following in vitro experiments is in vivo studies. However, due to ethical and financial problems, there are ongoing efforts to restrict animal studies with the regulations of 3Rs (replacement, reduction, refinement) testing approaches [1]. Besides, in vivo animal studies sometimes do not truly reflect the response to drugs in humans in terms of efficacy and toxicity due to the differences between species. The better in vitro cell culture method mimics in vivo physiological environment, the easier it is to accurately predict the clinical effect of the drug and the less likely the drug will fail later [2, 3]. Therefore, the number of studies with three-dimensional (3D) cell

Dania Movia and Adriele Prina-Mello (eds.), *Cancer Cell Culture: Methods and Protocols*,
Methods in Molecular Biology, vol. 2645, https://doi.org/10.1007/978-1-0716-3056-3_17,
© The Author(s), under exclusive license to Springer Science+Business Media, LLC, part of Springer Nature 2023

culture, which is more successful than two-dimensional (2D) cell culture in mimicking the in vivo environment, has been increasing in recent years [4, 5].

Among in vitro assays, high throughput cytotoxicity tests are widely used to evaluate the biocompatibility (safety) of the tested substances. These tests also determine the efficacy of cytotoxic drugs on tumor cell lines. WST-8 [2-(2-methoxy-4-nitrophenyl)-3-(4-nitrophenyl)- 5-(2,4-disulfophenyl)-2H-tetrazolium, monosodium salt] is known as the best option in terms of non-toxicity, sensitivity, and ease of handling among colorimetric tetrazolium-based cell viability assays such as MTT, XTT, and WST-1 [6]. Besides 2D in vitro models, the WST-8 assay has recently been accepted as a validated method for measuring the cell viability of 3D tissue models according to the Organisation for Economic Co-operation and Development (OECD) Test Guideline No. 492 [7].

Here, we established a modified protocol for measuring the cell viability of 3D prostate tumor spheroids using WST-8 assay. In this protocol, prostate tumor spheroids were formed on polyHEMA-coated non-adherent surfaces of round-bottom wells without the external addition of any extracellular matrix components. After applying treatments to the formed PC-3 prostate tumor spheroids, these spheroids were incubated with WST-8 reagent as in 2D monolayer cells. Following orange-color development, only the mixture of WST-8/media was transferred to a flat-bottom plate for spectroscopic measurement. Lastly, the percentage cell viability of treated spheroids was calculated relative to untreated spheroids based on their absorbance values. Reproducible results were confirmed with a low standard deviation of mean absorbance values obtained in three independent experiments. As demonstrated for PC-3 prostate tumor spheroids, this protocol can be applied to the spheroids formed with other tumor cells after completing the required optimizations.

2 Materials

Unless otherwise stated, prepare and store all materials at room temperature (20–25 °C).

2.1 PolyHEMA Microplate

1. 12% w/v poly(2-hydroxyethyl methacrylate) (PolyHEMA) stock solution: 25 mL 95% ethanol, 3 g polyHEMA powder (cell culture grade) (*see* **Note 1**)
2. 0.5% w/v polyHEMA working solution: 1 mL 12% w/v polyHEMA stock solution, 23 mL 95% ethanol (*see* **Note 2**)
3. Round (U)-bottom 96-well plate

4. Non-humidified (heat-only) incubator (optional, *see* **Note 3**)
5. Multichannel pipette
6. Pipetting reservoir

2.2 Formation of Spheroids

1. Cell line: PC-3 prostate cancer cell line (*see* **Note 4**).
2. Complete cell culture medium: DMEM/F12 1:1 medium, 5% heat-inactivated fetal bovine serum (FBS), 1% of 100 U/mL penicillin, 100 μg/mL streptomycin solution (*see* **Note 4**).
3. Phosphate buffered saline (PBS).
4. 0.25% trypsin-EDTA.
5. Centrifuge with microplate rotor suitable for 96-well plates.
6. Multichannel pipette.
7. Pipetting reservoir.
8. Sterile hood [Class II Biological Safety Cabinet (BSC)].
9. CO_2 Incubator.

2.3 Treatment of Spheroids and Determination of Their Viability

1. Microscope with a camera (*see* **Note 5**).
2. WST-8 (Cell Counting Kit-8, CCK-8) Cell Viability Assay (*see* **Note 6**). Store at 2–8 °C. For longer storage, store at −20 °C without repeated freeze-thaw cycles.
3. Drug: JIB-04.
4. Drug solvent: DMSO, cell culture grade.
5. Clear, flat-bottom 96-well plate.
6. Microplate reader (*see* **Note 7**).

3 Methods

Keep doing the routine monolayer (2D) cell culture of PC-3 prostate cancer cell lines while performing this protocol. Perform all procedures at room temperature (20–25 °C) unless otherwise specified. All steps of cytotoxicity assay with 3D tumor spheroids are summarized in Fig. 1.

3.1 Preparation of PolyHEMA-Coated U-Bottom 96-Well Plates

1. Because of the drying process, prepare the polyHEMA microplates at least a week before seeding the cells.
2. Add 60 μL of PolyHEMA working solution into each well of U-bottom 96-well plate using a multichannel pipette and a pipetting reservoir (*see* **Note 8**).
3. Drying step: Put the PolyHEMA-coated microplates with lids closed into a non-humidified incubator (37 °C) for 48 h to evaporate the ethanol (*see* **Note 3**).
4. Parafilm the PolyHEMA-coated 96-well plates with lids closed. Store at 2–8 °C until further use in cell culture (*see* **Note 9**).

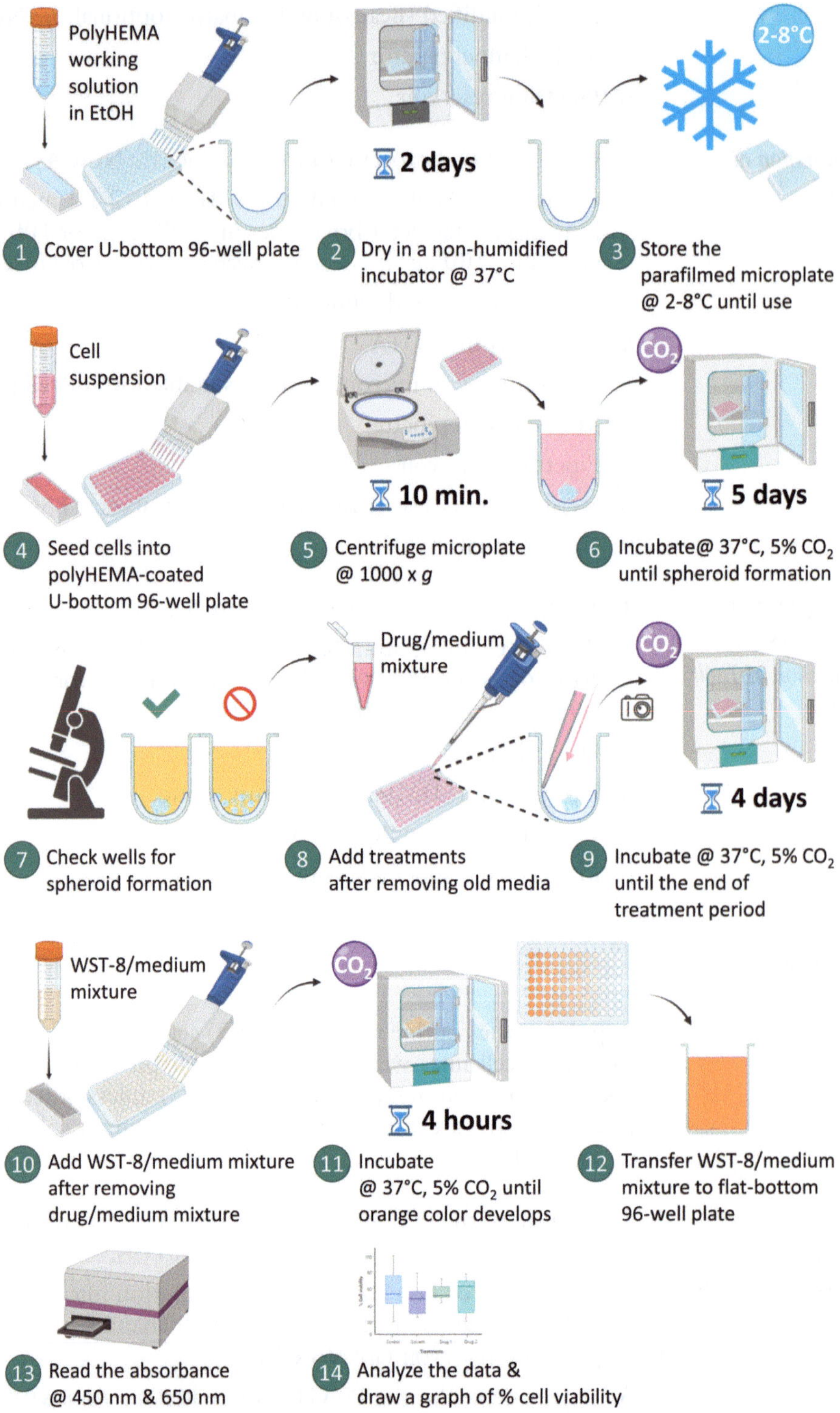

Fig. 1 Main steps in cytotoxicity assay with 3D tumor spheroids using polyHEMA non-adherent surface method. (This figure was created with BioRender.com)

3.2 Formation of Tumor Spheroids of PC-3 Prostate Cancer Cell Line

1. Before seeding the cells into the polyHEMA-coated microplate, bring it to room temperature (*see* **Note 10**).
2. Seed PC-3 cells (5×10^3 cells in 100 μL of complete cell culture medium per well) onto the PolyHEMA-coated microplate using a multichannel pipette and a pipetting reservoir (*see* **Notes 11–13**).
3. Immediately centrifuge the microplate harboring cells at 1000× *g* for 10 min (*see* **Note 14**).
4. Incubate at 37 °C with 5% CO_2 until evidence of spheroid formation (*see* **Note 15**).
5. Check the wells at certain intervals (i.e., every day after seeding) to determine the time for spheroid formation (*see* **Note 16**).

3.3 Drug Treatment

1. Determine regular shaped and similar sized spheroids using a microscope and use only these for the experiment (*see* **Notes 17** and **18**).
2. Freshly prepare all the drugs and mix them with the cell culture medium before treatment of spheroids (*see* **Note 19**).
3. Slowly remove the media from the wells using 200 μL pipette tips. Carefully aspirate at an angle not touching the center of the well where the spheroids are (Fig. 2a).
4. Quickly check the wells to be treated under the inverted microscope, to ensure that the spheroids are still intact/undamaged. Eliminate damaged spheroids, if any.

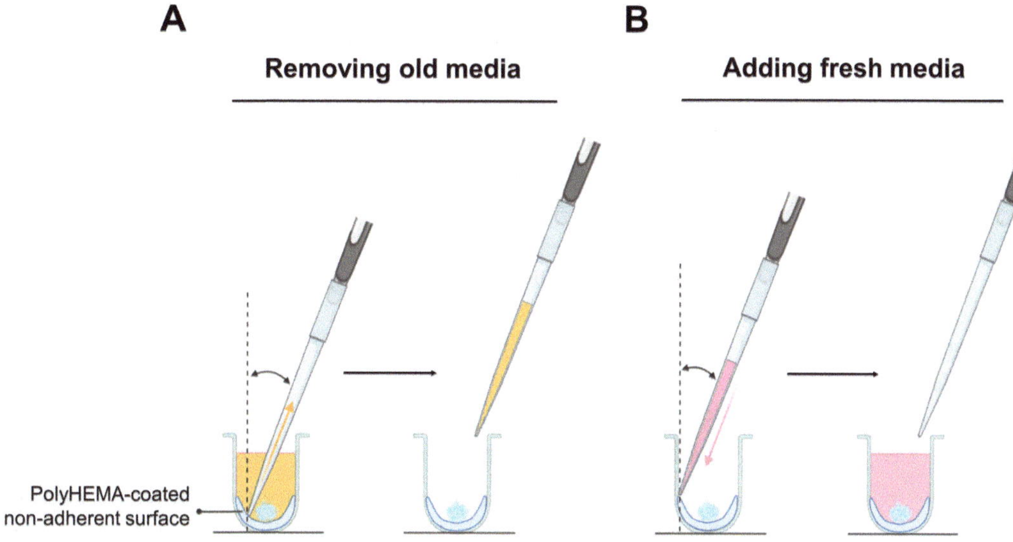

Fig. 2 Pipette use. The position of pipette tips and polyHEMA-coated U bottom wells while removing (**a**) and adding (**b**) media. (This figure was created with BioRender.com)

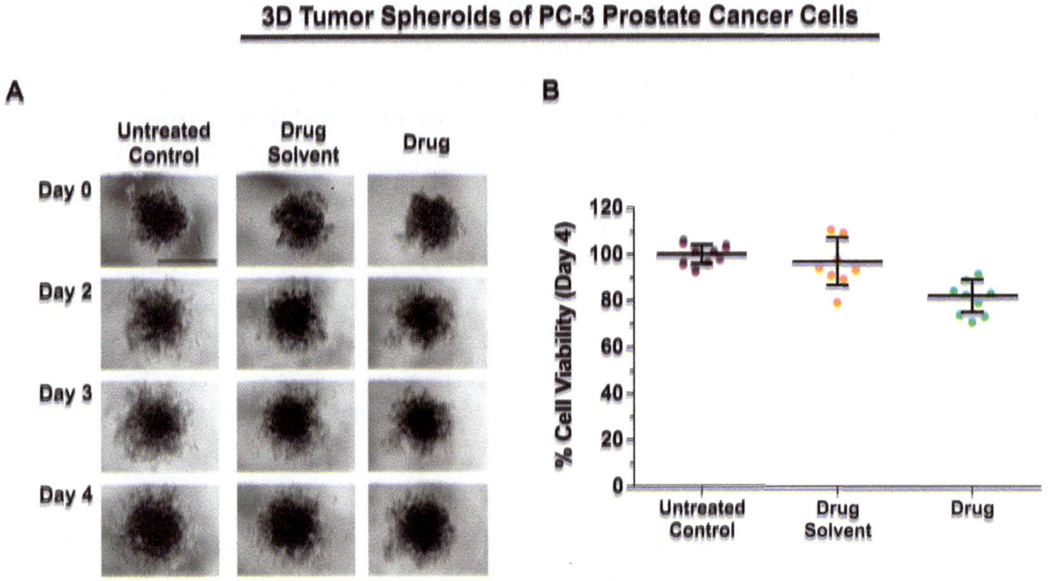

Fig. 3 Spheroid treatment. (**a**) Representative phase-contrast images of tumor spheroids at different time points after applying treatments (Scale bar: 1000 μm). (**b**) Graph showing the % cell viability of tumor spheroids relative to untreated control at the end of the 4-day treatment period. Dots represent the data collected from each well in three independent experiments, and error bars show their standard deviation ($n = 3$) (Adapted from Ref. [4] licensed under CC BY 4.0 and CC0 1.0)

5. Slowly add the mixture of drug/media into the wells harboring spheroids with desired shape and size. Carefully add the solution at an angle; the pipette tip must not touch the bottom and center of the well (Fig. 2b).
6. Soon after the treatment, take images of each well using a suitable microscope with a camera and record it as "day 0" or "0 h" (*see* **Note 20**).
7. Incubate the spheroids for the selected treatment time (*see* **Note 21**). If it is necessary to change the media with the fresh one at some point of the treatment, carefully follow **steps 3** and **4** described in Subheading 3.3 (*see* **Note 22**).
8. Acquire images of the spheroids at the determined time points, including the final end point (Fig. 3a).

3.4 Application of WST-8/Medium Mixture

1. Freshly prepare the mixture of WST-8/medium at a 1:10 ratio, respectively (*see* **Note 23**).
2. Slowly remove the mixtures of drug/medium from the wells without touching the spheroids (Fig. 2a).
3. Add 110 μL of WST-8 reagent mixture to each well, as shown in Fig. 2b.
4. Incubate the spheroids with the freshly prepared WST-8/medium mixture for 4 h in an incubator (37 °C, 5% CO_2) (*see* **Note 24**).

3.5 Measurement of the Absorbance

1. Before measuring the absorbance, check whether the treated spheroids were damaged during **steps 2** and **3** described in Subheading 3.4. If damage occurred, exclude these wells from the analysis.
2. At the end of the 4 h incubation period, transfer 100 μL of WST-8 Reagent mix from the polyHEMA-coated U-bottom 96-well plate to a flat-bottom 96-well plate (*see* **Note 25**).
3. Shake the microplate for 1 min and read the absorbance of each well at 450 nm and 650 nm (reference wavelength) using a microplate reader (*see* **Notes 26** and **27**).

3.6 Analysis of Percentage (%) Cell Viability of Spheroids

1. Use the below equation to calculate the percentage of the cell viability of treated spheroids relative to untreated spheroids based on their absorbance (cell viability of untreated control is assumed as 100%) (*see* **Notes 28** and **29**).

$$\%\text{Cell viability} = \frac{\text{Treated }(A_{450} - A_{650}) - \text{Blank }(A_{450} - A_{650})}{\text{Untreated }(A_{450} - A_{650}) - \text{Blank }(A_{450} - A_{650})} \times 100$$

A_{450}: Absorbance value at 450 nm wavelength.

A_{650}: Absorbance value at 650 nm wavelength.

2. Draw a graph with the data where the *x*-axis shows treatments, and the *y*-axis shows the % cell viability of spheroids (Fig. 3b).

4 Notes

1. While handling polyHEMA powder, wear protective equipment (dust mask type N95, eye shields, and gloves). In a hood, add ethanol in a sterile glass bottle with a sterile magnetic stirring bar (Open and use 95% ethanol only in the hood). Then add PolyHEMA powder to ethanol. Stir this mixture on a magnetic stirrer overnight at room temperature. Alternatively, you can stir it on a magnetic hot plate stirrer or on a shaking water bath at 55–65 °C until a homogenous/clear solution appears.
2. Before coating microplates, freshly prepare polyHEMA working solution in a 50 mL tube. Since parafilmed polyHEMA microplates can be stored indefinitely (*see* **step 4** in Subheading 3.1), several microplates can be covered with polyHEMA at the same time. The prepared 24 mL of 0.5% (w/v) polyHEMA working solution is sufficient to cover four 96-well plates. Vortex well for a homogenous solution.
3. To evaporate the ethanol at room temperature instead of using a non-humidified incubator at 37 °C, leave the PolyHEMA-coated microplate with its lid closed on a bench for 7–14 days or inside a sterile hood for 3 days [8].

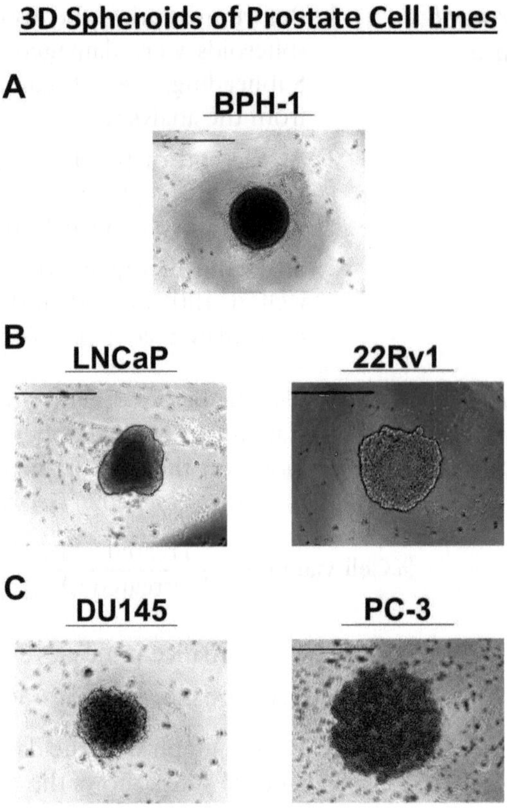

Fig. 4 Prostate tumor spheroids formed using polyHEMA-coated U bottom 96-well plates. (**a**) BPH-1 immortalized benign prostatic hyperplasia cell line (Day 3 after seeding). (**b**) Androgen-dependent LNCaP (Day 3 after seeding) and 22Rv1 (Day 1 after seeding) cell lines. (**c**) Androgen-independent DU145 (Day 3 after seeding) and PC-3 (Day 1 after seeding) cell lines. For all prostate cell lines, 10,000 cells were seeded in 200 μL of their complete media per well. (Scale bars: 1000 μm)

4. Here, a detailed protocol is given for PC-3 prostate cancer cell line. However, this method can also be applied for other prostate cell lines as well as other types of cancers after determining the optimal conditions (seeding number, type of cell culture medium, amount of FBS, incubation time, etc.) for each cell type. In our laboratory, we successfully established spheroids with BPH-1 benign prostatic hyperplasia cell line, androgen-dependent 22Rv1 and LNCaP prostate cancer cell lines, and also androgen-independent DU145 and PC-3 prostate cancer cell lines by seeding at 10^4 cells/well in 200 μL of their complete media using this protocol (Fig. 4).
5. We used an automated phase-contrast microscope (Lionheart Fx, BioTek) which is advantageous to take images of the same area of a well every time. However, one can also use an inverted

microscope with a camera. If 3D images of spheroids are required, a confocal fluorescence microscope with optical sectioning function might be the best.

6. For measuring the viability of 3D tumor spheroids, some other commercial 3D cell viability assay kits, such as colorimetric kits, including matrix dissociation solution for spheroids formed with extracellular matrix components and ATP-based luminescent kits with a solution for cell lysis, are also available.
7. For performing WST-8 assay with the best sensitivity, the microplate reader should have a 450 nm filter. However, a filter measuring the absorbance between 430–490 nm is also suitable for detection. To eliminate the background noise of each well, measure the reference absorbance with a 650 nm (or ≥600 nm) filter.
8. As ethanol is a volatile solvent, speed is important to have a constant concentration and enough volume of PolyHEMA working solution while covering the U-shaped bottoms of microplate wells. Therefore, it is strongly recommended to use a multichannel pipette to speed up the polyHEMA microplate preparation process.
9. We stored the parafilmed PolyHEMA microplates at 2–8 °C to reduce the risk of contamination. However, they can also be stored at room temperature (20–25 °C).
10. Since sterile technique and materials are used for preparing polyHEMA solutions, we do not perform a sterilization step. If you'd like to process sterilization before seeding the cells, sterilize the polyHEMA-coated U-bottom microplate with its lid off under UV light in a sterile hood for 30 min [8]. In other protocols, we also encountered washing of the polyHEMA-coated wells with physiological buffers such as PBS before the seeding step [9]. If both sterilization and washing processes are performed, be cautious about wrinkles/cracks that may appear on the polyHEMA-coated surface [10].
11. At the seeding step, the speed provided by using a multichannel pipette is important to avoid the attachment of cells to the well surfaces before centrifugation.
12. As the spheroids that do not meet the desired standards will be excluded, seed into more wells than you need. Therefore, before starting cell viability experiments, it is highly recommended to determine your yield of forming spheroids with desirable properties and then the required number of wells.
13. Add PBS to the outer wells of the 96-well plate due to the edge effect (i.e., evaporation of liquids is faster in the outer wells than in the inner wells of a microplate).

14. Modify the centrifuge force and time if cells do not form a cluster in the middle of the well. However, some types of cells may only form spheroids with the addition of extracellular matrix components such as fibronectin, laminin, collagen, and Matrigel®.
15. If necessary, change the media during spheroid formation, as shown in Fig. 2.
16. Many parameters (cell type, seeded cell number, components of cell culture medium, FBS ratio, incubation time, etc.) can affect the spheroid formation time. For example, PC-3 spheroids used for cytotoxicity assay were obtained after incubation for 5 days at 37 °C with 5% CO_2 (Fig. 3a, "Day 0" images). However, increasing the seeded cell number from 5000 cells/well to 10,000 cells/well and the percentage of FBS in the cell culture medium from 5% to 10% (v/v), and changing the media from DMEM/F12 (1:1) to Ham F12 resulted in the formation of similar PC-3 spheroids after 1-day incubation (Fig. 4c, right).
17. We selected the suitable-shaped and similar-sized spheroids by eye under a microscope. However, software-based methods are also available for this selection [11].
18. Mark the lid of the 96-well plate with a pen to distinguish the eliminated wells. Take a photo of the microplate with its marked lid or note the eliminated/selected wells in case the markings on the lid are erased.
19. Determine the components of cell culture media for treatment based on the type of the drug(s). For example, while treatment with therapeutic nucleic acids such as siRNA, the media with reduced FBS (5%) and without antibiotics can be used for an effective transfection [4].
20. To avoid culture contamination, acquire images with the plate lid closed.
21. While determining the treatment period for the first time, the change in the morphology of spheroids (increase or decrease in size, dissociation of the cluster, etc.) may give a clue about the required time for effective response at the applied concentration. Also, it would be beneficial to apply a wide range of drug concentrations for the first experiment, as spheroids are generally more resistant to drugs than the cells cultured as monolayers (2D) [4].
22. If WST-8 reagent will be directly added to the treatment media, prepare the wells of the blank with this fresh media since phenol red indicator in the media may affect the absorbance value.

23. If media change is not necessary, you can directly add WST-8 reagent to the treatment medium at a 1:10 ratio and continue from **step 4** in Subheading 3.4. However, tetrazolium-based reagents sometimes interfere with tested drugs or substances (Manganese-based nanoparticles with WST-1 [12], carbon nanomaterials with MTT [13], magnesium-based materials with both MTT and XTT [14]) and cause misleading results. Normally, reduction of tetrazolium-based reagents to orange-colored formazan by NAD(P)H occurs in live (healthy) cells, and this coloration is inhibited when there is an increase of damaged or dead cells in a cell population. Therefore, if there is any inconsistency between the cell viability and expected color at the end of the WST-8 assay, you should consider the interference of WST-8 with the tested material. To check the interference before setting up the viability assay; mix your drug, WST-8 reagent, and cell culture medium (without cells) and incubate. If color development occurs, change the treatment media, and add a washing step with PBS before **step 3** in Subheading 3.4 to remove the interfering drug. Alternatively, you can perform other cell viability assay(s) rather than tetrazolium salts.
24. The optimum incubation period of spheroids with the reagent can differ based on the viability assay, cell type, and seeding number. Therefore, it is recommended to perform the measurement every hour (at the end of 1, 2, 3, and 4 h) for the first experiment and follow the supplier's advice.
25. In our experience, carefully transferring 100 μL of the total volume (110 μL) of WST-8/media mixture into a new well of a flat-bottom microplate without taking any cells from the spheroids worked well for accurately measuring the absorbance. However, check the wells of the flat-bottom microplate after transferring the WST-8/medium mixture if any cells or parts of spheroids are mistakenly taken during transfer. If any, exclude these wells from the calculation.
26. If there are any bubbles in the wells, pop them using a thin needle or a 10 μL pipette tip before measuring the absorbance.
27. If the absorbance measurement could not be performed immediately, add 10 μL of HCl (0.1 M) or SDS (1% w/v) into each well. Then, cover the microplate with its lid closed using aluminum foil to protect it from light. Store this microplate at room temperature and measure the absorbance within 24 h [6].
28. Proper controls, such as blank (only the mixture of WST-8 and medium without spheroids), untreated control, solvent/vehicle control, should be included in the cell viability assays.

29. Some microplate readers automatically subtract the absorbance value set at the reference wavelength (A_{650}) from the absorbance value measured at the detection wavelength (A_{450}) and directly give one absorbance value as $A_{450-650}$ for each well. If this is the case, only subtract the blank's absorbance value from the absorbance of each sample as shown in the below equation:

$$\%\text{Cell viability} = \frac{\text{Treated } A_{450-650} - \text{Blank } A_{450-650}}{\text{Untreated } A_{450-650} - \text{Blank } A_{450-650}} \times 100$$

$A_{450-650}$: Absorbance value at 450 nm, which is automatically subtracted from its absorbance value at 650 nm by the microplate reader's software.

Acknowledgment

JIB-04 was a generous gift from Prof. Elisabeth Martinez from the Department of Pharmacology, University of Texas Southwestern Medical Center, USA.

References

1. Agency EM (2016) Guideline on the principles of regulatory acceptance of 3Rs (replacement, reduction, refinement) testing approaches
2. Langhans SA (2018) Three-dimensional in vitro cell culture models in drug discovery and drug repositioning. Front Pharmacol:9. https://doi.org/10.3389/fphar.2018.00006
3. Fitzgerald KA, Malhotra M, Curtin CM, FJ OB, CM OD. (2015) Life in 3D is never flat: 3D models to optimise drug delivery. J Control Release 215:39–54. https://doi.org/10.1016/j.jconrel.2015.07.020
4. Oner E, Kotmakci M, Baird A-M, Gray SG, Debelec Butuner B, Bozkurt E et al (2021) Development of EphA2 siRNA-loaded lipid nanoparticles and combination with a small-molecule histone demethylase inhibitor in prostate cancer cells and tumor spheroids. J Nanobiotechnol 19(1). https://doi.org/10.1186/s12951-021-00781-z
5. Han K, Pierce SE, Li A, Spees K, Anderson GR, Seoane JA et al (2020) CRISPR screens in cancer spheroids identify 3D growth-specific vulnerabilities. Nature 580(7801):136–141. https://doi.org/10.1038/s41586-020-2099-x
6. Dojindo Molecular Technologies I: Technical Manual of Cell Counting Kit-8. https://www.dojindo.eu.com/TechnicalManual/Manual_CK04.pdf. Accessed 16 Feb 2023
7. OECD. Test No. 492: Reconstructed human Cornea-like Epithelium (RhCE) test method for identifying chemicals not requiring classification and labelling for eye irritation or serious eye damage. 2019.
8. Phung YT, Barbone D, Broaddus VC, Ho M (2011) Rapid generation of in vitro multicellular spheroids for the study of monoclonal antibody therapy. J Cancer 2:507
9. Xu LH, Yang X, Bradham CA, Brenner DA, Baldwin AS Jr, Craven RJ et al (2000) The focal adhesion kinase suppresses transformation-associated, anchorage-independent apoptosis in human breast cancer cells. Involvement of death receptor-related signaling pathways. J Biol Chem 275(39):30597–30604. https://doi.org/10.1074/jbc.M910027199
10. Chuang Y-F, Wei M-K, Yang F, Lee S (2021) Water-driven surface wrinkling of poly (2-hydroxyethyl methacrylate) after ultraviolet irradiation. J Polym Res 28(10). https://doi.org/10.1007/s10965-021-02762-0
11. Zanoni M, Piccinini F, Arienti C, Zamagni A, Santi S, Polico R et al (2016) 3D tumor spheroid models for in vitro therapeutic screening: a systematic approach to enhance the biological

relevance of data obtained. Sci Rep 6(1): 19103. https://doi.org/10.1038/srep19103

12. Scarcello E, Lambremont A, Vanbever R, Jacques PJ, Lison D (2020) Mind your assays: misleading cytotoxicity with the WST-1 assay in the presence of manganese. PLoS One 15(4):e0231634. https://doi.org/10.1371/journal.pone.0231634

13. Monteiro-Riviere NA, Inman AO, Zhang LW (2009) Limitations and relative utility of screening assays to assess engineered nanoparticle toxicity in a human cell line. Toxicol Appl Pharmacol 234(2):222–235. https://doi.org/10.1016/j.taap.2008.09.030

14. Fischer J, Prosenc MH, Wolff M, Hort N, Willumeit R, Feyerabend F (2010) Interference of magnesium corrosion with tetrazolium-based cytotoxicity assays☆. Acta Biomaterialia 6(5):1813–1823. https://doi.org/10.1016/j.actbio.2009.10.020

Chapter 18

Analysis of Cancer Cell Line Secretomes: A Complementary Source of Disease-Specific Protein Biomarkers

Katie Dunphy, Despina Bazou, and Paul Dowling

Abstract

Various types of cancer cells enrich or condition the medium that they are cultured in by secreting or shedding proteins and small molecules. These secreted or shed factors are involved in key biological processes, including cellular communication, proliferation, and migration, and are represented by protein families, including cytokines, growth factors, and enzymes. The rapid development of high-resolution mass spectrometry and shotgun strategies for proteome analysis facilitates the identification of these factors in biological models and elucidation of their potential roles in pathophysiology. Hence, the following protocol provides details on how to prepare proteins present in conditioned media for mass spectrometry analysis.

Key words Biomarkers, Conditioned media, Culture, Metabolites, Proteins, Secretome

1 Introduction

Proteins, secreted/shed by cancer cell lines serve important roles in the regulation of many physiological and pathophysiological processes [1]. These bioactive molecules are of increasing interest to aid our understanding of disease and to identify potential biomarkers and therapeutic targets [2]. The group of proteins secreted/shed by the cell, termed "the secretome," is one of the cellular subproteomes and is readily analyzed by sensitive proteomics approaches, including mass spectrometry [3].

Protein biomarkers are increasingly needed in the clinic to predict outcomes and responses, effectively stratify cancer patient cohorts, and provide insights into the biology of cancer [4]. However, few protein biomarker candidates have been identified and pursued to support their clinical utility [5]. Serum/plasma-based biomarker discovery has proved to be problematic mainly due to the complexity and dynamic range of serum/plasma, coupled with

Dania Movia and Adriele Prina-Mello (eds.), *Cancer Cell Culture: Methods and Protocols*,
Methods in Molecular Biology, vol. 2645, https://doi.org/10.1007/978-1-0716-3056-3_18,
© The Author(s), under exclusive license to Springer Science+Business Media, LLC, part of Springer Nature 2023

the low abundance of many disease-specific proteins [6]. Proximal biofluids have proved to be a useful alternative in the quest to identify clinically valuable biomarkers; however, such biofluids also harbor limitations similar to serum/plasma, as exemplified by the high abundance of amylase in saliva [7].

Conditioned media from cancer cell lines and 2-/3-D cancer cell culture models has some advantages over clinical samples, including the relatively straightforward protocols around cell culture, unlimited source of mRNA and proteins, and the ability to rapidly prepare and store samples for subsequent analysis [8].

The role of the microenvironment in augmenting the presence of disease-specific biomarkers is a limitation of traditional 2D culturing on plastic. 3D cultures and coculturing of cells aim to mimic some components of the microenvironment with respect to producing more meaningful biomarker profiles [9]. The use of short-term primary cell cultures also reflects the more heterogeneous mix of cells derived directly from the living tissue [10].

Together with the physical environment, cancer cell lines depend on culture medium, a mixture of nutrients and growth factors at an optimized pH/osmolality, to maintain viability. Culture media is optimized to specific cell types, with cancer cell lines able to survive in a variety of media and supplements, while normal cells and stems cells need more bespoke formulations [11].

The technologies currently being utilized for discovery are now achieving the necessary sensitivity to allow evaluation of low abundance proteins and their respective modifications (e.g., phosphorylation/ubiquitination) that represent a potentially rich source of biomarkers [12–14].

The use of multiple cell lines for a particular disease will allow for adequate representation of disease heterogeneity in the discovery process [15–17]. It is also likely that future disease-specific biomarker panels will contain measurements from nucleic acids and metabolites, together with proteins. Conditioned media from cancer cell lines will serve as a useful reservoir for the identification of these diverse types of biomarkers during discovery phase analysis [18].

Generally, investigating cell line secretomes will provide a useful source of proteins in the search for clinically valuable biomarkers [19]. With the rapid improvement of new strategies for sample preparation, including filter-aided sample preparation (FASP) and analytical technologies [20, 21], this area of research has the potential to provide a continual flow of candidate biomarkers; and with the increasing complexity of culture models, will likely provide more effective information on disease diagnosis, prognosis, and therapeutic response (Fig. 1) [22].

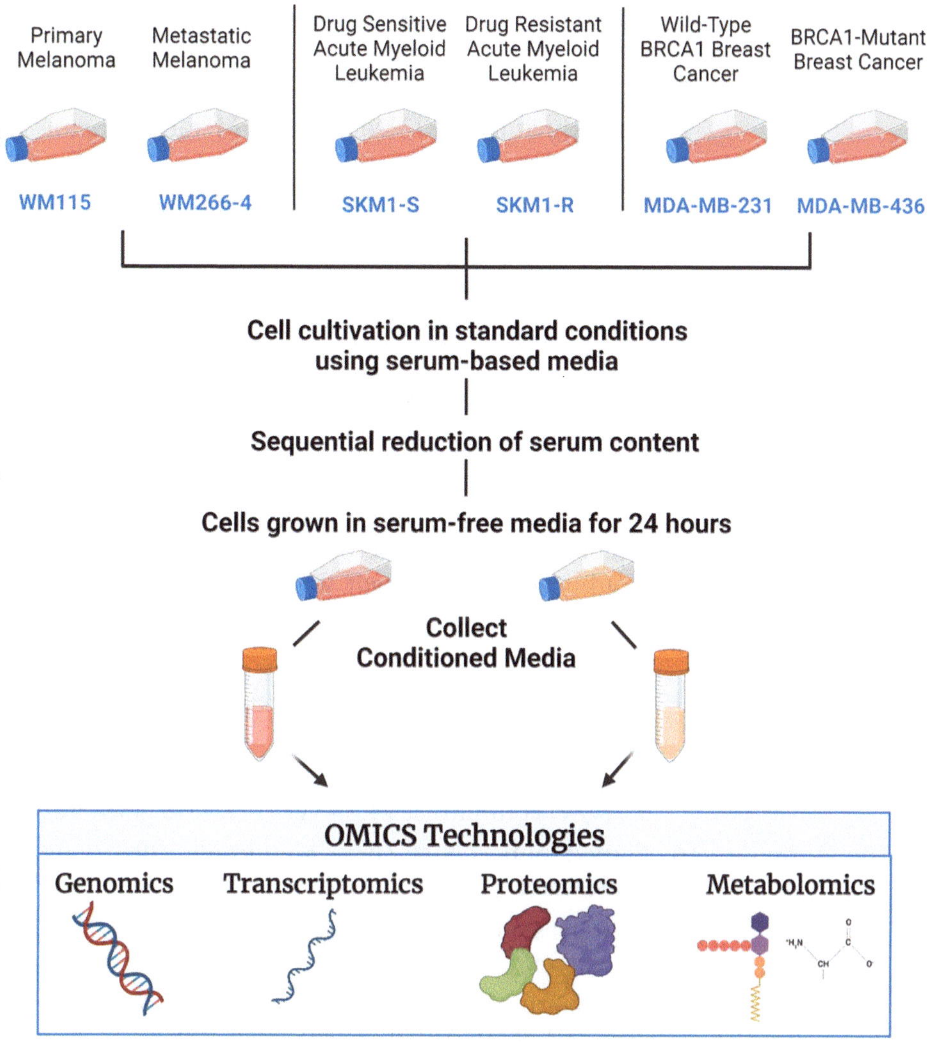

Fig. 1 Diagram showing workflow for the processing of conditioned medium. Secreted/shed proteins to be compared between cancer cell lines (for example, between the two isogenic colorectal cancer (CRC) cell lines, the primary SW480 cell line and its lymph node metastatic variant SW620), are initially cultured under normal conditions, exposed to reducing concentrations of serum and finally incubated for 24 h in serum-free media. The conditioned media is subsequently collected and analyzed using an appropriate OMICS platform. (Figure created using BioRender)

2 Materials

2.1 Solutions and Reagents

1. Phosphate buffered saline (PBS)
2. Recommended cell culture medium
3. Low serum-containing medium: 0.1–0.5% serum, recommended cell culture medium

4. Urea buffer (UA): 8 M urea, 0.1 M Tris/HCl (pH 8.5), LC-MS grade water
5. Iodoacetamide (IAA) solution: 0.05 M iodoacetamide, UA
6. Dithiothreitol (DTT) solution: 1 M dithiothreitol, LC-MS grade water
7. Resuspension buffer: 50 mM acetic acid, LC-MS grade water
8. Sequencing grade modified trypsin: reconstituted in resuspension buffer. Keep on ice before use
9. Ammonium bicarbonate: 50 mM ammonium bicarbonate, LC-MS grade water
10. 1% ProteaseMAX™ Surfactant (Promega): 100 μL ammonium bicarbonate, one vial of ProteaseMAX™ Surfactant (*see* **Note 1**)
11. Sodium chloride (NaCl): 0.5 M NaCl, LC-MS grade water
12. Ammonium bicarbonate buffer (ABC): 0.05 M Ammonium bicarbonate, LC-MS grade water
13. Sample buffer: 2% (v/v) trifluoroacetic acid (TFA), 20% (v/v) acetonitrile (ACN), 78% (v/v) LC-MS grade water
14. Activation solution: 50% (v/v) acetonitrile, 50% (v/v) LC-MS grade water
15. Equilibration/wash solution: 0.5% (v/v) trifluoroacetic acid, 5% (v/v) acetonitrile, 94.5% (v/v) LC-MS grade water
16. Elution buffer: 80% (v/v) acetonitrile, 20% (v/v) LC-MS grade water
17. Resuspension buffer: 2% (v/v) acetonitrile, 0.1% (v/v) trifluoroacetic acid, 97.9% (v/v) LC-MS grade water
18. Trapping buffer: 2% (v/v) acetonitrile, 0.1% (v/v) trifluoroacetic acid, 97.9% (v/v) LC-MS grade water
19. LC-Solvent A: 0.1% (v/v) formic acid, 99.9% (v/v) LC-MS grade water
20. LC-Solvent B: 80% (v/v) acetonitrile, 0.08% (v/v) formic acid, 19.92% (v/v) LC-MS grade water

2.2 Equipment

1. 3000 kDa MWCO centrifugal filter units (5–20 mL sample volume)
2. 660 nm Protein Assay Reagent Kit including Bovine Serum Albumin (BSA) standards (ThermoFisher Scientific) (*see* **Note 2**)
3. Microplate reader
4. Benchtop centrifuge
5. Thermo-mixer

6. Vivacon 500 centrifugal units – 30,000 MWCO (Sartorius) (*see* **Note 3**)
7. C18 reversed-phase resin spin columns
8. tC18 Sep-pak Cartridges
9. Dry incubator set to 37 °C/90 °C
10. SpeedVac vacuum concentrator
11. UltiMate 3000 HPLC (ThermoFisher Scientific) (*see* **Note 4**)
12. Orbitrap Fusion Tribrid (ThermoFisher Scientific) (*see* **Note 5**)

3 Method

3.1 Conditioned Media Preparation

1. Grow cells of interest with recommended medium in cell culture flasks to 60–70% confluency (*see* **Note 6**) at 37 °C.
2. Remove the growth medium, wash monolayer two-times (2×) with sterile PBS and then 2× with serum-free medium or low serum-containing medium. (*see* **Note 7**).
3. Add serum-free/low serum-containing medium and allow cells to condition the media for 24–48 h at 37 °C.
4. Collect the media from the cell culture flasks. Pipette or pour off the conditioned media from the flask directly into a 50 mL centrifuge tube.
5. Centrifuge for 5 min at 2000× *g* (4 °C) and gently harvest the supernatant, leaving a small volume of media behind (*see* **Note 8**).
6. Pre-wash a filter concentrator unit (3 kDa MWCO) with ABC for 10 min at 4000× *g* (4 °C) (fixed angle rotor) (*see* **Note 9**).
7. Fill the filter concentrator unit with up to 20 mL media volume ensuring the screw closure is fully sealed.
8. Centrifuge for 25 min at 4000× *g* (4 °C).
9. Empty the filtrate container and refill the concentrator with additional sample, if required.
10. Centrifuge for 25 min at 4000× *g* (4 °C) and repeat until the entire sample has been concentrated to the desired volume (e.g., 200–800 μL). Add 1 M DTT to each sample to give a final concentration of 0.05 M and incubate for 5 min (90 °C).
11. Remove the concentrate from the insert with a pipette and place in a tube placed on ice.
12. Proceed to the protein assay (*see* Subheading 3.2) or store the sample at −80 °C for use at a later date.

3.2 Protein Assay

1. Transfer 10 μL of Bovine Serum Albumin (BSA) standards (18.625, 37.25, 62.5, 125, 250, 500, 750, 1000, 1500 and 2000 μg/mL), blank (concentrated cell culture medium that

has not being incubated with cells) and test samples into individual microwells of a 96-well plate. All samples should be analyzed in triplicate.

2. Add 200 μL of the 660 nm Protein Assay Reagent into each well and mix well by pipetting up and down (avoid introducing bubbles during pipetting).
3. Incubate at room temperature for 5 min.
4. Vortex the plate gently and then quickly read optical density of the assay plate at 660 nm (*see* **Note 10**).
5. Subtract the average absorbances from the blank from the test samples/standards and plot a standard curve for determination of protein concentration of the test samples.

3.3 Filter Aided Sample Preparation (FASP) of Conditioned Media

1. Mix up to 30 μL of a sample (at a concentration of 10–250 μg, as determined by the protein assay) with 200 μL of UA in the filter unit (by pipetting it up and down 10 times) and centrifuge at 14,000× *g* for approximately 15 min (or until all the sample has moved through the filter unit).
2. Add 200 μL of UA to the filter unit and centrifuge at 15,000× *g* for 15 min at room temperature.
3. Discard the flow-through from the collection tube.
4. Add 100 μL of IAA solution, mix using a thermo-mixer for 1 min at room temperature (low mixing frequency) and incubate without mixing for 20 min in the dark.
5. Centrifuge the filter units at 15,000× *g* for 15 min at room temperature.
6. Add 100 μL of UA to the filter unit and centrifuge at 15,000× *g* for 15 min at room temperature. Repeat this step twice.
7. Add 100 μL of ABC to the filter unit and centrifuge at 15,000× *g* for 10 min at room temperature. Repeat this step twice.
8. Add 40 μL of ABC, with trypsin (enzyme to protein ratio 1:40 and 1 μL of 1% ProteaseMAX™ Surfactant) and mix using a thermo-mixer for 1 min at room temperature (low mixing frequency).
9. Incubate the units in a wet chamber at 37 °C overnight (approximately 18–20 h) (*see* **Note 11**).
10. Place the filter unit into a new collection tube carefully, and centrifuge the filter units at 15,000× *g* for 10 min at room temperature.
11. Add 60 μL of NaCl and centrifuge the filter units at 15,000× *g* for 10 min at room temperature. Repeat this step once.
12. Transfer 120 μL of collected sample into a new tube.

13. Acidify with sample buffer (1 μL for every 3 μL sample; for example, if 120 μL of sample was collected, add 40 μL of sample buffer) and desalt. For samples >100 μg, a tC18 Sep-pak (silica-based bonded phase with strong hydrophobicity) is recommended. For smaller samples, desalt using C18 reversed-phase resin spin columns (*see* Subheading 3.4).

3.4 Peptide Clean-Up

1. Place the C18 reversed-phase resin spin tubes into receiver tubes.
2. Add 200 μL of activation solution to rinse walls of the C18 reversed-phase resin spin tubes and to wet resin.
3. Centrifuge at 1500× *g* for 1 min at room temperature and discard flow-through solution.
4. Repeat **steps 2** and **3**.
5. Add 200 μL of Equilibration/Wash Solution.
6. Centrifuge at 1500× *g* for 1 min at room temperature and discard flow-through solution.
7. Repeat **steps 5** and **6**.
8. Pipette 100 μL of sample on top of resin bed (*see* **Note 12**).
9. Place the C18 reversed-phase resin spin tubes into receiver tubes.
10. Centrifuge at 1500× *g* for 1 min at room temperature.
11. To ensure complete binding, recover flow-through and repeat **steps 8–10** twice.
12. Place the C18 reversed-phase resin spin tubes into new receiver tubes.
13. Add 200 μL of Equilibration/Wash Solution to the C18 reversed-phase resin spin tube and centrifuge at 1500× *g* for 1 min at room temperature and discard flow-through solution.
14. Repeat **step 13**.
15. Place the C18 reversed-phase resin spin tubes in new receiver tubes.
16. Add 25 μL of Elution Buffer to top of the resin bed.
17. Centrifuge at 1500× *g* for 1 min at room temperature.
18. Repeat **steps 16** and **17** with the same receiver tubes.
19. Gently dry sample in a vacuum evaporator and resuspend the sample in 50 μL of resuspension buffer.

3.5 Mass Spectrometry Analysis

1. Using a liquid chromatography system (for example, the Thermo Scientific UltiMate 3000 UHPLC), load 500 ng of protein digest onto the trapping column at a flow rate of 25 μL/min with trapping buffer for 3 min, before being resolved onto an analytical column.

2. Elution of peptides is performed with the following binary gradient; LC Solvent A and LC Solvent B using 2 → 32% Solvent B for 75 min, 32 → 90% Solvent B for 5 min and holding at 90% for 5 min at a flow rate of 300 nL/min.
3. For peptide identification analysis, a data-dependent acquisition method is selected using a voltage of 2.0 kV and a capillary temperature of 320 °C.
4. Data-dependent acquisition with full scans in the 350–1750 m/z range is performed using a mass analysis (e.g., Thermo Scientific Orbitrap Fusion Tribrid) with a resolution of 120,000 (at m/z 200), a targeted automatic gain control (AGC) value of 4E+05 and a maximum injection time of 50 ms.
5. The number of selected precursor ions for fragmentation is determined by the top-speed acquisition algorithm.
6. Selected precursor ions are isolated by the quadrupole with an isolation width of 1.6 Da.
7. Ions with a charge state of 2+ to 6+ are detected (orbitrap), with a dynamic exclusion duration set to 60 s.
8. Precursor ions are fragmented using higher energy collision-induced dissociation (CID) with a normalized collision energy of 28%. Resulting MS/MS ions are measured in the linear ion trap.
9. The typical MS/MS scan conditions are as follows: a targeted AGC value of 2E+04 and a maximum fill time of 35 ms.

3.6 Protein Identification, Quantification, and Statistical Analysis

1. Identification and quantitation of proteins from conditioned media is performed using a variety of software including MaxQuant, Proteome Discoverer, and Progenesis QI for proteomics (*see* **Note 13**).
2. For subsequent analyses of proteomic datasets, a number of publicly available bioinformatics programs are available, including BLAST (https://blast.ncbi.nlm.nih.gov), PANTHER (http://www.pantherdb.org), STRING (https://string-db.org) and KEGG (https://www.genome.jp/kegg/).

4 Notes

1. We recommend the use of ProteaseMAX™ Surfactant as it enhances the enzymatic performance of trypsin.
2. We recommend the 660-nm Protein Assay as it is reproducible, quick, and more linear when compared with Coomassie-dye-based Bradford assay systems. The 660-nm Protein Assay is based on the binding of a polyhydroxybenzenesulfonephthalein-type dye and a transition

metal complex to proteins in acidic conditions, with the color produced increasing in proportion to increasing protein concentrations. Bovine Serum Albumin (BSA) is the standard reference for total protein quantitation by colorimetric assays.

3. The Vivacon 500 filter (Sartorius) membranes are compatible with high molarity urea solutions (e.g., 8 M) and allow fast liquid transfer. Thus, we recommend the use of these specific membranes.
4. Many LC configurations are available. The UltiMate 3000 HPLC series is used in this protocol because this system provides excellent chromatographic performance while supporting easy and consistent operation.
5. Many MS configurations are available. The Orbitrap Fusion Tribrid instrument is used in this protocol because it combines the best of quadrupole, ion trap, and Orbitrap mass analysis in a single system to provide exceptional depth of analysis.
6. The optimal confluency for cells at changing to serum-free media varies with each cell line, but upon reaching 60–70% confluence is most often used.
7. It is preferable to use serum- and phenol red-free media. It is known that phenol red mimics the action of some steroid hormones, particularly estrogen, and the use of phenol red-free media is particularly important when culturing breast or ovarian cancer cells concerning cell growth and viability. However, conditions will need to be optimized for each cell line to preserve the phenotype while minimizing cell death. This often results in the reduction, but not the complete removal, of serum. Typically, an adaptation protocol will involve reducing the serum concentration by approximately 50% with each subsequent culture passage until serum-free or low-serum levels are attained [23]. In experiments where serum is needed to minimize cell death, approaches like immunodepletion (albumin) can be employed to reduce the concentration of serum proteins identified in subsequent analysis.
8. The centrifugation step is needed to make sure there are no cells in the harvested media.
9. Select a filter concentrator unit with a low MWCO size (3000 or 5000 kDa) for maximum sample recovery.
10. If a 660 nm filter is unavailable, the assay plate can be read between 645 and 670 nm; however, this will result in a decrease in the linear range and in the assay sensitivity.
11. To prevent evaporation, it is necessary that incubations at 37 ° C are performed in a humidity chamber. A humidity chamber can be constructed using a plastic box with an air-tight lid. Place paper towels on the bottom of the box and wet

thoroughly with water. Place samples into a suitably sized tube rack that fits the size of box used, close securely, and place into an incubator at 37 °C.

12. Use a maximum of 30 μg of protein digest.
13. MaxQuant (www.maxquant.org) is a powerful, freely available software with a built-in search engine, Andromeda, used for the analysis of high-resolution mass spectrometry data. A database search can be performed in MaxQuant using a suitable reference proteome FASTA database (available online at www.uniprot.org). Proteome Discoverer (PD) is a commercial product of Thermo Fisher Scientific and provides a robust platform with both commercial and freely available tools to deal with most proteomics workflows, from labelled/label-free quantification, crosslink analysis, post-translational modification analysis and top-down proteomics. Progenesis QI for proteomics, by Waters, is a discovery analysis software that supports quantification and identification of proteins in complex samples using label-free mass spectrometry data.

References

1. Xue H, Lu B, Lai M (2008) The cancer secretome: a reservoir of biomarkers. J Transl Med 6:52. https://doi.org/10.1186/1479-5876-6-52
2. Dowling P, Clynes M (2011) Conditioned media from cell lines: a complementary model to clinical specimens for the discovery of disease-specific biomarkers. Proteomics 11(4): 794–804. https://doi.org/10.1002/pmic.201000530
3. Hathout Y (2007) Approaches to the study of the cell secretome. Expert Rev Proteomics 4(2):239–248. https://doi.org/10.1586/14789450.4.2.239
4. Papaleo E, Gromova I, Gromov P (2017) Gaining insights into cancer biology through exploration of the cancer secretome using proteomic and bioinformatic tools. Expert Rev Proteomics 14(11):1021–1035. https://doi.org/10.1080/14789450.2017.1387053
5. Hanash SM (2011) Why have protein biomarkers not reached the clinic? Genome Med 3(10):66. https://doi.org/10.1186/gm282
6. Geyer PE, Holdt LM, Teupser D, Mann M (2017) Revisiting biomarker discovery by plasma proteomics. Mol Syst Biol 13(9):942. https://doi.org/10.15252/msb.20156297
7. Teng PN, Bateman NW, Hood BL, Conrads TP (2010) Advances in proximal fluid proteomics for disease biomarker discovery. J Proteome Res 9(12):6091–6100. https://doi.org/10.1021/pr100904q
8. Kulasingam V, Diamandis EP (2008) Tissue culture-based breast cancer biomarker discovery platform. Int J Cancer 123(9):2007–2012. https://doi.org/10.1002/ijc.23844
9. Kapałczyńska M, Kolenda T, Przybyła W, Zajączkowska M, Teresiak A, Filas V, Ibbs M, Bliźniak R, Łuczewski Ł, Lamperska K (2018) 2D and 3D cell cultures – a comparison of different types of cancer cell cultures. Arch Med Sci 14(4):910–919. https://doi.org/10.5114/aoms.2016.63743
10. Pan C, Kumar C, Bohl S, Klingmueller U, Mann M (2009) Comparative proteomic phenotyping of cell lines and primary cells to assess preservation of cell type-specific functions. Mol Cell Proteomics 8(3):443–450. https://doi.org/10.1074/mcp.M800258-MCP200
11. Brandi J, Manfredi M, Speziali G, Gosetti F, Marengo E, Cecconi D (2018) Proteomic approaches to decipher cancer cell secretome. Semin Cell Dev Biol 78:93–101. https://doi.org/10.1016/j.semcdb.2017.06.030
12. Dayon L, Cominetti O, Affolter M (2022) Proteomics of human biological fluids for biomarker discoveries: technical advances and recent applications. Expert Rev Proteomics 19(2):131–151. https://doi.org/10.1080/14789450.2022.2070477

13. Nakamura R, Nakajima D, Sato H, Endo Y, Ohara O, Kawashima Y (2021) A simple method for in-depth proteome analysis of mammalian cell culture conditioned media containing fetal bovine serum. Int J Mol Sci 22(5). https://doi.org/10.3390/ijms22052565
14. Liu P, Weng Y, Sui Z, Wu Y, Meng X, Wu M, Jin H, Tan X, Zhang L, Zhang Y (2016) Quantitative secretomic analysis of pancreatic cancer cells in serum-containing conditioned medium. Sci Rep 6:37606. https://doi.org/10.1038/srep37606
15. Makawita S, Smith C, Batruch I, Zheng Y, Ruckert F, Grutzmann R, Pilarsky C, Gallinger S, Diamandis EP (2011) Integrated proteomic profiling of cell line conditioned media and pancreatic juice for the identification of pancreatic cancer biomarkers. Mol Cell Proteomics 10(10):M111 008599. https://doi.org/10.1074/mcp.M111.008599
16. Kulasingam V, Diamandis EP (2007) Proteomics analysis of conditioned media from three breast cancer cell lines: a mine for biomarkers and therapeutic targets. Mol Cell Proteomics 6(11):1997–2011. https://doi.org/10.1074/mcp.M600465-MCP200
17. Gunawardana CG, Kuk C, Smith CR, Batruch I, Soosaipillai A, Diamandis EP (2009) Comprehensive analysis of conditioned media from ovarian cancer cell lines identifies novel candidate markers of epithelial ovarian cancer. J Proteome Res 8(10):4705–4713. https://doi.org/10.1021/pr900411g
18. Fujita M, Imadome K, Somasundaram V, Kawanishi M, Karasawa K, Wink DA (2020) Metabolic characterization of aggressive breast cancer cells exhibiting invasive phenotype: impact of non-cytotoxic doses of 2-DG on diminishing invasiveness. BMC Cancer 20(1):929. https://doi.org/10.1186/s12885-020-07414-y
19. Li X, Liu H, Dun MD, Faulkner S, Liu X, Jiang CC, Hondermarck H (2022) Proteome and secretome analysis of pancreatic cancer cells. Proteomics:e2100320. https://doi.org/10.1002/pmic.202100320
20. Wiśniewski JR (2018) Filter-aided sample preparation for proteome analysis. Methods Mol Biol 1841:3–10. https://doi.org/10.1007/978-1-4939-8695-8_1
21. Ding Z, Wang N, Ji N, Chen ZS (2022) Proteomics technologies for cancer liquid biopsies. Mol Cancer 21(1):53. https://doi.org/10.1186/s12943-022-01526-8
22. Tanase C, Albulescu R, Neagu M (2016) Proteomic approaches for biomarker panels in cancer. J Immunoassay Immunochem 37(1):1–15. https://doi.org/10.1080/15321819.2015.1116009
23. Ozturk S, Kaseko G, Mahaworasilpa T, Coster HG (2003) Adaptation of cell lines to serum-free culture medium. Hybrid Hybridomics 22(4):267–272. https://doi.org/10.1089/153685903322329009

Correction to: Serum-free Production of Human Stem Cell-Derived Liver Spheres for Cancer Metastasis Research

Alvile Kasarinaite, James Drew, Mantas Jonaitis, Elaine Ma, Laura M. Machesky, and David C. Hay

Correction to:
Chapter 11 in: Dania Movia and Adriele Prina-Mello (eds.), *Cancer Cell Culture: Methods and Protocols*, Methods in Molecular Biology, vol. 2645, https://doi.org/10.1007/978-1-0716-3056-3_11

The original version of the chapter was inadvertently published with several errors. The chapter has now been corrected and approved by the author.

The updated original version of this chapter can be found at
https://doi.org/10.1007/978-1-0716-3056-3_11

Dania Movia and Adriele Prina-Mello (eds.), *Cancer Cell Culture: Methods and Protocols*,
Methods in Molecular Biology, vol. 2645, https://doi.org/10.1007/978-1-0716-3056-3_19,
© The Author(s), under exclusive license to Springer Science+Business Media, LLC, part of Springer Nature 2023

INDEX

Dania Movia and Adriele Prina-Mello (eds.), *Cancer Cell Culture: Methods and Protocols*,
Methods in Molecular Biology, vol. 2645, https://doi.org/10.1007/978-1-0716-3056-3,

© The Editor(s) (if applicable) and The Author(s), under exclusive license to Springer Science+Business Media, LLC, part of Springer Nature 2023

T

V

MIX
Papier aus verantwortungsvollen Quellen
Paper from responsible sources
FSC® C105338
FSC
www.fsc.org

If you have any concerns about our products,
you can contact us on
ProductSafety@springernature.com

In case Publisher is established outside the EU,
the EU authorized representative is:
Springer Nature Customer Service Center GmbH
Europaplatz 3, 69115 Heidelberg, Germany

Printed by Libri Plureos GmbH
in Hamburg, Germany